Urban Refugees

Urban refugees now account for over half the total number of refugees worldwide. Yet to date, far more research has been done on refugees living in camps and settlements set up expressly for them. This book provides crucial insights into the worldwide phenomenon of refugee flows into urban settings, repercussions for those seeking protection, and the agencies and organizations tasked to assist them. It provides a comparative exploration of refugees and asylum seekers in nine urban areas in Africa, Asia and Europe to examine issues such as status recognition, international and national actors, housing, education and integration. The book explores the relationship between refugee policies of international organizations and national governments and on the ground realities, and demonstrates both the diversity of circumstances in which refugees live and their struggle for recognition, protection and livelihoods.

Koichi Koizumi is Professor in the Faculty of International Relations at Daito Bunka University, Tokyo.

Gerhard Hoffstaedter is Lecturer in Anthropology in the School of Social Science at the University of Queensland.

Routledge research in place, space and politics series
Edited by Professor Clive Barnett
Professor of Geography and Social Theory, University of Exeter, UK

This series offers a forum for original and innovative research that explores the changing geographies of political life. The series engages with a series of key debates about innovative political forms and addresses key concepts of political analysis such as scale, territory and public space. It brings into focus emerging interdisciplinary conversations about the spaces through which power is exercised, legitimized and contested. Titles within the series range from empirical investigations to theoretical engagements and authors comprise of scholars working in overlapping fields including political geography, political theory, development studies, political sociology, international relations and urban politics.

Published

Urban Refugees
Challenges in protection, services and policy
Edited by Koichi Koizumi and Gerhard Hoffstaedter

Forthcoming

Space, Power and the Commons
The struggle for alternative futures
Edited by Samuel Kirwan, Julian Brigstocke and Leila Dawney

Nation Branding and Popular Geopolitics in the Post-Soviet Realm
Robert A. Saunders

Political Street Art
Communication, culture and resistance in Latin America
Holly Ryan

Geographies of Worth
Rethinking spaces of critical theory
Clive Barnett

Urban Refugees

Challenges in protection, services and policy

**Edited by Koichi Koizumi and
Gerhard Hoffstaedter**

Routledge
Taylor & Francis Group

LONDON AND NEW YORK

First published 2015 by Routledge

2 Park Square, Milton Park, Abingdon, Oxfordshire OX14 4RN
711 Third Avenue, New York, NY 10017

Routledge is an imprint of the Taylor & Francis Group, an informa business

First issued in paperback 2018

British Library Cataloguing in Publication Data
A catalogue record for this book is available from the British Library

Library of Congress Cataloging in Publication Data
Urban refugees : challenges in protection, services and policy / edited by
Koichi Koizumi, Gerhard Hoffstaedter.
    pages cm. – (Routledge research in place, space and politics)
    1. Refugees–Economic conditions. 2. Social integration. I. Koizumi,
    Koichi, editor. II. Hoffstaedter, Gerhard, editor.
    HV640.U673 2015
    362.8709173'2–dc23   2014041458

ISBN: 978-1-138-83980-9 (hbk)
ISBN: 978-1-138-54644-8 (pbk)

Typeset in Times New Roman
by Wearset Ltd, Tyne and Wear

Contents

PART II
A country's population on the move: Burma's refugees in Asia

Figures

Contributors

Linda Bartolomei is the Deputy Director of the Centre for Refugee Research (CRR) and the Coordinator of the Master of Development Studies at the University of New South Wales, Sydney. Since 2002 she has been engaged in a series of action research based projects with refugee and displaced communities across the world. A major focus of her work has been with women and girls who have survived rape and sexual violence in refugee and conflict situations.

Elizabeth Campbell holds a PhD and focused her research on urban refugees in Nairobi. She is an adjunct professor at Georgetown University Law Center and James Madison University where she teaches courses on forced migration, statelessness and global humanitarian issues. Since September 2011, Elizabeth has served as a Senior Humanitarian Policy Advisor in the Bureau of International Organizations at the Department of State in Washington, DC. Prior to her appointment, she was a Senior Advocate at the NGO Refugees International (RI) and the Director of the NGO consortium Refugee Council USA. Elizabeth has worked in Jerusalem and the West Bank, in Kakuma refugee camp in Kenya, and with UNHCR in Nairobi.

Nora Danielson holds an MPhil in Migration Studies from the University of Oxford, and is finalising doctoral research in anthropology at Oxford's Centre on Migration, Policy and Society. Her DPhil work, drawing from her experience contributing to the American University in Cairo's report on the Sudanese refugee demonstration of 2005, examines refugee protest and the politics of asylum governance, especially in urban, southern settings. Her research in Egypt over the course of the Arab Spring informed her recent article 'Revolution, its aftermath, and access to information for refugees in Cairo' (*Oxford Monitor of Forced Migration*). She is also interested in technology and the impact of communication about asylum for refugees, as explored in her recent working paper, 'Urban refugee protection in Cairo, Egypt: the role of information, communication and technology' (*New Issues in Refugee Research*). She has taught on anthropology, methodology and forced migration at the American University in Cairo, the University of Oxford and the University of East London.

Gerhard Hoffstaedter is a senior lecturer in Anthropology in the School of Social Science at the University of Queensland, researching religion, ethnicity and the state, international development and refugees in Southeast Asia. He has recently published *Modern Muslim Identities* with NIAS Press and is the author of *Religion and development: Australian Faith-Based Development Organisations*. A co-edited volume on *Human Security and Australian Foreign Policy* has just been published with Allen & Unwin. From 2014–2017 he holds a Discovery Early Career Researcher Award from the Australian Research Council to research refugee experiences in Malaysia.

Gail Hopkins' work focuses on migration, including forced migration, and integration, resettlement and associated social and economic impacts. She has conducted extensive qualitative field research on refugees in West Africa, Europe and North America under funding from ESRC, UNHCR and the British Academy, and currently combines independent field research and consultancy for organizations working on forced migration. She holds a PhD in Geography from the University of Sussex.

Koichi Koizumi is a professor in the Faculty of International Relations at Daito Bunka University, in Tokyo and specializes in Refugee and Forced Migration Studies. He focuses on the international community's response to humanitarian crises, with an emphasis on the socio-political dimensions of forced migration, sociological theories of forced migration, the evolution of the international refugee regime, and issues such as international refugee policy, international migration and asylum, the causes of migratory and refugee movements, refugee assistance and development process, post-conflict reintegration and reconstruction, refugee resettlement and livelihoods. He previously worked with UNHCR in the field, and is a former Visiting Research Fellow to the Refugee Studies Programme at the University of Oxford. He is the author of several books on refugees and forced migration, including *Who is a Refugee?* (San'ichi Shobo 1998), *Political Sociology of Forced Migration* (Keiso Shobo 2005) and *Globalization and Forced Migration* (Keiso Shobo 2009).

Eveliina Lyytinen completed an MSc in Human Geography from the University of Turku, Finland in 2006 and an MSc in Forced Migration from the Refugee Studies Centre, University of Oxford in 2008. Eveliina now works as a postdoctoral research fellow at the Department of Geography and Geology, University of Turku, Finland. Eveliina completed her doctoral studies (DPhil) in 2014 at the University of Oxford's Department of Geography. The overall aim of Eveliina's doctoral research, part of which is analysed in this chapter, examined refugees' experiences of the relationship between protection and insecurity in the urban setting of Kampala. Theoretically the focus of her research is to examine the connections between 'space', 'trust' and 'protection'. Empirically her research focuses on Congolese refugees living in Kampala.

Eileen Pittaway is the Director of the Centre for Refugee Research, University of New South Wales, Sydney, Australia, and Associate Professor in the

School of Social Sciences and International Studies, coordinating and teaching in the Master Programmes of *International Social Development*, and *Refugees and Forced Migration*. She has conducted research, provided training to refugees, UN and NGO staff in refugee camps and urban settings, acted as technical advisor to a number of projects, and evaluated humanitarian and development projects in Kenya, Thailand, Ethiopia, Bougainville, Egypt, India, Sri Lanka and Australia. She is currently working on applied research projects in Australia, India, Thailand and Kenya.

Barbara Sorgoni is an associate professor at the University of Turin. She holds an MA in Social Anthropology (University of Sussex); DPhil in Cultural Anthropology (Istituto Universitario Orientale, Naples). She has researched Italian colonialism in the Horn of Africa, in particular the history of Italian anthropology in the colonial context, customary and colonial laws, and interracial intimate and sexual relations between citizens and subjects (Sorgoni, B., 1998, *Parole e corpi*, Napoli, Liguori, 1998; Sorgoni B., 2001, *Etnografia e colonialismo*, Torino, Bollati Boringhieri). She is interested in the institutionalization of the asylum procedure in Europe, and on issues of credibility, truth and the bureaucratic production of asylum narratives (Sorgoni, B. ed., 2011a, *Etnografia dell'accoglienza. rifugiati e richiedenti asilo a Ravenna*, Roma, CISU; Sorgoni, B. ed., 2011b, 'Chiedere asilo in Europa. Confini, margini e soggettività', in *Lares*, LXXVII (1)).

Saburo Takizawa has worked for the United Nations, UNRWA, UNIDO and UNHCR, first at Headquarters and then as UNHCR Representative in Japan. He is currently Professor, Toyo Eiwa University and Project Professor, Human Security Programme, Graduate School of the University of Tokyo. He also serves as Chairman of the Board, UNHCR for Japan. Recent publications in English include 'Victims of Forced Displacement: Refugees and Internally Displaced Persons', in Morosawa, Dussich and Kirchhoff (eds), *Victimology and Human Security: New Horizons*, Wolf Legal Publishers, 2012; 'Refugees and Human Security: A Research Note on Japan's Refugee Policy', *Journal of the Graduate School of Toyo Eiwa University*, Vol. 7, 2011; 'Refugees and Victimology: Towards a more humane order', *United Nations Studies Series* No. 11, The Japan Association of the United Nations Studies, Vol. 11; 'A report on the local integration of Indo-Chinese refugees and displaced persons in Japan', UNHCR Representation in Japan, 2009.

Foreword

Barbara Harrell-Bond

Koichi Koizumi has put together this collection of papers in an effort to expose the situation of refugees living in urban areas in nine countries. Among the countries represented by the chapters of this book, there are six states that have ratified the 1951 Refugee Convention – Egypt, Gambia, Italy, Japan, Kenya and Uganda; and three states – India, Malaysia and Thailand – that have not. Theoretically, in states that have ratified the Convention, refugees should enjoy those rights that have been included in it.[1] While reading the chapters that follow, one might well ask how much difference it has made to the lived reality of the refugee, whether the Convention has been ratified or not. Even the fundamental right to protection against *refoulement*, which is now international customary law, is not always guaranteed.

The contributors to this volume were asked to highlight the needs of refugees living in urban areas and, it is good to see, all have emphasized refugee needs in terms of their *rights*. While many NGOs still conceive their efforts to provide for refugee needs as charity, increasingly academics as well as refugees themselves understand refugees' suffering as a result of their *rights* being violated (for example, see Danielson in this volume).[2] How can the humanitarian 'industry' be convinced to take a rights-based approach? Does the 'label' (refugee) obscure the fact that as human beings, wherever they are, refugees should enjoy human rights? A general rights-based approach to humanitarian assistance still seems a long way off.

Can one prioritize refugee needs? In any kind of emergency or disaster, whether the climate is hot or cold, we know that the first survival need of the human being is for shelter – before even water and food; that is why blankets should always be distributed first. But we can presume refugees living in an urban area are not in an 'emergency phase'.

In urban areas in Kenya and Uganda, we found that refugees' *greatest* need was for legal assistance[3] – to negotiate with landlords especially where there was no shared language, to protect them from exploitation in the work place, to assist them in gaining access to schools for their children, and so on, but above all we found their need for legal aid for help with their refugee status determination (RSD) adjudication, to be paramount. In Egypt, an independent study found that refugees considered their need for legal assistance with RSD as only second to their need for medical services.[4] Without legal assistance a refugee has little hope of being recognized, whether the State or UNHCR is the decision-maker.

It may surprise some readers that UNHCR is now the decision-maker for refugees' claims in more than 60 countries.[5] And, when UNHCR is the decision-maker, it can only grant refugee status under its 'mandate'. In countries that have not ratified the 1951 Convention (India, Malaysia and Thailand in this book), it is legitimate for UNHCR to carry out this *duty*, but it follows that it is also the duty of UNHCR is to find a third country that has ratified the Convention that will accept and resettle them. (Since few of the main countries of resettlement recognize mandate status, each refugee's claim must be re-examined by a representative of that country's government.) Why has UNHCR taken on this onerous role in countries that have ratified the Convention (Egypt and Kenya in this book) and have the obligation to protect refugees?

Many States that have ratified the 1951 Refugee Convention have failed to enact domestic legislation to regulate their refugee population.[6] In many countries that receive refugees en masse, it has been the norm that they have been recognized prima facie on the basis of their nationality. It has only been over the past decade that adjudicating refugee status on an individual basis in the global south has become the norm. Because RSD places a heavy toll on resources, there has been a growing tendency for countries to relinquish this responsibility and UNHCR has been forced to take up the slack. More alarming, when problems develop in relations between UNHCR and the host state, as in Kenya, the state may decide that responsibility for all the refugees on its territory lies with the 'international community' and stop recognizing refugees entirely.[7] In such a case, UNHCR has no alternative but to then make its own arrangements to conduct RSD adjudication.

There is a frightening absence of provision for legal assistance for refugees all over the world. In only a few countries is it considered to be a right and paid for by the state or by NGOs.[8] However, in the UK, to be paid, lawyers have to assess a claim as having 'more than a 50 percent chance' of being recognized.[9] In the US, the state does not provide refugees with legal assistance at all, although there are several pro bono university legal aid clinics where students prepare refugee clients for RSD adjudication and appeals under supervision. The European Union has issued a directive[10] requiring member states to provide free legal assistance to refugees who are appealing a first instance rejection of their asylum claim. In Cyprus, the refugee must appear – alone before a judge – to argue that his or her claim has more than a 50 per cent chance of winning. In one case, a lawyer who was assisting her rejected Cuban client (who did not speak Greek), briefed her on what to say. However, she did not understand the second question asked by the judge. He simply banged the gavel shouting, 'refused'. In Slovenia, there is a list of lawyers who have no training in refugee law, who are assigned the responsibility to defend refugee appeals.

It is not only lawyers who need training in refugee law.[11] Many who research refugee issues have not learned even the basic principles and thus make many errors in interpreting their findings. This is why 'Refugee Studies' at the Refugee Studies

Centre, University of Oxford, was constructed as an *interdisciplinary* course, with refugee law as a central feature. In fact, in 1982, the only places refugee law could be studied *as a course* was with James Hathaway, Osgoode School of Law, Toronto, and, a little later, at the Refugee Studies Centre.[12] Today refugee law is often taught as a component of immigration law; and a number of schools also now have specialized refugee law courses, but there are still far too few and none in the global south. The University of Michigan, under Hathaway's leadership, offers the most extensive refugee law curriculum in the world – not only a full introductory course, but four advanced seminars, a biennial Colloquium on Challenges in International Refugee Law, and student fellowships for top graduates to do refugee law 'on the ground'. It was only in 2015 that the degree course covering refugee law appeared – as an LLM in Immigration at Queen Mary, University of London.

As noted previously, in too few countries is legal assistance available to refugees; outside North America, some European countries, Australia and New Zealand, legal aid for refugees, as a specialty, is in its infancy. In 1997, when Guglielmo Verdirame initiated what became the Refugee Consortium of Kenya (RCK), and I helped to found the Refugee Law Project (RLP) at the Faculty of Law, Makerere University, Kampala, aside from one lawyer providing legal aid to Mozambicans in South Africa, as far as we know, there was no legal assistance for refugees in the entire global south.[13]

In an effort to contribute to rebalancing the rights versus charity approach to assisting refugees, since 1997 I have been promoting the establishment of legal aid NGOs, and encouraging the teaching of refugee law in universities.[14] The website for 'Information for Refugee Legal Aid'[15] is our effort to centralize widely scattered resources for providers of refugee legal assistance and for refugees themselves.[16] We also produce a monthly newsletter for legal aid providers.[17] Our 'Special Issues' pages are designed to assist lawyers with specially complicated claims.[18] Each page[19] includes the name of a 'resource person' who is a specialist in that area of refugee law and who has volunteered to give confidential help to lawyers with cases they are representing.[20] Refugee law is without doubt, *the* most complicated field of law, one that requires constant and continued learning, not only because of court decisions that can be used to argue the further development of law, but because states are becoming ever more restrictive, for example, in the application of the Exclusion Clauses.[21]

In 2011, UNHCR reported 54 countries that have produced 10,000 refugees or more.[22] This same source lists 199 states that produce refugees, including Canada, the United Kingdom, the USA and Sweden. There is no 'safe' country. These statistics suggest that lawyers are largely representing refugees of the same nationalities or from the same source countries. For example, lawyers in Hong Kong, Korea and Japan, like those in Israel, Zambia, Mexico and Brazil are all struggling to find information to defend cases from the Democratic Republic of Congo. Yet lawyers representing refugees tend to work in isolation; would it not be more effective if we all worked together?

Planning for this book coincided with the long-awaited announcement of UNHCR's new policy on urban refugees. This policy replaced the contentious 1997 'comprehensive' Urban Refugee Policy that assumed refugees belonged in camps and denied assistance or recognition to almost any who lived elsewhere. Although UNHCR has since launched a few 'feasibility studies' to test this new policy, only in the chapter on Kenya in this volume do we see some of the results. Sadly, these positive results have been overtaken by events since the chapter was written.

To get a glimpse of why it took 12 years to go public with a new urban policy, it is important to read the 1997 document.[23] It reveals the extent to which a significant number of UNHCR staff supported the continued *incarceration* of refugees, as well as the negative attitudes towards refugees who managed to escape camps.[24] Tens of thousands of dollars were paid to consultants who, one after another, attempted to devise a new urban policy that would overcome this resistance within UNHCR. Even in 2008, when what was considered *the* final document was on the High Commissioner's desk, rebellion in the ranks further delayed its publication until another revision emerged that the High Commissioner signed. What is less well known is that the non-governmental organizations (NGOs) that are contracted to implement assistance in camps and whose livelihoods were threatened by any deviation from the camp policy, strongly supported the 'pro-camp' faction within UNHCR.[25, 26] The chapters in this book will hopefully provide information for UNHCR to implement its urban policy with even more vigour.

Japan has always been a major contributor to UNHCR's budget, but this book marks the first significant academic work on refugees in English to come out of a Japanese university. It has been generously funded by the Japan Society for the Promotion of Science (Grant-in-Aid for Scientific Research (A) 2010–2013), the Ministry of Education, Culture, Sports, Science and Technology, Japan.

It is fitting that special mention is made of the contribution of Margaret Okole, who for 20 years has edited the articles for the *Journal of Refugee Studies*. Once the list of contributors was drawn up, she took on the task of finding referees for each manuscript, communicating their suggestions for revisions to the authors and finalizing the text which we now have before us.

Dr Barbara E. Harrell-Bond, OBE, Honorary Associate, Emerita Professor and Founding Director, Refugee Studies Centre, University of Oxford; Honorary Fellow, Lady Margaret Hall, University of Oxford; Director, Information for Refugee Legal Aid.

Notes

1 See: www.unhcr.org/4ec262df9.html
2 See also http://kanere.org/about-kanere/, an online newspaper produced by refugees at Kakuma refugee camp in Kenya.
3 G. Verdirame and B.E. Harrell-Bond, *Rights in exile: Janus-faced humanitarianism*, Berghahn Books, 2005.
4 N. Briant and A. Kennedy, 'An investigation of perceived needs and priorities held by African refugees in an urban setting in a first country of asylum', *Journal of Refugee Studies*, 17(4), 2004.
5 See http://rsdwatch.wordpress.com/where-unhcr-rsd-happens/. In five of the countries included in this book, UNHCR is the decision-maker.

6 See www.unhcr.org/4419921c2.html.
7 See Chapter 1, and especially pages 34–36 of Verdirame and Harrell-Bond, *Rights in Exile*. Before 1991, Kenya respected nearly all rights in the 1951 Convention. Today, after Campbell's chapter in this book was written, all refugees living in urban areas are threatened with encampment and the Somalis, with *refoulement*. See: http://frlan.tumblr.com/post/44315382902/refugees-in-kenya-a-briefing-on-recent-developments.
8 The UK has been perhaps the best example, but recent cuts in government expenditure seriously threaten legal aid to refugees. See, for example, 'Unkind cuts: UK refugee lawyers cite grave concerns over impending legal aid restrictions', http://ohrh.law.ox.ac.uk/?p=723.
9 Law firms are paid a fee for doing this initial assessment. We were alerted to a case of an Angolan couple who had twice been assessed by law firms and denied legal assistance. We had three days to prepare their case before their last appeal. One of my interns, a Greek lawyer unqualified to practice in the UK, accompanied them to court as 'McKenzie's friend'. They had a strong claim that was recognized.
10 Article 15: '2. In the event of a negative decision by a determining authority, Member States shall ensure that free legal assistance and/or representation be granted on request, subject to the provisions of paragraph 3' (COUNCIL DIRECTIVE 2005/85/EC of 1 December 2005 on minimum standards on procedures in Member States for granting and withdrawing refugee status).
11 Lawyers as well as social scientists also need to be taught the psychosocial consequences of the trauma of uprooting, torture, and of prolonged 'waiting in limbo' to which so many refugees are subjected, in order to properly represent or 'study' them. Sadly, this field of refugee studies has been eliminated from the curriculum of the Refugee Studies Centre since I left it in 1996. My intention in establishing a masters in refugee studies was to design it as a 'conversion' course, exposing students to refugee literature from several disciplines in order to prepare them to return to their own discipline for advanced research degrees.
12 When teaching International Public Law, professors may *mention* the 1951 Convention or exceptionally devote one lecture to it.
13 There are now – still only a handful – of NGOs providing legal aid in the global south. Most are members of the Southern Refugee Legal Aid Network or the Asia Pacific Refugee Rights Network (www.refugeelegalaidinformation.org/partners-2).
14 www.refugeelegalaidinformation.org/refugee-law-reader.
15 See www.refugeelegalaidinformation.org/.
16 The website is also for refugees, particularly the vast majority in the global south who do not have access to legal aid tailored for them. UNHCR addresses by country are at www.refugeelegalaidinformation.org/node/270. The Other Resources page (www.refugeelegalaidinformation.org/other-resources), lists newspapers developed by refugees, refugees' websites, and diaspora organizations, useful for searching for missing relatives. The *Pro Bono* legal assistance page (www.refugeelegalaidinformation.org/node/270#probono_table), serves to link refugees with legal support.
17 See: www.refugeelegalaidinformation.org/fahamu-refugee-legal-aid-newsletter.
18 www.refugeelegalaidinformation.org/special-issues-refugee-status-determination.
19 Special issues include: Apostasy or Conversion, Detention, The Exclusion Clause, Gang-based Asylum Claims, Gender-Related Persecution and Women's Claims to Asylum, Military Service Evasion, Palestinians who fall under the 1951 Convention, Unaccompanied and Separated Children, Sexual Orientation and Gender Identity, Statelessness, Accusations of Witchcraft, Trafficking, and Roma and groups outside nation-state organization.
20 Incidentally, these are all 'cutting edge' topics for doctoral research.
21 e.g. the introduction of material support to terrorist groups as a ground for exclusion and the expanding list of terrorist groups, initiatives of the US government that are rapidly 'aped' by other states.

22 See www.unhcr.org/4fd6f87f9.html.
23 The 1997 Comprehensive Policy on Refugees in Urban Areas can be found at www. refworld.org/cgi-bin/texis/vtx/rwmain?docid=41626fb64.
24 See www.refugees.org/resources/refugee-warehousing/ for a description of the campaign waged against camps, organized by the US Committee for Refugees and Immigrants.
25 In 2000, when UNHCR announced its plans to 'integrate' the Ugandan refugee camps' health and education services with those for citizens, the NGO responsible for implementing UNHCR's education programme in northern Uganda threatened to pull up stakes and move to South Sudan.
26 For discussion of the bureaucratic interests that refugee camps serve see: B.E. Harrell-Bond, E. Voutira and M. Leopold, 'Counting the refugees: Gifts, givers, patrons and clients', *Journal of Refugee Studies*, vol. 5, no. 3/4, 1992.

Between a rock and a hard place

Urban refugees in a global context

Gerhard Hoffstaedter

Urban refugees have, in the last 10 years, been slowly receiving the attention they deserve from policymakers, academics and service providers (see for example Sommers 2001; Parker 2002; Jacobsen 2006; Fábos and Kibreab 2007). This has largely been due to urbanization across the world changing the demographics in source, transit and destination countries, which means that finally the UN High Commissioner declared that 'almost half of the world's 10.5 million refugees now reside in cities and towns' (UNHCR 2009: 2). The numbers of urban refugees, as well as refugees in general, continue to rise and their movement to cities continues. Today 'more than half the refugees UNHCR serves now live in urban areas' (UNHCR 2013). UNHCR defines urban as

> a built-up area that accommodates large numbers of people living in close proximity to each other, and where the majority of people sustain themselves by means of formal and informal employment and the provision of goods and services.
>
> (UNHCR 2009)[1]

Concomitant with these changes has been a change in the make-up of refugee populations, with an increase in the numbers of women and children, either as part of family units or as individuals, making the urban refugee profile 'more like the "normal" distribution' of the population (Jacobsen 2006: 275). An unwelcome result of this demographic change has been the targeting of women. The Women's Refugee Commission (2011) states that: 'women, especially, are targeted by police and suffer verbal, physical and sexual abuse.' These recent developments have made the issue of urban refugees a pressing and critical concern.

In response to this the UNHCR issued an Urban Refugee Policy in 1997, which has undergone several revisions (UNHCR 1997). In 1997 the UNHCR urban refugee policy defined UNHCR's mandate as one of minimum engagement, 'based on the presumptions of state responsibility for protection and assistance, and refugee self-reliance' (UNHCR 1997: 2). The policy further stated that 'no assistance should be provided ... in urban areas ... where a UNHCR assistance programme exists in a rural camp or settlement' (UNHCR 1997: 2). Declaring this policy inadequate, refugee advocates mounted several campaigns for change. Following the massive response to urban refugees who had fled Iraq since 2003 to cities in neighbouring countries, UNHCR issued an update to its urban refugee policy in 2009. This explicitly made urban areas a 'legitimate place for refugees' (UNHCR 2009: 3) and changed the way protection spaces around the world were viewed and how UNHCR was going to resource its operations. In addition, the policy sought to engender new partnerships between UNHCR and local and national service providers and authorities in urban settings to provide sufficient support to refugees and strengthen these protection spaces. The latter has been a difficult task for UNHCR to achieve in many places. Sometimes interaction with host country authorities may be strained due to the legal situation, particularly if the host country is not a signatory to refugee conventions. Interactions with refugee communities, however, can also be strained due to suspicion, xenophobia, competing agendas and political constraints. Developing a global policy framework to deal with regional or local refugee population movements has proven difficult. Domestic host country concerns and the diversity (and complexity) of refugee populations make service provision and refugee advocacy a complex task. Thus there still exists a major disconnect between the UNHCR policy framework and the reality on the ground, as many of the chapters in this volume illustrate.

The immediate issue for most refugees is status, especially given that many host countries today have not ratified the 1951 Convention relating to the Status of Refugees or its 1967 Protocol. Many of the places where urban refugees seek (temporary) protection are non-signatory states that usually view refugees as illegal immigrants or grant them unofficial, temporary or conditional protection. This complicates the legal space asylum seekers inhabit and demands more resourcefulness on their part, whilst creating problems for those willing and able to help them. As such, many refugee service providers may themselves be deemed illegal under domestic regimes, or operate in a legal grey area helping unregistered, unwelcome or otherwise disadvantaged populations.

Status relies on documentation, which remains a major obstacle for refugees worldwide. Whether refugees are housed in Asylum Seekers' Reception Centres in Italy, literally stored away from Italian society in what Sorgoni (this volume) calls 'long-stay parking-places for human beings', or left to fend for themselves in the urban and peri-urban townships elsewhere, refugee lives are still best thought of as lives in limbo, as Pittaway (this volume) describes the experience of refugee life along the Thai–Burma border. Indeed, the life urban refugees find themselves in is somewhere between a rock and a hard place, i.e. they are wedged between equally bad outcomes for their protection, welfare and future.

Defining refugees and power relations

There is an enduring problem around defining who a refugee is, or rather should be, with UNHCR, national governments and civil society actors all varying in their assessment. UNHCR uses the well-known definition from the United Nations Convention Relating to the Status of Refugees (1951) that focuses on the fear of persecution and the fact of being outside one's country. This definition can be limiting and fear of persecution cannot be adequately measured or even qualitatively captured in a way that does justice to many refugee stories. Many chapters in this volume deal with people who have fled their countries of origin but remain, for one reason or another, outside the UNHCR definition. As a result, they are often even more marginalized as they are left without access to services and protection. One crucial issue is the process of registering as a refugee, which in many countries is a cumbersome, if not altogether impossible, task with restraints based on refugees' country of origin, means to access UNHCR and a range of other limiting factors detailed by the chapters presented here. Pittaway (this volume) draws our attention to the definition of a refugee proposed by the Assistant High Commissioner (Protection) for UNHCR, Ms Erika Feller. Feller's sentiment is that refugees are those people who have fled from persecution or fear of persecution. Those not yet registered as refugees are simply unregistered refugees. This definition sidesteps the often contentious difference between asylum seekers and refugees. The former includes those refugees not yet recognized and those who have had their refugee claims rejected. They are often excluded from UNHCR services and often remain in a legal grey zone, whilst recognized refugees have full access to UNHCR/host country services and protection. Whilst Feller's view can be a progressive step for researchers to look at the issue in a more holistic way and without differentiating along, often, indiscriminate lines, the issue remains especially problematic for those people who have fled persecution to countries that have not signed up to the Convention and thus have no legal recognition of refugees. Many of the following chapters grapple with the definitional conundrum in particular countries and the effects for specific refugee populations.

The same applies to other politically loaded and contentious labels, such as the country named Burma or Myanmar. In my field experience in Malaysia, many refugees I conducted interviews with called the country Myanmar, whilst Pittaway (this volume) reports the exact opposite for her field site in Thailand. This is a highly politicized issue for most refugee groups and where they strategically position themselves according to their beliefs and general usage patterns. Most Western governments continue to refer to the country as Burma. Some Western countries are reassessing this stance and code switching, as it were, due to the recent political developments in 'Burma'. Nonetheless, Burma captures the notion of a homeland fled, and a memory of a country once theirs and increasingly changing out of refugees' grasp and out of their imagination. (For consistency, we have taken an editorial decision to call the country Burma throughout this volume.) Thus, Burma will be used to denote the country, without political prejudice or motivation, in this book.

Beyond these definitional problems, the refugee stories and frameworks for their survival across the world have drawn out some striking similarities. There are a number of key issues faced by most urban refugees, whether they live in the global North or South, an Asian metropolis or an African township. The following chapters broadly encompass the issues facing urban refugees as well as detailing individual life stories that demonstrate how urban refugees find ways to make livelihoods, navigate the precariousness of their legal status, access to the UNHCR registration process and access to services such as health, education and shelter.

The case studies from industrialized countries such as Italy (Sorgoni) and Japan (Koizumi and Takizawa) demonstrate how power relations and dependence on the state problematize the relationships between host and refugee populations and can hinder integration processes. Similarly, the lack of work rights (if enforced) concomitantly increases dependence on service providers and puts additional pressure on refugee populations. Indeed the lack of work rights is a recurring issue for many urban refugees, as the stories stretching from Tokyo to Nairobi, from Kuala Lumpur to Cairo, vividly describe the exploitation of an often young, willing and able workforce of refugees in cities. The lack of work and/or work rights increases the reliance on transnational networks (mostly family members and friends living and working in the global North), but above all amongst themselves (most often co-nationals, intra-ethnic or religious groups) in new environments and in difficult situations. Rather than taking an individualistic attitude to survival, most refugees work together, establishing community organizations and helping each other. Acknowledging the contributions refugees make to a host society and to their own communities is captured in a refugee-centred approach (Jacobsen 2006) that is a welcome antidote to the dominant trope found in mainstream media in the many places that see refugees as 'problems' and more generally as a dangerous addition to the body politic.

In the urban and peri-urban areas refugees find places of temporary protection, where they are more reliant on their own networks, support groups and independent resilience, as UNHCR tends to offer only minimal support for the most vulnerable populations in terms of livelihood, accommodation and food, and limited services in terms of health and education. Unlike in a camp scenario, populations are scattered and often it is impossible for service providers to accurately predict and assess need, which leaves many urban refugees with added hardship. In 2008 a UNHCR report showed that '30 percent of the basic needs of refugees are not being met' (Voice of America 2008). The most sought-after resources for refugees are livelihoods, health and education for their children. The latter two are often costly and depend on refugees having access to some sort of income generation. Although many NGOs and UNHCR work towards offering some basic education in most urban settings, these programmes are never enough, often must be subsidized by the refugees and entail travel to and from schools, which is not only costly but may also be dangerous. The danger lies in being detected by authorities who may detain refugees or demand bribes. Even in host countries where education is fee-free for refugee children, the costs of travel, compulsory uniforms and study materials can be prohibitive.

Many refugees seek out the anonymity of the city because it allows them to escape capture by authorities and provides a means for survival. The city is a place of promise as well as danger and a space that refugees often have to learn to navigate for the first time in their lives. Hopkins' chapter (this volume) deals with the particular challenges for refugees of rural origin who seek refuge in an urban setting. While refugees may seek the anonymity of the city, some remain extremely visible in urban settings, such as black African refugees in Cairo and Kuala Lumpur, making their existence conspicuous and difficult. Such refugees are obvious targets for various forms of harassment and discrimination. However, there are also challenges for those trapped in a state of invisibility, as described by Hopkins (this volume), where refugees in The Gambia remain hidden from refugee networks and services. Indeed, the representation of refugees as victims, invisible and marginal (Harrell-Bond and Voutira 1992, 2007) that dominates most papers nevertheless, remains a contested field, as Lyytinen aptly explains in her exploration of Congolese refugees whose appearance is considered 'flash' by Kampala standards.

Discrimination, segregation, criminalization and securitization of the city have had a profound impact on refugees, their ability to move freely, find work and avoid arrest and detention by authorities as well as intolerance and violence from non-state actors. Most, if not all, developing countries see refugees as temporary residents, often forced to share their status with illegal immigrants and other people at the margins of society. It is also other people at the margins of the host society that refugees must compete with over scarce resources, such as paid work, living space and foodstuffs. On occasion this competition can become violent and cause further destabilization and threats for already vulnerable populations, but equally, on occasion, there can grow a shared understanding of their marginal situation, cooperation and support for each other.

Global contexts and local lives

While there exists no single urban refugee experience that stands for the global experience, the contributions to this volume demonstrate that there are certainly commonalities. Each chapter represents an attempt to explore the situation for urban refugees, or in some cases one particular refugee community that exists within a larger refugee population, in a specific locale, each one drawing to the fore the challenges unique to that host country. In most cases refugees find creative and imaginative ways of finding a space for themselves, their families and communities. Against the odds of living in limbo, they adapt, make spaces, homes and find livelihoods to sustain themselves. The host governments may support these spaces, as Takizawa, Koizumi and Sorgoni point out in this volume. In the case of Tokyo the presence of foreigners, especially refugees, prompted the municipal government to provide new multicultural spaces and support centres to cater to new and emergent needs. However, a lack of documentation, language and cultural issues persist, even in countries that provide resettlement and asylum protection such as Japan. In response to these limitations UNHCR has identified that 'the UN refugee agency needs to work in more innovative partnerships with municipalities, local community associations and others to adequately serve refugees in towns and cities' (UNHCR 2013). The effectiveness and commitment to this ideal are probed in several of the following chapters.

The book is divided into two sections. The first broadly focuses on African refugees spread across the content and making their way to Europe seeking protection. The second sections broadly focuses on refugees from Burma and their experiences in protection spaces across Asia.

Beginning with the situation for refugees in Africa is a chapter on Cairo, Egypt, home to one of the world's largest urban refugee populations. Danielson paints an absorbing portrait of a diverse refugee population seeking protection in a complex city. Danielson's chapter explores the challenges for refugees in Cairo through the lens of the 2005 mass demonstration by hundreds of Sudanese in Mustafa Mahmoud Park. She focuses on the experiences of those who participated in the demonstration and gives them a voice within the existing debates. This chapter also examines the situation for refugees in Cairo since the demonstration and the effects of the demonstration for the refugee population. Danielson's contribution importantly illuminates the position of refugees in Cairo from a refugee perspective. Echoing the urban refugee populations whose voices are heard throughout this volume, Danielson finds that Cairo's refugees identified 'employment, housing, healthcare and education' to be areas that needed to be addressed. Security, much sought after by the refugee interlocutors heard throughout this volume, is also identified by Cairo's refugees who demanded protection from harassment, abuse and arrest. Danielson recounts narratives from the demonstration, which include disturbing accounts of refugee life in Cairo – abuse, assault, harassment, disappearances and murder. These stories portray the anxiety and fear that characterize the refugee experience in Cairo.

Hopkins' chapter examines the urban refugee population in The Gambia originating from the Casamance region of southern Senegal. The Senegalese refugee population at the centre of Hopkins' research is comparatively small alongside the other refugee populations dealt with in this volume. While there are presently no refugee camps in The Gambia, there exists both an urban and rural refugee population in the country. Those who settle in rural areas are the chief recipients of refugee resources. The Gambia represents an interesting case study because unlike many other host governments dealt with in this volume, it promotes a policy of integration for refugees. In theory refugees are afforded the same rights as Gambian citizens. Those who register and receive a refugee identification card enjoy freedom of movement, the right to work, free health care and free education. Hopkins' research demonstrates, however, that poor dissemination of information and confusion amongst refugees in urban areas presents a major challenge to refugee access to education, health care and work. A range of other factors detailed by Hopkins also impact on the ability of urban refugees to take advantage of these services.

Lyytinen's chapter examines the significant population of Congolese refugees in Kampala, Uganda. The piece investigates the relationship between refugees and the organizations, which serve them. Lyytinen characterizes this relationship as marred by distrust. The chapter also examines the ethnic divisions and conflicts, which persist amongst the Congolese in Kampala and the mistrust that exists within the refugee population as a result. Lyytinen analyses the sense of insecurity felt by urban refugees as a consequence of this social and institutional mistrust.

Campbell's chapter on the position of urban refugees in Nairobi describes the advances made in the last decade, particularly in the areas of health and education. The year 2005 marked a new initiative and subsequent change in UNHCR's approach to urban refugees in Nairobi. New working relationships fostered by UNHCR with NGOs, refugee communities, and local authorities and organizations in Nairobi have led to significant advances in refugee access to health and educational services. Such improvements, however, have been achieved without any change in government policy, and Campbell importantly notes that such improvements need to be supported by government. The vast majority of refugees in Kenya live in the country's two main camps. For this reason, most of the resources are directed toward these refugee populations. One of the significant points to note from the Nairobi case study is that improvements have been implemented without any significant financial investment. Campbell's chapter provides an important case study of the impact of UNHCR's 2009 Policy on Refugee Protection and Solutions in Urban Areas. Official documentation, an issue for many of the refugee populations dealt with in this volume, is shown to be of critical importance for Nairobi's refugees. Employment in the city's informal sector is another feature of the refugee experience, which Nairobi's refugee community shares with their counterparts in Kuala Lumpur, Cairo, and The Gambia. Unlike the transitory refugee populations detailed in other chapters of this volume, Nairobi's urban refugee population is relatively settled, with most having lived in the city for 20 years or more.

Sorgoni's study of Ravenna is a detailed look into a local approach (funded by the government and EU) to welcome and support newly arrived refugees as well as the mediation that occurs between refugees, their cultural and legal translators and state officials. Sorgoni vividly describes the processes of objectification and bureaucratic subjugation that occur. The paternalistic attitude of state agencies, UNHCR and NGOs towards the subset of refugees who are on this managed programme provides a stark contrast with the experiences of many other refugees around the world who may struggle even to get an appointment with UNHCR or to submit an application for asylum – unfortunately a common occurrence in many overstretched and under-resourced UNHCR posts.

Similar issues plague the refugee experiences across Asia and the next section of the book focuses on the experiences of refugees from Burma as they seek protection and refuge across Asia's large cities with some more or less able to absorb the refugee populations.

Bartolomei describes the situation for refugees from Burma in Delhi, where they face a string of debilitating obstacles to survival. In her case studies and refugee stories Bartolomei constructs a damning picture of what life is like for these refugees in urban Delhi which goes a long way to explain why many decide to seek better protection elsewhere if they can afford the journey (many make the journey to Malaysia, where they have better chances to find a job and register with UNHCR for resettlement). The section on sexual violence and discrimination against women in particular is an important reminder of the gendered refugee experience and places a much-needed emphasis on women's stories. This is particularly so, because the refugee population has changed dramatically from being mainly young single men to include an equal proportion of families and young women and girls.

In Pittaway's chapter we are introduced to an interesting case of hidden refugees, uncounted and unknown in Thailand along the Thai–Burma border. Most refugees remain in camps dotted along the border with a small minority of urban refugees (around 2,000) registered in Bangkok. However, as Pittaway explains, there is a large contingent of refugees outside of the camps and away from Bangkok, living in regional towns and villages along the border. Refugees from Burma can only be registered in Thailand if they are in a camp; thus, these invisible refugee populations from Burma have no access to UNHCR protection and most end up working as cheap labour for local farms and businesses where they are highly vulnerable to exploitation and abuse. Pittaway also draws attention to the issue of gender and highlights specific matters such as sexual violence, but also the impact displacement, violence and the vulnerable situation they inhabit have on men and boys.

In my own chapter I note that the Malaysian state, has over time, given preference to Muslim refugees; however, a large contingent of Rohingya refugees have not been given preference, in fact they remain doubly discriminated against as policy shifts have seen them trapped long term in Malaysia without integration or resettlement as viable options. As in most other case studies, service provision is a major issue, complicated by the large caseload of refugees in Kuala Lumpur. Registration of refugees from Burma remains a critical point of contestation between UNHCR and refugee groups, as it is neither fast enough nor adequately resourced to be comprehensive enough. The entire refugee experience in Kuala Lumpur greatly depends on one's ethnic identity, as some Burmese refugee groups and their organizations have established a comprehensive service network, often with little UNHCR help.

In contrast to the earlier chapters which all concern self-settled refugees, Takizawa describes the recent Japanese pilot project to resettle Karen refugees from Thai refugee camps and the intricate ways that global, national and local levels of governance interplay in the refugee regime with often contradictory and surprising results. The programme had difficulty recruiting a limited intake from the refugee camps in its first two years and no takers at all in the third, due to negative reports back to friends and families from the first two groups. As Takizawa shows, policymakers were essentially motivated by an international political agenda and failed to understand these camp-based refugees' capacities and needs. Resettlement in and of itself clearly is no answer to the plight of refugees, if the refugees themselves cannot see a future or viable life in the third country. Japan's efforts to look good internationally thus have to be backed up by a sustained support network for refugees that adequately provides a new home and life for them. Takizawa also shows how a government resettlement programme can underestimate the capacity of refugees to know their own interests best; for example, to prefer a location where they have the support of co-ethnics (Tokyo) to the town designated for them, where they are isolated among a population whose language they can barely speak.

Beyond the resettlement programme, Koizumi's chapter presents the stories of asylum seekers who have made their way to Japan independently. The subjects of his study are likewise from Burma, often from an urban and even professional background, and have self-settled in a particular neighbourhood of Tokyo with a high foreign population. They range from illegally employed visa over-stayers to recognized refugees, and face the same obstacles as refugees in Kuala Lumpur, Nairobi or Cairo: lack of documentation, access to vital services and a livelihood, as well as language difficulties and an uncertain future. As in other cities around the world, refugees' main resource is each other. Koizumi makes clear the importance of intra-ethnic networking and support mechanisms for their survival and mental health, while not denying the divisions within the Burmese community.

At the outset I called the situation many refugees find themselves in as between a rock and a hard place, that is, they face an existential dilemma with little room for movement. As the contributions to this volume show, refugees must be resilient every day; when they are first displaced, in countries of first asylum and even in later onward movement. The reality for most urban refugees is thus a life in limbo; whether in countries, which have not ratified or even signed the Refugee Convention, or in signatory countries, which have formal procedures to cater for refugees, as the case studies of Japan and Italy attest. Between UNHCR's three permanent solutions of resettlement (generally an impossible dream), return to their own country (at what future date can only be guessed) and integration (dependent on the political will and resources of the host country), refugees have to 'make the best of it' and the following chapters describe in detail how refugees can achieve this in light of the many obstacles and countervailing interests and limitations. Indeed, life is fraught with a terrible anxiety and hopelessness where refugees are, as Pittaway describes it, 'caught between a snake and a tiger' (Pittaway, this volume).

Note

1 The distinction between camp dwellers and urban refugees has been complicated by, for instance, Doraï's exploration of the urbanization of Palestinian refugee camps (Doraï 2010).

References

Doraï, M.K. 2010. 'Palestinian refugee camps in Lebanon: Migration, mobility and the urbanization process.' In A. Knudsen and S. Hanafi (eds) *Palestinian refugees: Identity, space and place in the Levant*, pp. 67–80. London: Routledge.
Fábos, A. and Kibreab, G. 2007. 'Urban refugees: Introduction.' *Refuge*, 24: 3–10.
Harrell-Bond, B. and Voutira, E. 2007. 'In search of "invisible" actors: Barriers to access in refugee research'. *Journal of Refugee Studies* 20: 281–298.

Harrell-Bond, B.E. and Voutira, E. 1992. 'Anthropology and the study of refugees'. *Anthropology Today* 8: 6–10.

Jacobsen, K. 2006. 'Refugees and asylum seekers in urban areas: A livelihoods perspective'. *Journal of Refugee Studies* 19: 273–286.

Parker, A. 2002. *Hidden in plain view: Refugees living without protection in Nairobi and Kampala.* Human Rights Watch: London.

Sommers, M. 2001. *Fear in Bongoland: Burundi refugees in urban Tanzania.* New York/Oxford: Berghahn.

UNHCR 1997. *UNHCR Comprehensive Policy on Refugees in Urban Areas.* Available at: www.unhcr.org/refworld/docid/41626fb64.html (accessed 1 March 2013).

UNHCR 2009. *UNHCR Policy on Refugee Protection and Solutions in Urban Areas.* Available at: www.unhcr.org/refworld/docid/4ab8e7f72.html (accessed 1 March 2013).

UNHCR 2013. *Urban Refugees.* Available at: www.unhcr.org/pages/4b0e4cba6.html (accessed 1 March 2013).

Voice of America 2008. *UNHCR: Thousands of refugees in camps, urban areas left without basic aid.* Lanham, United States.

Women's Refugee Commission 2011. *Urban refugees '101'.* Available at: http://womensrefugeecommission.org/blog/1130-urban-refugees-101 (accessed 1 March 2013).

Part I

A continent on the move

Africa's urban refugees

Conflict and persecution continues to drive many people from their homes across Africa. However, Africa is not just a source of refugees; many places, countries and urban settings also offer a range of protection spaces. Refugees make their way to neighbouring countries and further afield to Europe. Chapters in this section focus on the varied and disparate experiences refugees face on their flight, in transit and destination countries as they seek asylum, are granted refugee status and make a life for themselves and their families in new places across the continent and beyond.

1 Demonstrable needs

Protest, politics and refugees in Cairo

Nora Danielson

Introduction

In the study of the situation of urban refugees, the city of Cairo is a compelling case. Positioned at the north of Africa, Egypt is host to refugees from war and political oppression in Syria, Sudan, South Sudan, Eritrea and Ethiopia. The largest city in the Middle East, it has also seen an influx of refugees from Iraq since the 2003 US-led invasion, and has long hosted Palestinians. Although considered by some to be a transit country, Egypt is a refugee host in its own right. With no refugee camps in the country, most refugees settle in its capital.

But Cairo, like Egypt as a whole, is a troubled refuge. Its troubles as such flashed to international headlines at the end of 2005, when at least 28 refugees were killed in the brutal forced eviction of a refugee protest. Although problems of asylum in Cairo have been well documented in academic and policy research, perhaps the most powerful call for attention to refugee needs in the city was raised by refugees themselves, over the course of that critical event of 2005 (Das 1995). For three months, camped in a traffic circle in a prominent Cairo neighbourhood, several thousand Sudanese had publicized their difficulties living in Cairo to staff of the nearby UNHCR office, to passers-by, to journalists and other visitors to the protest, and to the international community. The sit-in was not only an extraordinary event amongst urban refugees, but has also been called 'one of the most significant acts of public protest' in Mubarak-governed Egypt (Jones 2012: 16).

This chapter examines the basic needs of refugees in Cairo through an anthropological study of the recent politics of asylum in the city. First, it outlines the numbers of refugees in Cairo, the legal framework of asylum in Egypt, and the social, economic and geographic context. The second section presents analysis of the statements of the refugees and asylum seekers who staged the 2005 demonstration about the problems that had motivated their mass action. Third, I consider these needs in relation to changing research, politics and policy related to refugees in Cairo. The chapter thus contributes an overview of the state of asylum in Cairo, a refugee-centred assessment of it, and a consideration of how current dynamics may affect Cairo-based refugees' needs in the years to come.

This chapter draws from collaborative research conducted in Cairo in 2005–2006 as part of the team investigation undertaken for the report on the

2005 Sudanese protest published by the Forced Migration and Refugee Studies Program at the American University in Cairo, to which I contributed research, analysis, drafting and editing (Azzam 2006a). It is also based on fieldwork in Cairo 2010–2012 that included interviews and conversation with a range of refugees and asylum seekers, community-based organization leaders, service providers, and researchers, and participant observation at relevant workshops and events over that time.

Cairo as refuge: numbers, law and context

UNHCR reported 44,670 refugees and asylum seekers registered with its Cairo office in 2012 (UNHCR 2012: 1). This number represents some living in Alexandria and smaller towns in Egypt, and new asylum seekers who fled from the 2011 unrest in Libya to the border town of Saloum. The total does not include Palestinians living in Egypt, who do not receive assistance from UNHCR; people who have not sought recognition as refugees; people whose applications for refugee status have been rejected but continue living in Egypt; or some asylum seekers who have passed through the Sinai in an attempt to reach Israel. This chapter focuses on the needs of those who have sought refuge in Egypt's capital.

Approximately 56 per cent of the refugees in Egypt are from Sudan, 17 per cent from Iraq, 16 per cent from Somalia, 4 per cent from Eritrea, 4 per cent from Ethiopia and 3 per cent from other countries (UNHCR 2012: 1).[1] An initial study from 2003, prior to the large arrival of people from Iraq, found more than 30 first languages and dialects spoken by refugees of the six nationalities surveyed (Sudanese, Somali, Ethiopian, Eritrean, Burundian and Sierra Leonean), with many speaking two or more languages (Calvani 2003: 2–3, 9). The major dialects of Arabic relevant to refugee communities in Cairo (Sudanese, Fur, Egyptian and Iraqi) differ significantly, and are not always mutually well understood. Education and literacy levels in Cairo's refugee communities vary widely, and tend to be lower for women (Moghaieb 2009: 14–15).

Egypt is signatory to the 1951 Convention, and the 1969 Organization of African Unity Convention Governing the Specific Aspects of Refugee Problems in Africa. However, refugee protection in the country is limited by its reservations on the 1951 Convention, and a Memorandum of Understanding it signed with UNHCR in 1954 which further restricted refugee rights (Kagan 2011a). UNHCR carries out all refugee status determination procedures and registration of asylum seekers in Egypt. Its Cairo office works in partnership with local, national and international organizations to provide health, educational, livelihoods and legal assistance to asylum seekers and refugees in Egypt, available according to the varying eligibility guidelines of each service provider, and to arrange resettlement or voluntary return to countries of origin for a small minority of refugees.

Cairo's poverty, authoritarian governmental legacy, global position and current (2012) state of political upheaval significantly curtail the parameters for refugees and those who work with them. The government's position and policies on refugees has precluded local integration and places responsibilities on

UNHCR beyond its mandate, resulting in the creation of parallel systems and a cycle of dependency (Grabska 2005, 2006a; Gozdziak and Walter 2012: 5). Refugees are not in practice permitted to work, so must enter the tenuous and potentially exploitative informal market (Kagan 2011a: 18–19; Jureidini 2009).

The city of Cairo is an uneasy refuge: a troubled host, a place literally not easy to live in, and a source of anxiety for those who seek asylum in it. Cairo is the largest city in Africa and tenth largest urban area in the world in terms of both population and area (Demographia 2012), and is known for its overcrowded streets and traffic jams. Refugees and asylum seekers live throughout greater Cairo, and the city's refugee service providers are similarly dispersed. Transport to services can be costly and lengthy, and dangerous not just in terms of traffic accidents but interpersonal confrontations.

Refugee communities have long reported instances of discriminatory ill-treatment, harassment, arbitrary arrest and sometimes violent attacks in Cairo's streets, workplaces, hospitals, government institutions and police stations (Azzam 2006a: 15–16; Grabska 2006b: 296). As well as increasing daily fear and risk, patterns of harassment and violence contribute to mistrust in refugee communities of Egyptians and of institutions in Egypt. Cairo-based embassies of refugee-producing countries, most notably Eritrea, reportedly harass co-national refugees. During an interview, when I stated in passing that 'many refugees find life here hard', the refugee I was speaking with interrupted me: 'It's not some of them, it's all of them.'

Framing this hardship is the precarious state of Egypt itself: refugees live, as a service provider research participant stated, 'in a developing country with chronic problems where the nationals themselves struggle to live with dignity'. The outcome of the overthrow of the Mubarak regime has yet to be seen, and the bigger socio-economic picture remains bleak, especially for refugees. Livelihood difficulties are compounded by security problems. Legal recourse is frequently undermined by discriminatory police responses. Insecurity, unrest and crime have risen since the 2011 revolution, further increasing the risk and hassle of travel around the city (Jones 2012; Sadek 2012; Danielson 2012a, 2013: 37).

In theory, refugees who are neither able to return to their home countries nor to integrate in their present location are eligible to be resettled to a safe third country. Until 2004, a large proportion of refugees recognized by UNHCR in Cairo were resettled (Azzam 2006a: 8). However, following the 2004 Comprehensive Peace Agreement in southern Sudan, UNHCR began to give asylum seeker status rather than refugee status to new arrivals from Sudan, while waiting to see how the situation in southern Sudan would develop. As a result, southern Sudanese seeking asylum in Cairo were ineligible for resettlement except in exceptional cases (Kagan 2011a: 13).

The history of resettlement has created wide-reaching transnational social networks that contribute to the perpetuation of cycles of hope for resettlement (Crisp 1999; Fanjoy *et al.* 2005; Koser and Pinkerton 2002; Van Hear 2003). In light of continued instability in countries of origin – new South Sudan included – and increased insecurity in Egypt, resettlement remains most Cairo refugees' best hope. However, resettlement countries and UNHCR allot limited resettlement spots for refugees from Cairo – in recent years, fewer than 1,000 a year, with a low of 200 refugees resettled from Egypt in 2008 after lengthy and complex screening processes (Kagan 2011a: 27; Jacobsen *et al.* 2012: 15; Danielson 2013: 35). Refugees who hope for, expect, or have been selected for resettlement are unable effectively to plan their lives, due to not knowing if or when they will leave Egypt, which can cause great distress (Mahmoud 2009; Danielson 2013: 35).[2]

Demonstrating needs: the 2005 appeal by Sudanese refugees in Cairo

With this context laid out, the following section considers the needs of Cairo-based refugees through the rare assembly of refugee voices represented in the three-month sit-in by Sudanese in 2005. To do so, I first introduce previous relevant research, next outline the event itself, and finally analyse statements by its participants about the problems that motivated their actions.

Research on needs of refugees in Cairo before the 2005 demonstration

In 2005, many needs and problems in asylum in Cairo had been recognized and documented in academic and policy research. A 2001 review of the implementation of UNHCR's 1997 urban refugee policy showed the hardship for Cairo's refugees that had resulted from cuts to UNHCR's budget (Sperl 2001). Several researchers chronicled refugees' difficulties in Cairo slums (Ghazaleh 2002, 2003; Le Houérou 2004). A large study found African refugees in Cairo to need increased assistance with medical services, legal aid, food, shelter, and education (Briant and Kennedy 2004). Further research found narratives of 'social and cultural breakdown' amongst Sudanese refugees in Cairo (Coker 2004a, 2004b).

Papers produced for the Forced Migration and Refugee Studies Program (FMRS; now the Center for Migration and Refugee Studies) at the American University in Cairo suggested the need for changes in UNHCR's procedures for refugee status determination in Egypt (Kagan 2002), poor nutrition of southern Sudanese migrants in Cairo (Ainsworth 2003), the lack of educational opportunities in Cairo for refugee children (Afifi 2003), and issues with the protection of unaccompanied child refugees (Maxwell and El-Hilaly 2004). FMRS researchers also studied the high hopes in Sudanese refugees' expectations of resettlement (Fanjoy *et al.* 2005), the marginalization of Sudanese people whose applications for asylum had been rejected (Grabska 2005), and refugees' limited access to health care (Eidenier 2006). Other papers noted the limited livelihood pathways for Cairo-based refugees (SUDIA 2003; al-Sharmani 2004). The barriers facing refugees in Cairo in 2005 were thus both extensive and well documented (see Shafie 2004 for a comprehensive listing).

The 2005 Sudanese demonstration at Mustafa Mahmoud

Given this documentation of problems in asylum in Cairo, the 2005 demonstration 'should have come as no surprise' (Azzam 2006a: 17). Though perhaps not a surprise, the event was extraordinary. It started on 29 September 2005 in Mustafa Mahmoud park, the large, green-gardened traffic circle that served as waiting room and screening grounds for nearby UNHCR and was named after Mustafa Mahmoud mosque across the street. A group of 12 Sudanese refugees – mostly young men, who had initially come together as 'a reading group to inform themselves and their fellows of refugee rights' (Harrell-Bond 2008: 226) and called themselves *suut al-ajeen al-sudaniyeen bil qahira*, in English *The Voice of Sudanese Refugees in Cairo* (sometimes referred to in their documents as simply *Refugee's Voice* (Rowe 2007: 4) – had created a petition, email

Figure 1.1 Banners and clothes hang on the park fence, seen here backdropped by Cairo office and apartment buildings, 26 November 2005. Photo by Youssef Assad (Assad 2005).

address, and public statements. On 29 September they delivered their petition, signed by others, to UNHCR and, with some already waiting in the park, began their protest. The group disseminated a list of requests to local and international media, including postings on the website *Sudanese Online*.

As word spread amongst the Sudanese refugee community in Cairo, the occupation grew into a mass sit-in. Estimates of the number of participants vary, from 800 to 4,000. Fewer people were in the park during the day as many worked and returned in the evenings, and numbers continually increased over the course of the protest, with some moving completely into the park, together with all of their belongings (Azzam 2006a: 15). The great majority of participants were either asylum seekers or recognized refugees, as shown at the protest's end, when UNHCR identified all but 169 of the 2,174 protesters removed to detention as people 'of concern' to the office (Azzam 2006a: 43). The only quantitative survey of the park's population, a non-random survey of 149 of its participants (Schafer 2005a), found that a majority of those surveyed were men, 60 per cent of whom were unmarried, and mostly between 20 and 35 years of age. Eighty per cent of women surveyed were married, and many had their children with them at the protest. The sample population was half Muslim and half Christian, and originated from all regions and tribes of Sudan. Thirty-six tribes were represented, the most common of which were Dinka, Nuba and Nuer (Schafer 2005a: 9).

Through the distributed list of demands, banners hung around the park, statements to media and visitors, and meetings with UNHCR and other stakeholders, protesters explained their problems and their hope that they might be solved in various ways, as this section will examine closely. Formal negotiations between protest leaders and UNHCR started on 3 October, and ran more than 40 hours over eight subsequent meetings (UNHCR Regional Office Cairo 2006: 14–15). However, the goals of each party were at odds, and those in the park were largely distrustful of the office. Although protest representatives and UNHCR came to an agreement on 17 December that was supposed both to address the protesters' problems and result in the protest's end, it was rejected by the majority in the park.

The sit-in continued until the evening of 29 December 2005, when approximately 4,000 police circled the park. Authorities and protesters negotiated for several hours on the terms under which protesters would leave the park – protesters wanted guarantees that they would be taken to safe camps and not deported (Azzam 2006a: 35). Over the course of the night police sprayed protesters with water cannons for prolonged periods on several occasions, prompting protesters to protect children under blankets and plastic sheeting in the centre of the park. At around 5 a.m., the police surrounding the protest closed in, beating protesters with batons and removing them one by one whilst others were trampled or smothered in the melee (Azzam 2006a: 35). At least 28 people were killed and many more injured, and the survivors were bussed to detention centres around the city, with some families still remaining separated weeks later. Relatives of protesters killed in the protest break-up were blocked from transporting their bodies to Sudan for burial. Most protesters also endured significant loss of personal belongings, including identity cards, education records and family photographs (see photos at Lewis 2006b).

The demonstration at Mustafa Mahmoud has been studied and remembered in various ways. Reports in its immediate aftermath examined what went wrong and who was responsible (Azzam 2006a; UNHCR RO-Cairo 2006). Students in the graduate diploma in forced migration studies at the American University in Cairo investigated the views of its participants (Gomez *et al.* 2005a; Schafer 2005a, 2006) and of other stakeholders (Gomez *et al.* 2005b; Salih 2006), examined the protest as an instance of political resistance in exile (Lewis 2006a, 2007), and considered issues of masculinity and generation amongst its leadership (Rowe 2006, 2007). Literature since has largely studied the protest as an instance of one theme or another: as refugees 'claiming rights and seeking justice' (Harrell-Bond 2008: 222), as refugee mobilization around the need for 'meaningful protection' (Grabska 2008: 71), as 'an act of global political society' (Moulin and Nyers 2007), and as a way of understanding the rhetorical machinations of the Egyptian government (Giri 2007). More widely, the event has come to be considered a landmark case of the 'quest for recognition as political subjects' by asylum seekers and refugees (Sigona 2014: 370). In Cairo's refugee communities, the protest is remembered, first and foremost, as a massacre. This chapter focuses primarily on the voices of its participants themselves, with inductive analysis of the problems they raised, in order to restore attention to them and their actions, and gain a refugee-centred insight into the broader politics of the situation as a whole.

'Solve our problems': a refugee-framed view of refugee needs

Although overshadowed by the brutality and loss of the forced eviction, the 2005 demonstration was more than just a tragedy (Azzam 2006a). Condensed into three months and a single city block, the protest comprised a spectacular concentration of creative and political expression, and represented unprecedented collective organization by urban refugees. The objections that protesters raised through the protest offer a singular view into a refugee-initiated presentation of experiences of asylum in Cairo.

For the purposes of this chapter, I will focus on statements by protesters in official demands, banners, and interviews that expressly articulated their needs and explained the impetus for their actions. This analysis draws from materials produced by protesters and collected for the FMRS report on the event by the research team that produced it, of which I was a part (Azzam 2006a), interviews with participants during and after the event by myself and others (Schafer 2005a, 2006; Azzam 2006a, 2006b), and my visits to the protest and participation observation at related events. This section thus examines a selection of protester explanations of an event that, through its very staging, sought to bring attention to their words and grievances. As one woman participant explained in an interview shortly after the protest's end,

We hoped to attract the attention of the office [UNHCR], because there were so many cases with problems that didn't have individual causes. We also hoped that maybe we would attract the attention of the media, of the world, that then they would see our problems.

(Azzam 2006b)

The phrase 'Solve our problem or send us to another country' was painted on a protest banner (Schafer 2005b) and indeed, in sources consulted, protesters repeatedly framed the demonstration as the confluence of and reaction to an intersecting set of problems. Participants especially raised the need for solutions to livelihood, housing and service barriers; insecurity; and problems in dealing with asylum bureaucracy. With an effort not to separate these intersecting themes artificially, the following sections examine them in turn.

The need for improved access to livelihoods, housing and services

Protesters at Mustafa Mahmoud park called for improved access to the basic structures of support – employment, housing, health care and education – necessary for urban survival. Such needs were addressed briefly in the paper list titled 'Requests', printed in English and Arabic, given to me and a companion at the protest on 10 October 2005 by one of the event's organizers. One of the points dealt with the difficulty of local integration into life in Egypt, and another asked for special assistance for the elderly, disabled, widowed, vulnerable women and unaccompanied child refugees and asylum seekers.

If the protest organizers' lists delivered an official set of problems to stakeholders and others interested, the banners they prepared delivered messages to visitors and passers-by, and, when photographed, to a wider audience. Created by protest organizers and participants using fabric, coloured paper, pens and sometimes printed photographs or photocopied images, the banners were written in both Arabic and English, and fastened and hung on the fencing in a ring around Mustafa Mahmoud park, facing outwards. The need for improved livelihood access was illustrated in multiple banners. One written in Arabic stated: 'We die of hunger is better than getting killed from unfairness and poverty' (translated from Arabic by Dina al Shafie; see Assad 2005). This statement ties together poverty with unfairness, suggesting bigger complaints than a mere lack of work or money – perhaps an allusion to discrimination, or legal barriers. The banner also makes explicit the problem of poverty. The points from the list of demands also took form on fabric banners that stated 'We object local integration' (in English), and 'We need care for the old people and the children' (translated from Arabic by Dina al Shafie; see Assad 2005).

The banners that hung on the park's fence throughout the protest formed a material and narrative circle that united and conveyed the messages of the protesters who lived within its periphery. Yet this circle contained a multiplicity of voices. Statements collected from protesters by researchers during and

Figure 1.2 A man washes plates near drying laundry, cooking canisters and belongings along the fence of Mustafa Mahmoud park, 26 November 2005. Photo by Youssef Assad (Assad 2005).

immediately after the demonstrations give a richer view into the kinds of experiences raised in shorthand in the lists and banners.

The interconnected nature of the problems represented by the protest is set out in the statement of a man who described his participation in reference to local integration, one of the three 'durable solutions' for refugees emphasized by UNHCR. A middle-aged man from Omdurman, Sudan's largest city, the protest participant was a small business owner before being imprisoned and tortured for his political activity. He fled to Cairo in 1992, where he was given refugee status. In his statement to the Nadim Centre, in Arabic and translated by the centre into English, he said he had been accepted for resettlement to the United States in 2000, but was still awaiting progress in 2005. He connected the difficulty of local integration with problems of access to housing, poverty, work, education and financial aid, asking:

> How can they talk of local integration if we have no place to live? We are not allowed to work. There are no work opportunities in the first place. We cannot even secure the food. I have seven children. I try to educate some of them in cheap private school because we are not allowed to put our children in public schools. Sometimes I receive irregular aid from Caritas. This stopped totally since June 2004. They said 'we have no money'. I decided to join the protest. Maybe this would help.
>
> (El Nadim Center 2006)

This multifaceted explanation shows how barriers to these necessities of urban subsistence influenced this man's participation in the Mustafa Mahmoud demonstration, impeding both his own livelihood and local integration, and his children's. The problems of education he mentions were a common theme in interviews conducted during the sit-in by FMRS researcher Stacy Schafer, who, with the assistance of interpreters, surveyed 150 of the demonstrators and held 15 unstructured group interviews at the demonstration in mid-November 2005. She writes that 'every single mother [she] interviewed emphasized that either she could not afford her children's schools or that the schools her children went to were inadequate' (Schafer 2005a: 11).

The struggle for economic survival in Egypt was an overarching problem resurfacing in interviews with protesters. Of the 'long string of problems that people listed when asked why they were here' at Mustafa Mahmoud park, Shafer writes, 'lack of employment was often the first' (2005a: 8–9). Her research in the park found that this lack was not simply a matter of there not being enough jobs to go around. Rather, multiple factors contributed to problems with work:

> [Some protesters] tell me about the factories they were working in 15 hours a day for 10 LE [around 1 GBP] per day, or poor treatment by employers who pay them less than their Egyptian counterparts and force them to work longer. Many quit their quasi-jobs in the weeks or months before the protest citing problems with employers. When asked if they could return to these jobs, all were firmly opposed. One man explained that after working 20-hour shifts, his employer would only pay him half his earnings; the refugee was warned that if he quit he would be turned over to the police for working illegally and 'stealing' from the employer. However, most of the people I talked to had no job to complain about. Some could occasionally find work cleaning houses or doing manual labor, but a number insisted they could not get any job.
>
> (Schafer 2005a: 8–9)

In these explanations, the difficulties of finding work that might be expected in an economically weak country like Egypt were aggravated by discriminatory and abusive practices by bosses who exploited refugees' vulnerable legal and economic positions.

The theme of exploitation ran through interviews with those who were working or had worked in Egypt. Of the protest participants who spoke with Schafer about their employment experiences in Cairo, 'nearly all spoke of the harassment and ill treatment they encountered daily' at their jobs (2006: 9). Two men interviewed in Arabic (later transcribed and translated) on 12 January 2006 by FMRS researchers cited similar problems as their reason for protesting. A man who had been in Egypt since early 2004 and was awaiting the outcome of his asylum application explained: 'There is no work here for refugees. We want to go anywhere in the world where we will be protected.' As if to explain, a second man, a recognized refugee who had arrived in Egypt in 2001, followed this by saying: 'Here the Egyptians exploit us. There is no sense of humanity. They believe we don't deserve work.' The first answer tied together a lack of work with the need for protection, and the second framed that shortage in socio-cultural terms, based on exploitation and discrimination by Egyptians. Indeed, those interviewed described a vicious circle of livelihood problems: being effectively barred from legal employment by Egypt's reservations on the 1951 Convention, seeking jobs where there are few, and facing further problems from discriminatory and exploitative employers against whom they had little recourse given their precarious legal and economic positions.

If work led people to protest in frustration at a sense of unfairness and ill treatment, related problems with accommodation more immediately impacted some protesters' decision to join. Many of those who Schafer spoke with 'joined the protest because they had lost their jobs and could no longer pay their apartment's rent' (2006: 8). A 58-year-old woman who came to Cairo from Omdurman in early 2005 cited this problem in describing her reasons for protesting. When asked why she wanted to join the protest, she gave a tri-partite answer: 'I wanted to join to solve my protest. The UNHCR still did not see my case. I don't have means for the rent' (Azzam 2006b).

Indeed, poverty, employment, housing, health care and education were repeatedly cited in both interviews and public statements as needs that protesters hoped to address through their actions in Mustafa Mahmoud park. However, the examples above also illustrate that such barriers were presented as being enmeshed with security and legal problems. Employment, housing, healthcare and education can be said to be fundamental aspects of security in urban environments, just as legal recourse is. However, as the following section illustrates, a sense of safety and security was elusive for the Cairo protesters, and comprised one of their greatest needs.

The need for improved security

Protesters at Mustafa Mahmoud described harassment, physical and sexual abuse, organ theft, arbitrary detention, disappearances and death as pervasive in the experiences and fears of Sudanese in Cairo – in the city's workplaces, public areas, prisons and hospitals. The list of statements distributed by protest organizers to UNHCR, the media and local stakeholders raised the need for security in multiple forms. The list condemned the arbitrary arrest and detention of Sudanese refugees, and asked for UNHCR's assistance with those imprisoned or mistreated by Egyptian authorities. One point on the list called for UNHCR to protect refugees from Sudanese government personnel. Another point requested a search for Sudanese refugees and asylum seekers who had gone missing since their arrival in Egypt – one version of the list placed the number of missing at 500, citing the Red Cross (Damanga Coalition for Freedom and Democracy 2005).

Banners hung at Mustafa Mahmoud during the demonstration also served as visual representations of bodily and legal insecurity. One banner was written in English with the call 'Save our children from Sexual Abuse' (Schafer 2005b). The banner specified neither perpetrator, form nor location of abuse; demonstrating a generalized alarm in its appeal for aid and protection. Such an appeal took different form in three banners that showed pictures of three Sudanese people who were said to have been killed, detained and disappeared, respectively.

On one banner, the caption of a photograph of a man read, 'Who killed him? And why was he killed? Name: Benjamin Bill Tooh.' Linking together refugee flight with problems in exile, a caption below another man's photograph read, 'Running away from Sudanese detentions to the Egyptian prison.' Another banner asked: 'Where has Adroub vanished!? Refugee – member of the Beja Union – disappeared in unknown reasons – where is he now???' (Schafer 2005b). By asking where the pictured people were, why they had gone, and who was responsible, these banners conveyed a message not just of dismay at the physical violations of those pictured, but also of frustration at a lack of knowledge or control in the search for legal recourse. The use of photographs personalized the problem of disappearance raised in the organizers' list; the faces of the missing and arrested looked out at and confronted passers-by.

Figure 1.3 Protesters and police sit along the protest's periphery, 26 November 2005.
Photo by Youssef Assad (Assad 2005).

The need for legal recourse to the problem of detention was conveyed in a newspaper article, photocopied and attached to a sign posted along the park's periphery. The headline of the article read: 'From a Sudanese youth to the Egyptian police station: I want my right.' (Schafer 2005b) The hand-written sign placed responsibility for the detention of the youth pictured upon UNHCR, charging one of the Cairo office's senior staff by name as having 'lied in stating that you will protect us …' (Translation from Arabic by Dina al Shafie; Schafer 2005b.) Such a statement suggests that a one-time belief in UNHCR protection has since been betrayed, through events related to the detention by the Egyptian police of the young man pictured.

Linking protest banners and interviews, a story of disappearance was described by a woman who lost her daughter earlier in 2005. The woman's daughter, as researcher Stacy Schafer explained, 'is Fatima, the beautiful girl whose picture can be seen on a number of the banners lining the garden. She was killed last June while working for an Egyptian family in Alexandria' (2005: 17). When the mother tried to investigate what had happened, she discovered her daughter's body had been defiled: she eventually 'found her body in a hospital (morgue), where her organs had allegedly been removed' (ibid.). Fatima's mother then 'sought answers from the family [who she had worked for] as to what happened to her daughter, but they only offered her money in exchange for her silence', suggesting that the family had knowledge about the circumstances surrounding the death, or perhaps had killed her and sold her organs themselves (ibid.). The bereaved mother would not keep silent. However, when she 'went to the police and UNHCR, both said they could not help her', and her attempts to solve and find justice in her daughter's death were thwarted (ibid.).

Furthermore, at the time of the protest at Mustafa Mahmoud, Schafer reported that 'Fatima's mother believes this family is now trying to kill her and she refuses to leave the perimeters of the demonstration' (ibid.). Her participation in the park was a direct result of the insecurity she felt; as she told Schafer, she is in the park because she 'wants UNHCR to protect her and to investigate her daughter's death' (ibid.). This account describes a nightmare situation that links several themes that arose in protester statements: workplace abuse; disappearances; organ removal; no recourse with police and none with UNHCR; and throughout them all, a sense of pervasive mistreatment, injustice and insecurity. In a comment on Fatima's mother's story, Schafer wrote that 'it is accounts like these that lay heavily on the minds of Sudanese refugees. Whether such fears are valid or well-founded or not, such concerns and rumors pervade their thoughts' (2005: 17).

A similar scenario was described by a man interviewed on 14 January 2006 for the FMRS report by the pair of FMRS researchers mentioned above. The interview took place in a small apartment in the Cairo suburb of Nasr City, with three men in their late 20s and early 30s. One had been in Cairo for three years, one for four, and one for 12 years. In discussing what led them to join the demonstration at Mustafa Mahmoud, one man said, 'I feel afraid, I don't feel safe' (Azzam 2006b). Another added:

I know friends and neighbors who were slaughtered and we don't know who did this. I know this Sudanese woman who worked for an Egyptian family in the North coast. She died. Her employer claimed that she died in an accident in the swimming pool, but the autopsy showed that she was hit in the head.

(Azzam 2006b)

Like Fatima's mother's, his account of a covered-up murder suggests an Egypt where justice is on the side of the more powerful (employers, Egyptians), where refugees can be murdered with impunity, and where just going to work can be a dangerous act.

Another account of insecurity was told by Natara, a 36-year-old woman who I interviewed for the FMRS report almost a month after the forced eviction, on 28 January 2006. She was from Darfur and had arrived in Cairo earlier in 2005. In our interview, which was interpreted from Sudanese Arabic into English by an FMRS-affiliated Sudanese volunteer who himself had participated in the demonstration, she said, 'The major reason that I joined the protest was that I was afraid of Egyptians in general, especially after my troubles with the police in Maadi, afraid for me and for my children, my family' (Azzam 2006b). Natara was the mother of eight children, who were in Cairo with her, along with her husband, mother and other relatives. She added that they shared her sense of insecurity: 'When I had my accident in Maadi, all my other family members were afraid too' (Azzam 2006b).

Natara unfolded this story: her troubles began when she was attacked in the relatively affluent Cairo neighbourhood of Maadi. She said, 'In September I was beaten up by some Egyptians in Maadi who took all my documents' (ibid.). These documents included the UNHCR-issued yellow card that proved her legal status as an asylum seeker. She struggled to get legal recognition for the crime: 'I tried to file a report at the police station, but they wouldn't let me' (ibid.). However, she persisted and was ultimately successful: 'Finally after several tries they agreed and I got a police report' (ibid.). Police report in hand, she went to the Cairo office of UNHCR to apply for a replacement for her stolen yellow card. There, however, her problems continued, as the final section of this chapter will pick up.

In summary, protestors described the need for security through accounts of mysterious disappearances, unpunished murders and assaults, a lack of recourse, a failure of refugee ID cards to afford protection, arbitrary detention and police harassment. The stories told reveal protesters' views of both bodily safety and access to justice as things systemically denied to them in Cairo. Not only did the government of their host country fail to provide legal recourse but also sometimes its police perpetrated abuses themselves. The experiences described, and the sense of their pervasiveness amongst Sudanese refugee communities resulted in fear, frustration and disillusionment, motivating those interviewed to stage their protest at UNHCR. But in several stories, issues with refugee status, identity cards, and UNHCR formed part of the problem as well, as the following section examines further.

The need for improved asylum bureaucracy

The third major set of needs raised in the Mustafa Mahmoud demonstration related to protesters' frustrations with dealings with processes and treatment at organizations governing asylum and resettlement. Calls for changes to asylum procedures dominated the publicly-distributed list of protest demands. Protest leadership called on UNHCR to cease what they perceived as pressure from the office to repatriate to Sudan, in light of continued instability in the country. The organizers' list of points also requested improvements in UNHCR's refugee status determination (RSD) procedures in several ways.[3] If solutions to the problems facing Sudanese refugees could not be found in Egypt, the list asked that they be moved to another country 'where they will be more effectively handled' and 'where their refugee status would be determined in a more clear and transparent way' (or, in another version of the list, to a country 'where there is no discrimination') (Damanga Coalition for Freedom and Democracy 2005).

Criticisms of refugee assistance were also a refrain on protest banners. One banner made the charge through the English-language slogan 'We are the victims of mismanagement', suggesting that those who were supposed to take responsibility for refugees were failing to do so. Another banner asked, 'Why do we have to give up on the UNHCR?' (Translation from Arabic by Dina al Shafie). A series of political cartoons affixed to posterboards at the park suggested that these complaints

Figure 1.4 A car passes protesters at Mustafa Mahmoud park, 26 November 2005. Photo by Youssef Assad (Assad 2005).

might be linked to wider criticisms of the United Nation's role in geopolitical conflict and aid, presumably in Darfur, South Sudan, or other wars on the African continent. The next depicts a man sitting on the ground as if waiting, outside a door marked 'UN'. In another, an emaciated man and cow eat from the olive leaves surrounding the UN logo's globe. The banner stating 'No no for the political trading in our problem' (translated from Arabic by Dina al Shafie) similarly points to the importance of the wider political context, with its suggestion that some were gaining politically from the problems of Sudanese refugees.

In protester statements, livelihood and physical insecurities intertwined with difficulties with interactions with UNHCR in Cairo. In late September 2005, Natara, the woman quoted earlier who had been beaten up and managed to file a police report, approached UNHCR in Cairo to apply for a new yellow card, since hers had been taken when she was assaulted. However, she was refused entry to the office and badly treated by the security staff outside. She said: 'one officer shouted "Go, go, stay away" at me and the other people who were trying to get appointments, and the police shouted at us' (Azzam 2006b). On that day Natara was unable to get an appointment. She returned the next day, but again she was refused entry and was unable to make an appointment. She continued to return to the office, as she explained in her interview, with no luck, for more than a week: 'All together I went to the offices every day for eight days trying to get an appointment; I also wrote a written request but nobody responded' (ibid.).

Finally, on one night, still before the protest had begun, she slept in Mustafa Mahmoud park so she could be there early, as she had heard from others that UNHCR 'only let in a limited number of people a day' (ibid.) and that early arrivals would be at an advantage. Finally she was successful in her quest for an appointment. The day of her appointment she went to UNHCR, but this, her eighth trip to the office, coincided with the beginning of the protest. She explained:

> I saw that there was a group of people in the park, including two women with their children. I asked them what they were doing there, and they told me they were protesting. I was already having problems with my flat, so that night I came back with my eight children and husband and mother, and went and joined the protest and stayed at the park.
>
> (Ibid.)

And thus Natara and her family joined the protest and would stay for its duration. The narrative she told encompassed a manifold sense of insecurity: an actual assault, trouble with police when she tried to file a report, mistreatment outside the UNHCR office, lack of access to the office when she needed a replacement copy of the card that would prove her identity and legal status in Egypt, and problems with her flat.

Another account of problems of access to UNHCR was told by a woman I interviewed on 29 January 2006. The woman had worked in Khartoum before fleeing Sudan to Cairo in 2003. Within three months of her arrival, UNHCR had issued her a blue card, signifying her refugee status. Facing further troubles in Cairo due to medical problems, she was in the pipeline for resettlement to a third country. However, over the year and nine months that passed after she received her blue card, she was unable to access further assistance or information about her case.

When she visited the office, she explained, she 'could not even find an officer to talk to, just a security man, and I couldn't even talk to him for five minutes' (Azzam 2006b). In exasperation, she joined with others to try to make it easier to register questions with the office: 'We tried to suggest that they put a suggestion box in the [Mustafa Mahmoud] garden where we could write down our inquiries and give it to them, but they never did' (ibid.). The woman made trips across Cairo, made more difficult by her medical problems, on a weekly basis, for more than a year. By fall 2005, she said, she 'was disappointed in the office, because I had gone repeatedly for the year, putting in inquiries into my situation, and they never helped me' (ibid.).

Nonetheless, she continued her weekly visits to the UNHCR office in hopes that the situation might change. On one such visit in early October 2005, it did:

> Three days of the start of the demonstration, I went to the office and saw all the people there in the park. I asked what everyone was doing there and they said they all have a lot of problems, and so I went there and joined the protest. I was upset with UNHCR and decided to stay. I stayed until it was broken up.
>
> (Azzam 2006b)

In addition to her problems with access, she noted that she 'was also protesting the illegal ways in which UNHCR processed the applications' (ibid.). In her account, the relationship between problems with UNHCR and her participation in the protest is both paramount and pronounced. Despite her initial recognition by the office, her inability to gain further help or audience with its staff made her feel as though she had 'lost her rights', and her strategy to access them took more radical form.

Through their occupation, protesters at Mustafa Mahmoud raised the need for more effective legal recourse, accessible information, more transparent and fair procedures, and professional treatment in the processes of refugee registration, protection and assistance. In protesters' descriptions, within a context with problematic access to livelihoods, education, and housing and where police were more of a menace than a reassurance, lack of access to and responsiveness from the office that was supposed to provide protection created insecurity on top of insecurity.

Publicly-released lists, banners, and statements by protesters highlighted a range of needs amongst refugees and asylum seekers in Cairo. This chapter has reviewed indicative examples that showed some of the ways that protesters presented the difficulties that had precipitated their actions: problems of poverty, work, housing, health care, education; of harassment, discrimination, arbitrary

arrest, disappearances and injustice; of procedure and treatment in the legal processes around asylum. The demonstration at Mustafa Mahmoud was the source
of much debate and contention, with protesters levelling strong accusations
against UNHCR in particular, and protesters denounced as simply seeking economic gain or resettlement (Danielson 2008). Yet, as research before and since
has illustrated, the event also brought attention to legitimate needs in asylum in
Cairo in 2005. To what extent are these needs still relevant?

Changing Cairo, changing needs?

In the time since the demonstration at Mustafa Mahmoud park in Cairo, much
has changed. The situations of sending countries, of refugee populations in
Cairo, and of Egypt itself are in some ways quite transformed. Yet the basic
needs of Egypt's urban refugees are largely the same as those raised by the
Sudanese demonstrators in 2005. This section examines this relationship. First, it
reviews major political and policy changes related to refugees in Cairo. Second,
it considers the problems raised by the Mustafa Mahmoud demonstration in light
of these changes and more recent research. Through this study, the chapter con-
cludes with a longer view of the needs of urban refugees in Egypt, both in its
past and for its future.

Changes affecting asylum in Cairo

Changes affecting asylum in Cairo have come from many sides since the demonstration that tried to trigger change – although generally not in ways that its
participants sought. The biggest immediate change following the occupation of
Mustafa Mahmoud was its brutal eviction, which, in addition to at least 28
deaths and hundreds of injuries, resulted in the temporary detention of all
involved, the loss of belongings of the many who had moved their lives into the
park, and enduring psychosocial effects on participants, assistance workers, and
wider refugee communities in Cairo (Azzam 2006a: 39–54; Lewis 2011: 82;
UNHCR RO-Cairo 2006: 24–25).

Not long after this eviction came a sharp rise in migration from Cairo across
the Sinai into Israel by Sudanese, both asylum seekers and recognized refugees.
This movement, often seen as a direct outcome of the protest, grew from less
than 10 Sudanese arriving in Israel yearly from 2000–2005, to 270 in 2006, to
5,502 in 2009 (Human Rights Watch 2008; Afeef 2009: 9; Yacobi 2011: 52;
UNHCR 2011a). From 2007 such migration started to grow amongst Eritreans –
fewer than 10 Eritreans arrived in Israel yearly from 2000–2007, but 4,426
arrived in 2008, and 11,852 arrived in 2009 (UNHCR 2011a) – although
research has found this flow largely comes not from Cairo but through the south
of Egypt from Sudan and Ethiopia (Afeef 2009: 10). Indeed, as Eritrean migration to Israel rose, it declined amongst refugees from Cairo – unsurprisingly,
given Egypt's practice of shooting migrants crossing its border into Israel, rising
kidnappings and extortion of refugees in the Sinai itself, and reports of difficult

conditions from those who did make it to Israel (Furst-Nichols and Jacobsen 2011; Greenwood 2012; Human Rights Watch 2008, 2010).

In Cairo, the end of the Mustafa Mahmoud protest also marked the rise in collective organizing by some Sudanese youth into groups, sometimes identified as 'gangs', united by social solidarity and the cultural practices of American hip hop style (Lewis 2009, 2011; Forcier 2009). This activity, by several hundred youth, attracted attention not only for its visible affiliation with global, rather than Sudanese or Egyptian, youth culture, but also for a rise in violence between its members, including the stabbing to death of one young man outside the 2007 World Refugee Day celebration in Cairo (Forcier 2009: 6). However, due in part to peacemaking efforts by group members and service providers, such violence has largely subsided in the years since.

A more major shift in the population of refugees in Cairo came with the arrival of many asylum-seeking Iraqis, starting in 2006 (Pascucci 2012; Yoshikawa 2007). By 2011 Iraqis made up almost 20 per cent of those of concern to UNHCR in Egypt, and were largely recognized as refugees by the office (UNHCR 2011b). The new arrivals, although more affluent than many refugees in Cairo, brought particular psychosocial needs; a new challenge for service providers (Pascucci 2012: 50). Although some have since been resettled or have returned to Iraq, most remain living in Egypt, albeit in a state of uncertainty about their futures (Minnick and Nashaat 2009; El-Shaarawi 2012).

The 2011 formation of South Sudan, following decades of civil war and a referendum for independence in which many Cairo-based refugees participated, signalled a renewed possibility of return for many Sudanese refugees, who in 2012 made up more than half of Cairo's refugee and asylum seekers (UNHCR Egypt 2012). However, in light of limited infrastructure and violence and conflicts in multiple areas, many Sudanese refugees who originated from the south continue to stay in Cairo, waiting until the situation in their new country is less volatile (IRIN 2008; Ahmed 2009; King 2012; UNHCR 2012).

Certainly the most well-known recent change to the situation in Cairo is the 2011 revolution that, starting on 25 January, brought to a world stage some of the very problems raised by the Sudanese protesters in 2005 – arbitrary arrests and detention, lack of legal recourse, poverty – and ultimately brought down the Mubarak government. After a period of instability and army rule, the election of new president Mohamed Morsi represents the possibility of a more democratic future for Egypt and those who live in it.

In the short term, however, the revolution meant a host of new or worsened problems for refugees living in Cairo: immediately, loss of jobs and violence in the streets, and heightened insecurity and xenophobia in the months that followed (IRIN 2011; Lammers 2011; Jones 2012; Sadek 2012; Danielson 2012b). After Mubarak's deposing, refugees of multiple nationalities staged demonstrations at UNHCR Cairo, protesting what they felt was their abandonment due to the office's closure during the revolution (Viney 2011; Farag 2011; Jones 2012; Danielson 2012b). Although the Arab Spring resulted in increased resettlement spots for refugees in Egypt, the majority of these were allocated to those who fled Libya to the Egyptian border town of Saloum – including many sub-Saharan Africans who have since staged their own protests for rights and resettlement after finding themselves in legal limbo there almost two years after arrival (FIDH 2011; Chasek-Macfoy 2012; Danielson 2012a).

From the protest break-up to the rise in onward flight to Israel and youth violence in Cairo, from the influx of Iraqi refugees to the independence of South Sudan, and from the overthrow of the Mubarak regime to ongoing political struggles in Egypt, much has changed since Sudanese refugees occupied the Mustafa Mahmoud park. Despite these changes and developments, however, many of the fundamental problems raised by protesters at Mustafa Mahmoud remain essentially the same, and some have become worse.

Changing needs? Asylum in contemporary Cairo

Pre-2005 research and the requests of the 2005 Sudanese protesters showed how the lives and experiences of refugees in Cairo were constrained by barriers to livelihoods, housing, education and health care, by a range of security problems, and by difficulties in interactions with UNHCR. Although some of these needs have been somewhat alleviated due to the organizational efforts of both service providers and refugees, the changes outlined have worsened others, most notably security, and the new government has not shifted the civil practices and legal restrictions that impinge upon refugees' lives.

First, the need for improved refugee access to non-exploitative jobs, fairly priced housing, and services such as health and education remains hampered by legal, economic and social constraints. Due to these restrictions, most refugees continue work in the informal sector, where Egyptian labour law does not apply – occupations 'with few protections' (Buscher and Heller 2010: 21; see also Rizvi and Buscher 2012: 9). As a result most have 'little hope of finding dignified and safe employment' in the city (Rizvi and Buscher 2012: 1). Recent research with refugees in Cairo has also documented 'immense challenges' for refugees trying to make a living in the city: not just high unemployment rates and legal restrictions, but also 'financial, cultural, and – for non-Arabic speakers – linguistic barriers' (Gozdziak and Walter 2012: 17).

Some refugees lost their jobs during the revolution when employers left the country, and others have since, as the economic situation has worsened for all in Egypt (Jacobsen *et al.* 2012: 28). Access to education, especially the Egyptian public school system, also remains hampered, by 'complicated bureaucratic procedures, overcrowded schools and xenophobic attitudes' (Ensor 2010: 26). A 2011 survey of 565 Sudanese heads of households, 70 per cent of whom were registered refugees or asylum seekers, found that, in the previous year, 85 per cent of respondents did not have sufficient money to pay rent; 31 per cent were unable to send their children to school; and 28 per cent had been 'denied or unable to pay' for health care (Jacobsen *et al.* 2012: 36). The goal of local integration for refugees in Cairo remains 'nearly always impossible' (Kagan 2011a: 5).

Second, the need for security for refugees in Cairo has grown, especially following the revolution. Criminality, xenophobia, sexual harassment and street violence have risen as policing has weakened (*Egypt Independent* 2012; Jones 2012: 16; Jacobsen *et al.* 2012: 21, 28; Rizvi and Buscher 2012: 6). Refugees also remain subject to harassment or arbitrary arrest by authorities. A Cairo-based organization that provides legal aid to refugees and asylum seekers reported a 20 per cent increase in refugee complaints of 'arbitrary arrest and detention, acts of violence and acts of discrimination' over 2011 (Jones 2012: 16). The 2011 survey of 565 Sudanese heads of households found that in the past year, 83 per cent had experienced harassment in Cairo's streets; 40 per cent had been subject to robbery; 36 per cent had been physically assaulted; 24 per cent had been harassed by authorities; and 18 per cent had been arrested or detained by police (Jacobsen *et al.* 2012: 36).

The problem of legal recourse raised by the protesters at Mustafa Mahmoud continues; as the 2011 survey reports, refugees in Cairo continue to

> face difficulty or are reluctant to report incidents, and non-Arabic speaking refugees such as Eritreans and Ethiopians are less able to defend themselves against accusations or to report crimes or harassment. Refugees report that the police in post-revolution Egypt are less helpful than before.
>
> (Jacobsen *et al.* 2012: 39)

Despite the protection that recognition by UNHCR should offer, refugees or asylum seekers who are detained by police still can 'not be confident that [they] would be able to meet with UNHCR, nor that UNHCR's pleas on [their] behalf would be heeded' (Kagan 2011a: 5). Egyptian authorities' increasing practices of *refoulement* and 'prolonged arbitrary detention' over the later 2000s has resulted in what some have called a 'protection crisis' (Kagan 2011a: 4).

Insecurity at work has also continued to be a problem, due to exploitation amongst employers, corruption in police forces, and refugees' precarious legal positions. Many participants in the 2011 survey 'spoke of their livelihood vulnerability, i.e. that they feared being raided by the police and having to pay bribes to keep their businesses and stay out of jail' (Jacobsen *et al.* 2012: 28). The many who find employment as domestic workers face more insidious dangers, with 'little recourse': in a 2009 survey of 633 foreign-born Cairo-based domestic workers, 35 per cent of Sudanese respondents said that their employers physically abused them, and 15 per cent of Sudanese respondents reported 'sexual harassment' which ranged from verbal harassment and unwanted touching to attempted and actual rape (Jureidini 2009: 87).

The allegations of organ trafficking raised by some protesters at Mustafa Mahmoud have also since been documented: a 2011 study estimated the existence of 'at least hundreds of Sudanese victims of organ trafficking in Egypt' as well as 'numerous others' of survivors originating from Eritrea, Ethiopia, Somalia, Iraq, Syria and Jordan – cases in which organs have been removed 'either by inducing consent, coercion, or outright theft' (COFS 2011: 1, 2). The need for improved security for refugees in Egypt is greater than ever. There have been renewed protests at the offices of UNHCR in 2011 and 2012 by refugees and asylum seekers asking for resettlement or secure segregated housing. That self-settled, urban refugees should be essentially requesting encampment has been called 'a sign of the fear felt' in refugee communities in post-revolution Cairo (Jones 2012: 17).

Third, the need for improvements in the bureaucracy around refugee status and resettlement – as documented in earlier research and demanded by the 2005 protesters – has received attention and effort. However, it remains constrained by institutional funding priorities and practices, and especially by the Egyptian government's restrictions on refugee rights and increasing use of *refoulement* and detention against refugees (Kagan 2011a: 4). Since the 2005 protest, changes to refugee services and organizing have helped improve the meeting of refugee needs. Service providers have increased inter-agency coordination and communication, as well as greater provision of information for refugees about available services and legal processes. A new organization provides psychosocial support and outreach to refugees in Cairo. Recent years have also seen the growth of community-based organizations in refugee communities, through which refugees may help themselves and each other (Danielson 2012b: 18–19; Jones 2012: 16).

Recent research has shown, however, that 'refugees and asylum seekers lack information about most of the vocational training programs targeting them' (Moghaieb 2009: 30) as well as more generally about rights, services, protection, and legal processes (Danielson 2012b: 11). An over-reliance on site-specific information provision, requiring refugees and asylum seekers to make often frustrating journeys across the city, tends to be the norm amongst service providers, albeit with notable exceptions, such as one service provider pioneering the use of text messaging (Danielson 2012b: 36). Information, updates and time estimates regarding resettlement opportunities are especially lacking, to considerable psychological effect (Mahmoud 2009: 3, 95; Danielson 2012b: 11). Outreach to refugees, and efforts to ease their access to needed information, continue to require greater attention at the institutional level.

Not only has the political sea change of 2011 not alleviated the restricted rights and access of Cairo-based refugees, in many ways it has made their situations worse, with fewer livelihood opportunities, more insecurity, and further strain on the functioning of service providers. The growth in numbers of Cairo's refugee population and the arrival of new communities, notably Iraqi and more recently Syrian, have – without significant corresponding funding – also heightened needs.

Refugee needs in Cairo: from history to future

The non-governmental and governmental institutions best positioned to meet the needs of refugees in Cairo are hindered, albeit in very different ways. The history of UNHCR's urban refugee policy and relationship with refugees in Cairo has had 'lingering effects' (Gozdziak and Walter 2012: 5). Recent research has pointed to the 'negative impact of the 1997 UNHCR [policy on urban refugees]' – a policy called 'probably one of the most controversial documents ever produced by the agency', since revised in 2009 (Gozdziak and Walter 2012: 10; Marfleet 2007: 38). Both the reduction of resettlement since 2004 and the violent end to the 2005 protest, which some community organizations alleged UNHCR was involved in planning, also worsened refugee trust of UNHCR Cairo (Kagan 2011a: 4–5; Marfleet 2007: 42). Furthermore, because the arrival of many new asylum seekers and refugees in the city has not been met with a 'proportional increase in funding and durable solutions', existing assistance 'shows signs of strain' (Gozdziak and Walter 2012: 5, 10).

Although the blame for such strain may lie with donors and UN priorities at a global level, the lack of interest by the Egyptian government, leadership, and, to some extent, civil society in refugee needs in Cairo also limits and distorts efforts to meet them. Despite the change in leadership, Egypt's policy towards refugees continues to be characterized as one of 'benign neglect' (Kagan 2011b: 3; Jones 2012: 16), with both 'restrictive legal framework and lack of wide-spread implementation of existing rights' increasing refugees' reliance on humanitarian assistance (Gozdziak and Walter 2012: 10). This problem is not just structural but is common amongst Egypt's major political parties, whose 'ignorance and indifference' towards refugees has been called 'the biggest danger to refugees in Egypt', both presently and for the future (Jones 2012: 17).

Despite the recent boom in civil society activity, its actors, too, have tended to overlook the rights and needs of refugees in Egypt (Sadek 2012: 2, Samy 2009). Civil society's post-revolutionary growth has also created staffing problems for refugee service providers, who must increasingly 'compet[e] for staff with mainstream civil society organizations with higher political profiles and often offering higher salaries' (Jones 2012: 17). NGOs have also been put on alert by recent government raids on foreign-funded organizations and increased xenophobia (Jones 2012: 16). In the absence of possibilities for inclusion into Egyptian institutions and regulated employment sectors, refugees continue to rely largely on a system of parallel services provided only to refugees, which 'inadvertently contribute[s] to refugees' separation from mainstream Egyptian society' (Gozdziak and Walter 2012: 5).

With reference to urban refugees in Cairo, it has been said that although 'for centuries the city was a place of sanctuary; for people of the global South it is increasingly a place of danger' (Marfleet 2007: 43). Despite the considerable and complex circumstances impinging on the lives of Egypt's urban refugees, however, and the needs that this chapter has highlighted, recent reports have drawn attention both to where problems lie, and what might be done about them. Although beyond the scope of this chapter, for those interested in possible remedies, they are worth review (Buscher and Heller 2010; Danielson 2012b; Gozdziak and Walter 2012; Grabska 2006a; Jacobsen *et al.* 2012; Kagan 2011a, 2011b; Rizvi and Buscher 2012; Sadek 2012).

This chapter has taken a historical and anthropological approach to examining the basic needs of refugees in Cairo, using the 2005 protest by Sudanese refugees as a lens. After introducing the demographic, legal, economic and social framework of asylum in Cairo, I outlined previous research on Cairo-based refugees' needs and the events of the 2005 protest. The chapter then examined how participants in the protest presented the problems that had triggered their action: needs of access to safe livelihoods, housing and services, of safety and legal recourse, and of better treatment in accessing refugee status and resettlement. Finally, the chapter reviewed changes that have come in the years since, and assessed recent research to see to what extent the needs raised in 2005 had changed or been addressed, finding the situation much the same and in some respects worsened.

In a July 2011 interview with an Egyptian who has long provided legal assistance to refugees in Cairo, I asked if anything good had come from the Sudanese demonstration of 2005. He answered that its only possible benefit was that it might still be useful as 'propaganda' or 'show': 'to use it as background to 2011. 'If you don't make anything for refugees, maybe it will happen again.' Because every day we have more refugees. We can use the Mustafa Mahmoud incident to help refugees today.' Whether the event will muster such power remains to be seen. But at the very least, the protesters contributed an important view into their problems in Cairo and, more broadly, of urban asylum in the global South – forcing conversation with their demands at the centre. As misguided and ultimately tragic as the demonstration at Mustafa Mahmoud may have been, the voices and actions of the Mustafa Mahmoud protesters brought these problems to life and sought our attention to them. Their needs mostly persist for refugees in Cairo, and they ought not be forgotten.

Notes

1 At the time of publication, these populations had further shifted with the arrival of people fleeing the Syrian crisis. For UNHCR Egypt's recent statistics and planning figures, see its website: www.unhcr.org/pages/49e486356.html
2 This contextual section draws from research carried out with support from UNHCR's Policy Development and Evaluation Service, and is partially revised from from a resulting *New Issues in Refugee Research* working paper (Danielson 2012b).
3 These included requests to conduct RSD on an individual basis, rather than on the basis of region of origin, ethnicity, or nationality; to stop using the 'Four Freedoms' agreement between Sudan and Egypt as justification for the repatriation or local integration of Sudanese refugees in Egypt; to review and reopen the files of Sudanese people whose asylum claims had been rejected by UNHCR, and to allow people denied refugee status to appeal their cases (for the full list, see Azzam 2006a: 62–65).

References

Afeef, K.F. 2009. 'A promised land for refugees? Asylum and migration in Israel'. *New Issues in Refugee Research* 183.

Afifi, W. 2003. 'Field report: Preliminary investigation of educational opportunities for refugee children in Egypt'. Forced Migration and Refugee Studies report, American University in Cairo. Available at: www.aucegypt.edu/GAPP/cmrs/reports/Documents/wesal.pdf (accessed 14 April 2011).

Ahmed, Y. 2009. 'The prospects of assisted voluntary return among the Sudanese population in greater Cairo'. Center for Migration and Refugee Studies paper, American University in Cairo. Available at: www.aucegypt.edu/GAPP/cmrs/reports/Documents/AVR_Final_March10.pdf (accessed 4 April 2012).

Ainsworth, P. 2003. 'Changing dietary practices amongst Southern Sudanese forced migrants living in Cairo'. Forced Migration and Refugee Studies report, American University in Cairo.

Al-Sharmani, M. 2004. 'Refugee livelihoods: Livelihood and diasporic identity constructions of Somali refugees in Cairo'. *New Issues in Refugee Research* 104.

Assad, Y. 2005. 'Photo album, Sudanese protest against the UNHCR in Cairo, 26 November 2005'. YoussefAssad Flickr webpage. Available at: www.flickr.com/photos/50229137@N00/sets/1456762/with/67498003/ (accessed 15 September 2006).

Azzam, F. (ed.) 2006a. 'A tragedy of failures and false expectations: Report on the events surrounding the three-month sit-in and forced removal of Sudanese refugees in Cairo, September–December 2005'. Forced Migration and Refugee Studies report, American University in Cairo. Available at: www.aucegypt.edu/GAPP/cmrs/reports/Documents/Report_Edited_v.pdf (Accessed 14 March 2011).

Azzam, F. (ed.) 2006b. 'Background material and original sources for the FMRS report on the events surrounding the three-month sit-in and forced removal of Sudanese refugees in Cairo'. Unpublished archival collection, CMRS Grey Files KE/LSU-66 M Azz, American University in Cairo library.

Briant, N. and Kennedy, A. 2004. 'An investigation of the perceived needs and priorities held by African refugees in an urban setting in a first country of asylum'. *Journal of Refugee Studies* 17(4): 437–459.

Buscher, D. and Heller, L. 2010. 'Desperate lives: urban refugee women in Malaysia and Egypt'. *Forced Migration Review* 34: 20–21.

Calvani, D. 2003. 'Initial overview of the linguistic diversity of refugee communities in Cairo'. Forced Migration and Refugee Studies Working Paper, American University in Cairo.

Chasek-MacFoy, N. 2012. 'Refugees protest legal limbo in Egyptian border camp' *Daily News Egypt*. Available at: www.dailynewsegypt.com/2012/09/01/refugees-protest-legal-limbo-in-egyptian-border-camp/ (accessed 18 November 2012).

COFS 2011. 'Sudanese victims of organ trafficking in Egypt: A preliminary evidence-based, victim-centered report'. Coalition for Organ-Failure Solutions. Available at: www.cofs.org/english_report_summary_dec_11_2011.pdf (accessed 12 July 2012).

Coker, E. 2004a. 'Dislocated identity and the fragmented body: Discourses of resistance among southern Sudanese refugees in Cairo'. *Journal of Refugee Studies* 17(4): 401–419.

Coker, E. 2004b. '"Traveling pains": embodied metaphors of suffering among southern Sudanese refugees in Cairo'. *Culture, Medicine and Psychiatry* 28(1): 15–39.

Crisp, J. 1999. 'Policy challenges of the new diasporas: Migrant networks and their impact on asylum flows and regimes'. *New Issues in Refugee Research* 7.

Damanga Coalition for Freedom and Democracy 2005. 'Darfur refugees protest UN policies in Cairo, Egypt'. Damanga website. Available at: www.damanga.org/refugee_protest.html (accessed 22 January 2006).

Danielson, N. 2008. 'A contested demonstration: Resistance, negotiation and transformation in the Cairo Sudanese refugee protest of 2005'. Unpublished Masters thesis, University of Oxford.

Danielson, N. 2012a. 'Field report: Revolution, its aftermath, and access to information for refugees in Cairo'. *Oxford Monitor of Forced Migration* 2(2): 57–63.

Danielson, N. 2012b. 'Urban refugee protection in Cairo: The role of communication, information and technology'. *New Issues in Refugee Research* 236.

Danielson, N. 2013. 'Channels of protection: communication, technology, and asylum in Cairo, Egypt'. *Refuge* 29(1): 31–42.

Das, V. 1995. *Critical events: An anthropological perspective on contemporary India.* Delhi: Oxford University Press.

Demographia 2012. *World urban areas and population projections,* 8th edition available at: www.demographia.com (accessed 2 April 2012).

Egypt Independent 2012. 'Several arrested as sexual harassment surges in Cairo'. *Egypt Independent.* Available at: www.egyptindependent.com/news/several-arrested-sexual-harassment-surges-cairo (accessed 22 August 2012).

Eidenier, E. 2006. 'Providing health care information to refugees in Cairo: Questions of access and integration'. Panel paper on access to health care, American University in Cairo. Available at: www.aucegypt.edu/GAPP/cmrs/reports/Documents/Eidenier.pdf

El Nadim Center 2006. 'Testimony 9', Sudanese testimonies to El Nadim Center. On file with author.

El-Shaarawi, N. 2012. 'Living in an uncertain future: An ethnography of displacement, health, psychosocial well-being and the search for durable solutions among Iraqi refugees in Egypt'. PhD dissertation, Case Western Reserve University. Available at: http://etd.ohiolink.edu/send-pdf.cgi/ElShaarawi%20Nadia.pdf?case1325709084anddl=y (accessed 15 May 2012).

Ensor, M. 2010. 'Education and self-reliance in Egypt'. *Forced Migration Review* 34: 25–26.

Fanjoy, M., Ingraham, H., Khoury, C. and Osman, A. 2005. 'Expectations and experiences of resettlement: Sudanese refugees' perspectives on their journeys from Egypt to Australia, Canada and the United States'. Forced Migration and Refugee Studies report, American University in Cairo. Available at: www.aucegypt.edu/GAPP/cmrs/reports/Documents/resettlement-final-edited___000.pdf (accessed 14 March 2011).

Farag, M. 2011. 'Cairo refugees and the revolution'. Student Action for Refugees presentation, 23 March, American University in Cairo, fieldnotes on file.

Fiddian-Qasmiyeh, E., Loescher, G., Long, K. and Sigona, N. (eds) 2014. *The Oxford Handbook of Refugee and Forced Migration Studies.* Oxford: Oxford University Press.

FIDH 2011. 'Exiles from Libya flee to Egypt: double tragedy for sub-Saharan Africans'. International Federation for Human Rights (FIDH) report 565a. Available at: www.fidh.org/IMG/pdf/libyeegypt565ang.pdf (accessed 4 August 2011).

Forcier, N. 2009. 'Divided at the margins: A study of young southern Sudanese refugee men in Cairo, Egypt'. Center for Migration and Refugee Studies paper, American University in Cairo. Available at: www.aucegypt.edu/GAPP/cmrs/reports/Documents/Divided_at_the_Margins_Forcier_2010_Final_for_Publication.doc (accessed 14 March 2011).

Furst-Nichols, R. and Jacobsen, K. 2011. 'African refugees in Israel'. *Forced Migration Review* 37: 55–56.

Ghazaleh, P. 2002. 'Two miles into limbo: Displaced Sudanese in a Cairo slum'. *Middle East Report* 225: 2–7.

Ghazaleh, P. 2003. '"In closed file" limbo: Displaced Sudanese in a Cairo slum'. *Forced Migration Review.* Available at: www.fmreview.org/FMRpdfs/FMR16fmr16.8.pdf (accessed 4 September 2011).

Giri, M. 2007. 'On contagion: Sudanese refugees, HIV/AIDS, and the social order in Egypt'. *Égypte/Monde arabe* 3(4). Available at: http://ema.revues.org/index1769.html (accessed 15 March 2011).

Gomez, A., Lewis, T., Rowe, M., Salih, A.K., Sander, L., Schafer, S. and Smith, H. 2005a. 'Sudanese refugees in Cairo: We'll wait here, we'll die here'. *Pambazuka News*.

Gomez, A., Lewis, T., Rowe, M., Schafer, S., Sander, L. and Smith, H. 2005b. 'UNHCR responds to the Cairo refugee sit-in: an official response'. *Pambazuka News*.

Gozdziak, E. and Walter, A. 2012. 'Urban refugees in Cairo'. Institute for the Study of International Migration. Available at: http://issuu.com/georgetownsfs/docs/urban_refugees_in_cairo/1 (accessed 22 November 2012).

Grabska, K. 2005. 'Living on the margins: Analysis of the livelihood strategies of Sudanese refugees with closed files in Egypt'. Forced Migration and Refugee Studies Working Paper, American University in Cairo.

Grabska, K. 2006a. 'The lessons of Cairo: A system of diffuse responsibility, with blame shared by all', RSD Watch. Available at: http://rsdwatch.wordpress.com/2006/06/16/forum-the-lessons-of-cairo-a-system-of-diffuse-responsibility-with-blame-shared-by-all/ (accessed 12 May 2010).

Grabska, K. 2006b. 'Marginalization in urban spaces of the global south: Urban refugees in Cairo'. *Journal of Refugee Studies* 19(3): 287–307.

Grabska, K. 2008. 'Brothers or poor cousins? Rights, policies and the well-being of refugees in Egypt', in K. Grabska and L. Mehta (eds), *Forced Displacement: Why Rights Matter*, London: Palgrave Macmillan, pp. 71–92.

Greenwood, P. 2012. 'Egyptian authorities look the other way as Bedouin kidnap refugees'. *Guardian*. Available at: www.guardian.co.uk/world/2012/feb/14/egypt-bedouin-kidnap-refugees-israel (accessed 4 August 2012).

Harrell-Bond, B. 2008. 'Protests against the UNHCR to achieve rights: Some reflections', in K. Grabska and L. Mehta (eds), *Forced Displacement: Why Rights Matter*, London: Palgrave Macmillan, pp. 222–243.

Human Rights Watch 2008. 'Sinai perils: Risks to migrants, refugees and asylum seekers in Egypt and Israel'. Available at: www.hrw.org/sites/default/files/reports/egypt-1108webwcover.pdf (accessed 13 March 2011).

Human Rights Watch 2010. 'Egypt: Guards kill 3 migrants on border with Israel'. Human Rights Watch website. Available at: www.hrw.org/en/news/2010/03/31/egypt-guard-kills-3-migrants-border-israel (accessed 2 April 2010).

IRIN 2008. 'Egypt–Sudan: Sudanese refugees face dilemma of return'. IRIN Humanitarian News and Analysis. Available at: www.irinnews.org/Report/78641/EGYPT-SUDAN-Sudanese-refugees-face-dilemma-of-return (accessed 18 November 2012).

IRIN 2011. 'Egypt: Refugees hit by discrimination, violence amid heightened nationalism'. IRIN Humanitarian News and Analysis. Available at: www.irinnews.org/printreport.aspx?reportid=94294 (accessed 19 April 2012).

Jacobsen, K., Ayoub, M. and Johnson, A. 2012. 'Remittances to transit countries: The impact of Sudanese refugee livelihoods in Cairo'. *Cairo Studies on Migration and Refugees* 3. Available at: www.aucegypt.edu/GAPP/cmrs/reports/Documents/paper per cent20No. per cent203.pdf (accessed 30 October 2012).

Jones, M. 2012. 'We are not all Egyptian'. *Forced Migration Review* 39: 16–17.

Jureidini, R. 2009. 'Irregular workers in Egypt: Migrant and refugee domestic workers'. *International Journal on Multicultural Societies* 11(1): 75–90.

Kagan, M. 2002. 'Assessment of refugee status determination procedure at UNHCR's Cairo office 2001–2002'. Forced Migration and Refugee Studies Working Paper, American University in Cairo.

Kagan, M. 2011a. 'Shared responsibility in a new Egypt: A strategy for refugee protection'. Center for Migration and Refugee Studies paper, American University in Cairo.

Kagan, M. 2011b. '"We live in a country of UNHCR": The UN surrogate state and refugee policy in the Middle East'. *New Issues in Refugee Research* 201.

King, N. 2012. 'Refugees fear return home'. Deutsche Welle. Available at: www.dw.de/refugees-fear-return-home/a-15924240 (accessed 18 November 2012).

Koser, K. and Pinkerton, C. 2002. 'The social networks of asylum seekers and the dissemination of information about countries of asylum'. Migration Research Unit, University College London. Available at: www.homeoffice.gov.uk/rds/pdfs2/socialnetwork.pdf (accessed 20 January 2013).

Lammers, E. 2011. 'Lives in transition'. *The Broker*. Available at: www.thebrokeronline.eu/Magazine/articles/Lives-in-transition (accessed 3 July 2011).

Le Houérou, F. 2004. 'Living with your neighbor: Forced migrants and their hosts in an informal area of Cairo, Arba wa Nus'. Forced Migration and Refugee Studies workshop paper, American University in Cairo.

Lewis, T. 2006a. 'Nothing left to lose? An examination of the dynamics and recent history of refugee resistance and protest'. Paper presented at the 4th Annual Forced Migration Post-graduate Student Conference, University of East London, 18–19 March. Available at: www.aucegypt.edu/GAPP/cmrs/reports/Documents/Lewis.pdf (accessed 15 September 2006).

Lewis, T. 2006b. 'Photo album, Sudanese Protest'. Sudanese Protest Flickr webpage. Available at: www.flickr.com/photos/37076007@N00/ (accessed 15 September 2006).

Lewis, T. 2007. 'Nothing left to lose? Protest, political action, and resistance strategies in exile: A case study of Sudanese Refugees in Cairo'. Unpublished dissertation, MSc in Forced Migration, Refugee Studies Centre, University of Oxford.

Lewis, T. 2011. '"Come, we kill what is called 'persecution life'"': Sudanese refugee youth gangs in Cairo'. *Oxford Monitor of Forced Migration* 1(1): 78–92.

Mahmoud, H. 2009. 'Disrupted lives and shattered dreams: Culture, identity, and coping pathways among Sudanese refugees in Cairo'. PhD dissertation, Faculty of Politics, Psychology, Sociology, and International Studies, University of Cambridge.

Marfleet, P. 2007. '"Forgotten," "hidden": Predicaments of the urban refugee'. *Refuge* 24(1): 36–45.

Maxwell, L. and El-Hilaly, A. 2004. 'Separated refugee children in Cairo: A rights-based analysis'. Forced Migration and Refugee Studies Working Paper, American University in Cairo.

Minnick, E. and Nashaat, N. 2009. '"Stuck" in Egypt: Iraqi refugees' perceptions of their prospects for resettlement to third countries and return to Iraq'. Center for Migration and Refugee Studies working paper, American University in Cairo, available at: www.aucegypt.edu/GAPP/cmrs/reports/Documents/CMRS_Reporton_Return_and_Resettlement_For_Web.pdf (accessed 2 March 2010).

Moghaieb, H. 2009. 'Strengthening Livelihood Capacities of Refugees and Asylum Seekers in Egypt'. UNHCR Survey Report: 30.

Moulin, C. and Nyers, P. 2007. '"We live in a country of UNHCR": Refugee protests and global political society'. *International Political Sociology* 1: 356–372.

Pascucci, E. 2012. 'Migration, identity, and social mobility among Iraqis in Egypt'. *Refuge* 28(1): 49–58.

Rizvi, Z. and Buscher, D. 2012. 'Shifting sands: Risk and resilience among refugee youth in Cairo'. Women's Refugee Commission. Available at: http://womensrefugeecommission.org/resources/doc_download/858-shifting-sands-risk-and-resilience-among-refugee-youth-in-cairo (accessed 20 November 2012).

Rowe, M. 2006. 'Performance and representation: Masculinity and leadership at the Cairo refugee demonstration'. Paper presented at the 4th Annual Forced Migration Post-graduate Student Conference, University of East London, 18–19 March. Available at: www.auc-egypt.edu/GAPP/cmrs/reports/Documents/Rowe.pdf (accessed 14 March 2011).

Rowe, M. 2007. 'The experience of protest: Masculinity and agency among young male Sudanese refugees in Cairo'. MA thesis, sociology and anthropology, American University in Cairo.

Sadek, S. 2012. 'Where are the refugees from Egypt's revolution?' *Egypt Independent.* Available at: www.egyptindependent.com/opinion/where-are-refugees-egypt's-revolution (accessed 6 September 2012).

Salih, A.K. 2006. 'Sudanese demonstration in Cairo: Different stands and different opinions'. Paper presented at the 4th Annual Forced Migration Post-graduate Student Conference, University of East London, 18–19 March. Available at: www.aucegypt.edu/GAPP/cmrs/reports/Documents/Assad.pdf (accessed 14 March 2011).

Samy, S. 2009. 'The impact of civil society on refugee politics in Egypt'. CARIM research report 2009/07. Available at: http://cadmus.eui.eu/dspace/bitstream/1814/11414/1/CARIM_RR_2009_07.pdf (accessed 15 March 2011).

Schafer, S. 2005a. 'Sudanese demonstrators in Mohandiseen: Who they are and why they're here'. Paper submitted in partial fulfillment of the requirements for the Graduate Diploma in Forced Migration and Refugee Studies, American University in Cairo.

Schafer, S. 2005b. 'Photo album, Sudanese refugees protest UN policies in Cairo, Egypt, 13 October 2005'. Fahamu and *Pambazuka News.* Flickr webpage. Available at: www.flickr.com/photos/93729649@N00/sets/1157901/ (accessed 15 September 2006).

Schafer, S. 2006. 'Solace and security at the Cairo refugee demonstration'. Paper presented at the 4th Annual Forced Migration Postgraduate Student Conference, University of East London, 18–19 March. Available at: www.aucegypt.edu/GAPP/cmrs/reports/Documents/schafer.pdf (accessed 14 March 2011).

Shafie, S. 2004. 'Egypt research guide'. *Forced Migration Online.* Available at: www.forcedmigration.org/guides/fmo029/ (accessed 14 April 2012).

Sigona, N. 2014. 'The politics of refugee voices: representations, narratives, and memories'. In *The Oxford Handbook of Refugee and Forced Migration Studies*, 369–385.

Sperl, S. 2001. 'Evaluation of UNHCR's policy on refugees in urban areas: A case study review of Cairo'. UNHCR Evaluation and Policy Analysis Unit 7.

SUDIA 2003. 'Developing pathways into work for Sudanese refugees: A study of the labour market and employment needs of Sudanese refugee communities in Cairo'. Sudanese Development Initiative EETS Program report.

UNHCR 2011a. 'UNHCR Statistical Online Population Database'. Available at: www.unhcr.org/pages/4a013eb06.html (accessed 12 April 2012).

UNHCR 2011b. 'UNHCR Egypt fact sheet (February 2011)'. Available at: www.unhcr.org/4d8216359.html (accessed 15 April 2011).

UNHCR 2012. 'UNHCR Egypt fact sheet January 2012'. Available at: www.unhcr.org/4f4c956c9.pdf (accessed 4 April 2012).

UNHCR RO-CAIRO 2006. 'Response to FMRS report: "A tragedy of failures and false expectations"'. June (on file).

Van Hear, N. 2003. 'From durable solutions to transnational relations: Home and exile among refugee diasporas'. *New Issues in Refugee Research* 83.

Viney, S. and Beach, A. 2011. 'Sudanese protests force UNHCR to close Cairo office'. *Al Masry Al Youm* English Edition. Available at: www.almasryalyoum.com/node/383443 (accessed 2 April 2011).

Yoshikawa, L. 2010. 'Iraqi refugees in Egypt'. *Forced Migration Review* 29: 54.

2 Casamance refugees in urban locations of The Gambia

Gail Hopkins

Introduction

Refugees from the Casamance region of southern Senegal are living in urban locations in The Gambia due to protracted, albeit low-level, conflict in Casamance which continues to present security challenges and periodic refugee influxes from Casamance across its northern border into The Gambia.

The Casamance refugee presence in The Gambia can be split into those who arrived before 2006 and those who arrived after. Pre-2006 arrivals are mostly from around Ziguinchor and were urban dwellers, whereas post-2006 refugees are from rural communities. In urban areas of The Gambia there is a mix of pre- and post-2006 refugees, but with post-2006 arrivals being the majority.[1] The post-2006 arrivals are linked to the Government of Senegal increasing its military presence in Casamance in an attempt to halt opposition activity in 2005/2006. The year 2006 is a point in time that is identified by The Gambian Immigration Department (GID) and by UN agencies such as United Nations High Commission for Refugees (UNHCR) and World Food Programme (WFP) as the moment when the current influx of Casamance refugees began arriving. It is also the point in time at which refugee identity cards were first issued to Casamance refugees in The Gambia by GID. It is for these reasons that the focus here is on post-2006 Casamance refugees.

Arrivals of Casamance refugees since 2006 comprise mostly those who have made rural to rural moves, fleeing Casamance border villages and settling in Gambian border villages.[2] Those who have made rural to urban moves often did so via Gambian border villages as a first port of call. Those who made the rural to urban move represent approximately 20 per cent of the total Casamance refugee population in The Gambia (conversation with GID, February, 2012).

Aims

This chapter explores why some post-2006 Casamance refugees relocate to urban areas, their motivation to do so and what keeps them there despite the clear hardships they endure. The chapter also explores how access to assistance is impacted by refugees' move to urban areas and the challenges faced by local and international agencies in supporting urban refugees. Showing there to be a considerable extent to which urban Casamance refugees become invisible to agencies, implementing partners and to the Government of The Gambia, the chapter discusses how this occurs, why it matters and the impact this has on urban refugee settlement and goals of self-reliance and basic needs provision, thus connecting with UNHCR objectives and with the current UNHCR policy toward refugees in urban areas (UNHCR 2009). As Kibreab pointed out in 1996, the 'invisibility' of rural refugees due to their remote location was replaced with the invisibility of urban refugees who became difficult to identify readily within the general urban melee and who were mostly assumed to have easy access to support. The unfamiliarity of the urban context and the challenges refugees face there prompts Sommers (1999: 24) to state: 'refugees who migrate to urban areas are actually a particularly vulnerable kind of urban migrant'. The situation Kibreab described persists today and it is important that refugees' invisibility within urban situations is recognized and understood so that strategies for adequate assistance and protection may be developed.

Experiences of urban Casamance refugees are presented here to reflect on the effectiveness of assistance and entitlements and to highlight barriers to 'success' of refugees achieving self-reliance and integration in a context of existing poverty. What is meant by self-reliance, integration and effectiveness is highly debatable and these terms have to be considered in relation to the context of The Gambia. UNHCR defines self-reliance as 'the social and economic ability of an individual, household or community to meet basic needs (including protection, food, water, shelter, personal safety, health and education) in a sustainable manner and with dignity' (UNHCR 2011: 15). The traditional approach to self-reliance by humanitarian organizations has largely been from a technical perspective, focusing on the effective design and implementation of initiatives such as income generation, micro-finance, agriculture and vocational training (Crisp 2003: 3). UNHCR follows a 'graduation approach', from grant-based capacity-building for the poorest households to targeting economically active individuals with existing skills to targeting individuals with a high capacity for success and integration (UNHCR 2011: 15). Such an approach may be over complex for rural-origin refugees and may require extensive outreach activities and substantial funding which may not be practical in smaller refugee situations such as Casamance and The Gambia.

For the purposes of this chapter, self-reliance is held to mean the ability of the refugees to support themselves and their dependants to a level that was previously achieved in Casamance, which was almost completely a subsistence farming lifestyle. This view of self-reliance is adopted because most post-2006 Casamance refugees were previously adequately supporting themselves prior to flight (refugee conversations, 2010–2012) and it is reasonable for them to wish to regain that position. Effectiveness of assistance here is held to mean the degree to which self-reliance and a rebuilding of life and life-chances are enabled by the policies of the Government of The Gambia and UNHCR, and the degree to which differences in individual needs (related to, for example, differing education and literacy levels or former employment skills) are recognized and met. 'Success' is therefore clearly subjective and variable according to the individual, and expectations or aspirations are likely to differ. Effectiveness or success here is not held to mean *numbers* of people assisted or *quantities* of food distributed, it is about (as confirmed by refugees themselves) the impact on lives and the appropriateness of assistance, whether that assistance is food, training, health or other interventions.

This research complements existing literature on the Casamance conflict itself (for example, Evans 2000, 2002, 2004, 2007; Home Office 2010; de Jong and Gasser 2005; Sonko 2004) and the internal and inter-agency reports, assessments, and evaluations which inform the focus and extent of assistance and interventions to post-2006 Casamance refugees in The Gambia.

The Gambia and Senegal

Geographically, The Gambia is a 300 kilometre strip of land following the river Gambia inland from the Atlantic coast of West Africa. Running through the middle of Senegal, The Gambia at its widest point is 48 kilometres and cuts Senegal almost completely in two, with the Casamance region of Senegal to the south (see Figure 2.1). With an area of 11,295 square kilometres (CIA 2012) The Gambia is West Africa's smallest state. From a population of 1.8 million (CIA 2012), 52 per cent live in urban areas.

Historically, The Gambia was a British colony until independence in 1965. The British established a settlement on James Island from 1661 as part of the slave trade. In 1816, the British bought Banjul from the Portuguese as a new base from which to enforce the Act for the Abolition of the Slave Trade of 1807. Throughout the nineteenth century, tribal conflict led various chiefs to seek protection from the British, and the British administration gradually extended inland

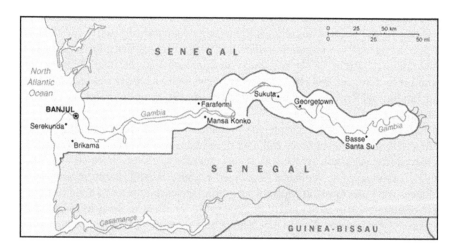

Figure 2.1 Map of The Gambia (source: CIA World Factbook: www.cia.gov/library/publications/the-world-factbook/graphics/maps/ga-map.gif).

from Banjul, until in 1889 an agreement on British rule was reached between the British Crown and the French who had colonized surrounding Senegal. This somewhat friendly colonialism of a state often considered historically an enclave of Senegal, could explain why, today, the British are generally welcomed in The Gambia as brothers and sisters and not as former oppressors.

Since independence in 1965, The Gambia has been a republic and follows a mixture of English common law, customary law and Islamic law. President Yahya Jammeh has been head of state since 1994, initially as chairman of the junta which took control in 1994, and latterly as president since 1996. He was overwhelmingly re-elected in November 2011 for a further term.

The Casamance conflict

The geographic separation of the southern Casamance region of Senegal from the north of Senegal and the capital, Dakar, where the Senegalese government is located, has produced some long-term tensions concerning marginalization of the south which are situated historically in an unkept promise for Casamance's own independence made at the time of independence in 1960 by the then President, Léopold Senghor.[3] At that time, a general feeling existed amongst Casamançais that Dakar had exploited the south's resources without a corresponding financial, material and infrastructural input for the south. As a result, a desire for independence from Senegal developed in Casamance, expressed at first by popular protest during the early 1980s and then by guerrilla war from the late 1980s and early 1990s (Evans 2002), continuing to date with intermittent incursions along the Casamance/Gambian border in the Foni districts, particularly since 2006. Discussions with refugee respondents demonstrate that feelings of exploitation and marginalization continue.

Not all the Casamance population wanted independence, but the general dissatisfaction and frustration regarding under-resourcing of the south generated support for the formation of the separatist Mouvement des Forces Démocratiques de Casamance (MFDC) in 1982, leading to armed conflict between MFDC and Senegalese military forces. Since this time, there have been numerous attempts at brokering peace deals;[4] the more recent attempts supported by President Jammeh of The Gambia and President Wade of Senegal, and mediated by Guinea Bissau and ECOWAS. These attempts failed to produce long-term peace and stability. Furthermore, over the years opposition forces fragmented into two main groups that further split into smaller splinter groups. The prospect of opposition forces speaking with one voice in any peace negotiation is therefore a fundamental challenge to lasting peace and the possibility of future refugee repatriation to Casamance.

The conflict in Casamance is the longest running conflict in Africa (Evans 2004), and has been characterized by sporadic fighting between MFDC and the Senegalese armed forces, by factional fighting, by MFDC-led attacks on villages considered to oppose MFDC aims for independence, and by Senegalese military-led raids on villages considered to harbour MFDC members. A review of

Gambian newspapers from 1982 to 2010[5] revealed peaks and troughs in fighting and in attacks on civilians and Casamance villages. Over a period of years, newspapers reported fighting during the various peace talks as a result of divisions within the MFDC itself and MFDC leadership battles. In 2006, the Government of Senegal increased its military presence in the region in an attempt to eradicate opposition activity. After an initial lull, this presence actually served to increase instability and spates of conflict.

The situation from 2006 to the current date is that there are a series of 'border bases' on the Casamance side, control of which alternates between the Senegalese military and one or other of the two main opposition forces. The majority of these bases were formerly homes of Casamance villagers. Especially since 2006, clashes occur not only between the military and opposition groups, but also between the two main opposition groups.

Since the increased presence in 2006 of the Senegalese military, the previous pattern of refugee flight and return – movement across the border to The Gambia to escape the fighting followed by a move back to Casamance again when the fighting stopped – has been largely replaced by a permanent movement into The Gambia. Although a minority of refugees – mostly men – do still return to their Casamance villages to gather fruit, farm or to attempt to rescue belongings, episodic flight and return can no longer be called a pattern. The permanence of refugee flight to The Gambia since 2006 has been recognized by UNHCR and the Government of The Gambia by the issuance of refugee identity cards to Casamançais in that year.

Refugees are reluctant to return to Casamance and are fearful of opposition groups and Senegalese forces, who occupy some villages and compounds having forcibly removed villagers by threat to their life, and appropriation or destruction of houses, land, crops and livestock which formed the core of village livelihoods. During one period of fieldwork, in January and February 2012, an instance occurred of incendiary bombs being dropped by Senegalese planes to disperse opposition forces and villages were destroyed by fire as a consequence (conversation, International Committee of the Red Cross (ICRC), February 2012). Refugee return is additionally prevented by fear of inadvertent association with opposition forces and consequent violence or punishment by the Senegalese military or other opposition groups.

Although still the most fertile area of Senegal, farming and livelihoods in Casamance have been disrupted by conflict-associated activities such as forced abandonment of villages, animals and crops, and by the laying of landmines.[6] The sense of the hopelessness of return is apparent when many refugees report that, even if the region was safe, their land has been spoiled and their homes and villages have been destroyed and would need complete rebuilding: an overwhelmingly daunting task for them.

The peaks and troughs of activity saw hostilities increase in 2009 (see Home Office 2010: 16)[7] and throughout 2010 with increased fighting in the border area in October/November 2010 (*Daily Observer* 2010). This activity continued throughout the period of fieldwork in 2010–2011, with fighting occurring most nights along the border running west from Gikess to Sibanor. This generated an influx of approximately 450 new refugees during the period mid-January to mid-February, 2011. At the end of February 2011, the Senegalese armed forces sealed the border between Gikess and Siwol in order to flush out opposition forces by preventing them escaping across the border to The Gambia. One result of this strategy was that Casamance civilians were trapped in the forest, having fled their homes trying to reach safety in The Gambia. Some managed to walk for several days through the forest to Gikess, entering The Gambia at that point. The border remained sealed for approximately three weeks, after which time other refugees arrived along the border in a highly distressed and malnourished condition, with some individuals missing (conversation with Gambia Food and Nutrition Association (GAFNA) field representative, 18 March 2011). Through this period and beyond, according to GID border records and conversations with UNHCR, by mid-March 2011 the number of new refugees had reached 700. By the end of 2011, 1,100 new refugees had arrived from Casamance (personal conversation, GID, March 2012). In January and February, 2012, the most recent period of fieldwork, approximately a further 650 refugees arrived.

Although the wish for independence has been accepted by many Casamançais as unattainable, and even expressed as undesirable by one MFDC leader, Commander Diatta (*Observer*, 9 August 2001), and although many Casamançais no longer support the MFDC, independence remains an underlying goal for many, and opposition forces continue their activities against Senegalese armed forces and against each other. As a result, the region remains unstable and unpredictable at the time of writing (December 2012).

Urban Casamance refugees in The Gambia

Gamblan refugee policy

The Gambia is a signatory to the 1951 Convention relating to the status of refugees and its 1967 Protocol, and to the 1969 Organization of African Unity (OAU) Convention Governing the Specific Aspects of Refugee Problems in Africa. Whilst UNHCR is present in The Gambia, it is GID which is legally responsible for the protection of refugees, and UNHCR and GID work closely together to this end. In 2008, The Gambia passed the Refugee Act (Government of The Gambia 2008), which sought to set out the parameters by which refugee populations would be managed. Prior to the Act there was no specific legislation relating to refugees and the Government of The Gambia managed previous refugee populations such as Sierra Leoneans, Liberians, and pre-2006 Casamançais according to the above Conventions. The major component of the Refugee Act (2008) was to establish The Gambia Commission for Refugees to take responsibility for coordinating and managing policy matters relating to refugee affairs in The Gambia.

The Act specifies who is and who is not a refugee but is somewhat simplistic in defining rights of refugees and is therefore open to interpretation and inconsistency

in implementation. In Sections 41 and 42, which refer to rights of refugees, the Act states only:

41 (1) A person who is duly recognized as refugee is entitled to:
  (a) be issued a Refugee Identity Card;
  (b) be issued a Convention Travel Document, if the refugee needs to travel;
  (c) engage in wage-earning employment or self-employment;
  (d) freedom to move freely within, and to settle anywhere in, The Gambia; and
  (e) access to social amenities available in The Gambia.

42 In cases where refugees arrive in large numbers, the Secretary of State may by an Order published in the *Gazette* designate refugee camps to house the refugees.

In particular, specifics of employment law and taxation related to employing non-Gambians (which are discussed later) are not sufficiently detailed within the Act and present areas of potential confusion which impact on refugees and their ability to regain self-reliance. As at late 2012, new legislation relating to refugees and asylum is being drawn up by GID and the Ministry of the Interior to support and implement the existing Act.

The 1979 ECOWAS Protocol Relating to the Free Movement of Persons, Residence and Establishment, may underlie the simplicity of The Gambia's approach to refugee management and the absence until 2008 of specific legislation. Freedom of movement across ECOWAS borders has, in practical terms, negated the need for provision relating to migration by ECOWAS citizens. The Protocol states 'Community citizens have the right to enter, reside and establish in the territory of Member States' with a valid travel document. The Protocol became fully effective in three phases, completing the third phase in 1994. The recent recognition by the Government of The Gambia of the need for specific refugee legislation marks an acceptance of the particularity of situations of forced migration.

Refugee camps existed in The Gambia until 2005 at Kwinella, Bambali, Sifoe, Kitti and Basse. They were not tented camps, but of brick construction resembling a housing estate (conversation GAFNA, June 2009): 'five star refugee camp' as one NGO staff member joked. Around 750 Casamance refugees lived in these camps in the late 1980s to 1990s, most in Basse. The camp at Basse also housed around 200 refugee arrivals from Sierra Leone from 1997 to 2003, and Basse and Bambali camp housed some Liberian refugees from 1999 to 2003.[8] Since refugees were not compelled to remain in camps, all camps closed when refugee numbers dropped due to decreased hostilities and due to refugees' gradual move to other rural or urban locations. There are now no refugee camps in The Gambia and there are no designated 'refugee' areas.

UNHCR and GID considered reopening camps at Sifoe and Kitti in 2006 at the time of the increased tensions, but were reluctant to place refugees near these

border areas in case the existence of camps near the border encouraged incursions from armed opposition forces into The Gambia or created the possibility for those forces to live in the camp under the guise of refugees, giving them a point from which to attack Senegalese military bases across the border (conversations, GID November 2010, and UNHCR November 2010). The Government and UNHCR suggested sending Casamance refugees to Bambali in 2006, but as this is on the North Bank, refugees refused on the basis they would be too far removed from family and villages of origin (conversation, UNHCR November 2011).

Crisp (2004: 2) points out that during the 1960s–1980s, many African countries receiving refugees gave them land upon which to live and farm, with an expected outcome of local integration. This position towards refugees altered in the 1990s–2000s and, as Fábos and Kibreab (2007: 3) highlight, it is now typical amongst countries of the south to exercise spatial segregation of refugees, which 'is seen as an important instrument of preventing refugees' integration into host societies by prolonging their refugee status.' Citing Sudan and Egypt as examples of countries that actively regulate the presence of refugees in urban locations, they add that this strategy of *preventing* integration is undermined if refugees are allowed to settle freely in urban areas. Crisp (2003: 3) comments that many of the world's refugees are unable to re-establish independent livelihoods because they cannot exercise rights to which they are entitled under international human rights and international law, and have limited freedom of movement, no access to land or the labour market, and lack legal status, residence rights and documentation.

The Gambia differs in this respect and continues to practice integration as a preferred method of managing refugee influxes, and registered refugees in The Gambia receive all that Crisp cites as lacking in other cases. By supporting free movement, The Gambia allows refugees to wait for a durable solution to their plight (Conway 2004). Rather than attempting to solve the 'problem' of refugees, The Gambia to a large extent leaves refugees to solve their own problems related to displacement by giving them some basic support and freedom of movement. The Gambia does not regulate refugees' movement at all once they hold a refugee identity card, whether in rural or urban locations, and they are free to settle and work where they wish and to move unhindered within The Gambia. Without the card, they are treated as any other migrant and are required to purchase an annual residence permit and alien card. Even if through language or ethnicity it is clear to immigration and police officers that the individual is a refugee, identity documents are required to enable free movement. Many urban respondents did not have refugee identification and had acquired by some means Gambian documents. Most did not realize the benefits of having a refugee card, such as free education and health care.

Following the argument of Fábos and Kibreab (2007), in the Gambian context therefore, one of the major obstacles to refugee integration is removed. With a policy of integration, the Government of The Gambia provides an example of a southern receiving country where refugee segregation is neither enforced nor encouraged and where registered refugees enjoy the same legal, civil and human rights as Gambian nationals (see Harrell-Bond 2010: 13). As far as enshrining these rights in law, the Refugee Act implies equal civil rights such as rights to work, to move freely and to social amenities in The Gambia but currently goes no further than this. Refugees, like other immigrants to The Gambia, are permitted to work in the informal sector and to buy land and property. However, if they wish to work in the formal sector or to start a formal business, they must acquire the same work or business permits as other migrants. This is a subject of much heated discussion amongst refugees who say this amounts to having no right to work.

The liberal government policy toward migrants and refugees in The Gambia can perhaps be explained as simply one of practicality and logistics: it is less onerous for the government to allow registered refugees free movement and the right to work than to designate and maintain refugee areas. This stance extends to all refugees, but in the case of Casamance refugees it perhaps also has roots in the close relationship between Senegal and The Gambia and the porous borders that facilitate easy access to nearby villages and extended families existing on either side of the border. Some settlements straddle the border and in the past have passed from Gambian to Senegalese control and back again. Cross-border marriages are a frequent occurrence and on several occasions I met Gambian women who had married Casamance men in nearby villages across the border. Having moved to Casamance upon marriage, they had fled back to The Gambia with their husband and family when conflict occurred. Due to the geographic situation of The Gambia within the larger boundaries of Senegal, family, tribal and social networks have developed almost as though no border exists. Porous borders and the fact that many of the current refugees have lived and traded cross-border for many years may partly explain a policy of integration in the case of Casamance refugees, since Gambians and Casamançais have much in common already. In some sense, it could be argued, moving across the border is simply a permanent enactment of a previously oft-repeated transient lifestyle, the difference clearly being that possessions and livelihoods are left behind permanently in a context of refugee flight, creating emotional, financial and material hardship and vulnerability.

This does not explain the liberal policy toward other refugee groups such as Sierra Leoneans and Liberians, and the case of Gambian policy toward refugees is not easily explained in these cases except as a survival of the mid-twentieth century approach to refugees that Crisp (2004) points out, and a liberal attitude inherited from the British.

Assistance and programming

UNHCR and other UN agencies are present in The Gambia and focus most of their attention on rural areas, as do other humanitarian organizations in the country. Assistance to post-2006 Casamance refugees has included food assistance and material items such as sleeping mats and mosquito nets. Food assistance ended in summer 2010, apart from emergency interventions for the January/February 2011 influx, for example by ICRC in March 2011 and by UNHCR in May 2011. Current, long-term assistance for all registered refugees takes the form of educational costs, free health care, and waiving of annual residence and alien card cost. Skills training such as soap-making and tie-and-dye for women is provided in rural and urban areas, in addition to some vocational training for both men and women in urban areas such as hairdressing for women, and mechanic, electrician and plumbing training for men. IT skills courses ran for three months before funding ended. There have been some capacity building projects for rural communities such as the setting up of a bakery in Sibanor that initially foundered due to lack of business knowledge (conversation with Alkalo,[9] bakery cooperative members and GRC, April 2011) but has now been restarted. Access to micro-credit to start small businesses was reported by refugee respondents to be available only to Gambian nationals, but a GAFNA micro-loan initiative specifically for refugees established in cooperation with the National Cooperative Credit Union of The Gambia (NACCUG) existed from 2010 until June 2011, when it was withdrawn due to widespread non-repayment by refugees. Those refugees who were aware of this facility expressed dissatisfaction because frequently only part of the requested loan was made which impeded the success of the proposed project. The result was that the business did not start and the loan was spent on basic needs resulting in non-repayment. The refugee identification card issued by GID provides access to all these forms of assistance.

Refugee dependance on Gambian host households or family members brings additional responsibility for the receiving household and from the post-2006 influx onwards, poverty and vulnerability were seen to increase amongst receiving rural households in terms of food security, health and housing (GAFNA/GRC conversations June 2009). As a result, UNHCR and WFP intervened to provide support in rural areas, not only to refugees but to host households in recognition of their own consequent vulnerability. Distributions occurred only in rural locations with none in urban areas. Whole families were required to travel to the village of registration to enable collection of their full supply based on presence of family members. Urban-based refugees and all family members who had initially registered in the village were required to travel for their supply. The time and cost of the journey meant that many urban refugees did not do so.

Assistance was never intended to be long term (conversation with WFP, April 2011) and during 2010 there was a significant scaling down of donor assistance to Casamance refugees in The Gambia. Since then, food aid and ongoing material assistance has been phased out in favour of livelihood programming and skills training from UNHCR via GAFNA. Although occasional emergency responses assist highly vulnerable new arrivals in rural locations, the withdrawal of regular assistance has impacted refugees and hosts alike, particularly in the area of food security and particularly in urban areas (conversations with local NGOs and refugees, May–July 2010). Distribution to rural refugees of seeds, farming tools and animals has become an annual occurrence as harvests fail to produce good crops, and is drawing UNHCR into a maintenance pattern. There is no comparable assistance for urban refugees. Urban refugees have rights to skills training, which is discussed later, but which presents refugees with barriers relating to access.

Location and numbers of refugees

Casamance refugees are the largest group of refugees in The Gambia: 10,847 persons of concern from Senegal resided in The Gambia in August 2014 according to UNHCR data. Other much smaller refugee populations are from Sierra Leone (135) and Côte d'Ivoire (252). As at August 2014 the total population of persons of concern was identified by UNHCR as 11,439.[10] Sierra Leoneans are no longer considered refugees by the Government of The Gambia or UNHCR but still identify themselves as such. Similarly, cessation[11] was declared for Liberian refugees on 30 June 2012, at which time they were offered assisted voluntary repatriation or permanent local integration in The Gambia with work and residence permits being issued free of charge for the first two years.

According to GID registration data, there were 7,890 registered refugees from Casamance in The Gambia at the end of 2007 (UNHCR–WFP 2009). The number had risen to 8,241 in March 2010 (UNHCR 2010). Of this, the number of registered refugees in rural areas of the Foni districts (running east from Kafuta on the map below) had fallen slightly (7,290/6,494), whilst the number in urban areas had risen (600/1,747).

Refugee data are gathered by GID and GAFNA officers who constantly log new refugee arrivals at border villages. Between mid-January 2011 and mid-March 2011 approximately 700 new refugees arrived in the Foni border areas. By the end of 2011, 1,100 new refugees had arrived. During January and February 2012, there were around 650 new Casamance arrivals. The increase from March 2010 (8,241) to August 2014 (10,847) demonstrates the continued generation of new influx refugees from the Casamance region. The current split between urban and rural presence is not readily available.

Refugee figures are under-recorded because some refugees remain unregistered due to their absence from the village at the time of the rural registration exercise, such as those visiting family, those who were in the bush gathering firewood, those travelling for health care. These figures also exclude those who go direct to Gambian urban areas, as there is no system to log them unless they pass through a rural registration point or settle temporarily in a border village, or unless they present themselves at the GID office in Banjul for registration. Official figures for registered Casamance refugees include those who arrived from the early 1980s until the present time, but records make it difficult to state accurately how many are pre-2006 and how many are post-2006. In fact, pre-2006 refugees are considered as integrated by GID; however, they are issued with refugee identity cards and therefore form part of the total. Estimates by local donors and NGOs of the total number of Casamance refugees (registered and unregistered) living in The Gambia was thought to be between 11,000 and 12,000 at the time of the most recent fieldwork (conversations with GAFNA and

Map of Foni District of The Gambia: Road Network & Border Settlements

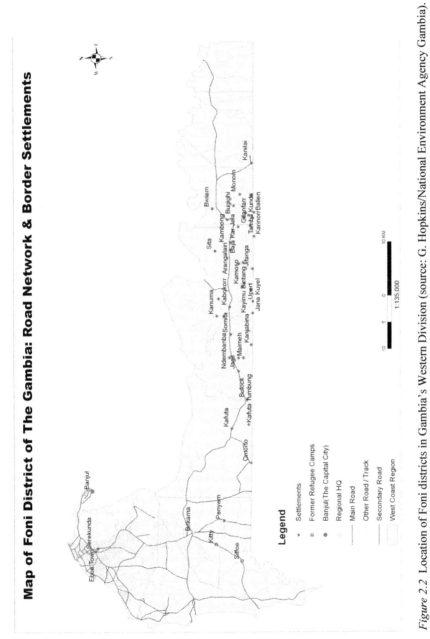

Figure 2.2 Location of Foni districts in Gambia's Western Division (source: G. Hopkins/National Environment Agency Gambia).

Gambian Red Cross (GRC), February, 2012). This estimate is likely to have increased to around 13,000–14,000 at the time of writing given the level of new influx refugees in that period.

Usually represented in reports and evaluations by aid agencies as subsistence farmers by origin, this is true of almost all post-2006 Casamance refugees with only a few exceptions, whether they are currently residing in rural or urban areas of The Gambia. The characteristics of the urban refugee population is therefore largely of farming background, with limited, if any, formal education or literacy skills, particularly amongst women and girls, and lacking in urban-relevant work skills. Those who have attended school have often had, and continue to have, interrupted education due previously to the conflict and currently to financial constraints, and therefore have not attained a high academic standard.

Settlement

The main pattern of settlement for post-2006 refugees arriving in The Gambia is to stay with host families at border villages, or with urban or rural family members if they can locate them (Hopkins 2011). Stages of settlement usually entail moving several times until refugees find family or until they are able to secure independent housing of some type. As the majority originate from Casamance villages, many, initially at least, choose to remain in villages on the Gambian side of the border. For these, movement is therefore rural to rural. For others, a rural to urban move may follow to find work or to improve housing conditions. A smaller number of refugees flee directly from rural Casamance to urban areas, bypassing the host villages. A handful of these may be urban dwellers originally but refugees usually choose this route when they have known family members in urban centres, albeit extended family. The result of this pattern of settlement is a highly mobile refugee population, particularly in the very early stages of flight but also throughout the first few years as family connections are discovered.

Most Casamance refugees now located in urban areas reside within 30 minutes travelling distance of Serrekunda, with the capital Banjul in one direction and Brikama in the other direction (see Figures 2.3 and 2.4). Serrekunda and surrounding areas are the most populated areas, rather than Banjul which is a commercial area and port. There are no Casamance refugees currently registered as living in Banjul and respondents had no knowledge of anyone living there.

Urban refugees organize themselves to a limited degree, with refugee leaders acting as conduits for information between refugees and international organizations, their local partners, and GID; however, the effectiveness of this is significantly curtailed by distance, cost of transport and mobile phone credit. This results in many refugees either having no contact with a refugee leader or maintaining only occasional contact with a rural leader in their village of first arrival. The consequence of this is that information does not effectively reach many urban Casamance refugees, and those involved in their assistance face barriers to maintaining effective contact.

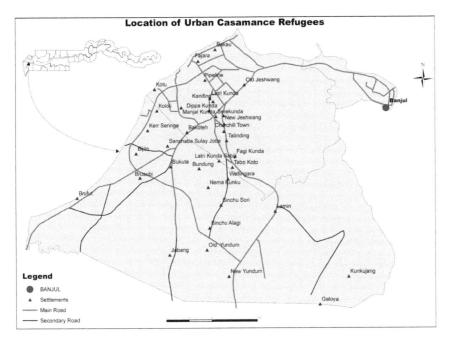

Figure 2.3 Location of urban Casamance refugees (source: G. Hopkins/National Environment Agency Gambia).

Methodology

This research began in June–July 2010 with a pilot study that revealed the ongoing but under-reported impact of the Casamance conflict on refugees and on The Gambia as a host state. Fieldwork conducted from November 2010 to February 2011 focused on Casamance refugees in rural areas of The Gambia (see Hopkins 2011). Further fieldwork in May–June 2011 focused on those in urban areas. This chapter is therefore based on data gathered during each of these phases but on urban interviews conducted in May–June 2011.

Interviews were held with Casamance refugees settled in urban areas of The Gambia and with their refugee leaders. In addition, discussions were held with staff members of international and national agencies working with refugees, such as UNHCR, WFP, GRC, GAFNA and GID from June 2010 to the time of writing.

The locations and numbers of urban Casamance refugees are as reported in the UNHCR Assessment March 2010, and are further informed by conversations with UNHCR Gambia and GAFNA. The assessment document was used to select areas for interviews and access to respondents was through refugee leaders, snowballing and through family members previously met in rural areas. The rationale for selection was to cover a wide geographic area of urban locations rather than focusing

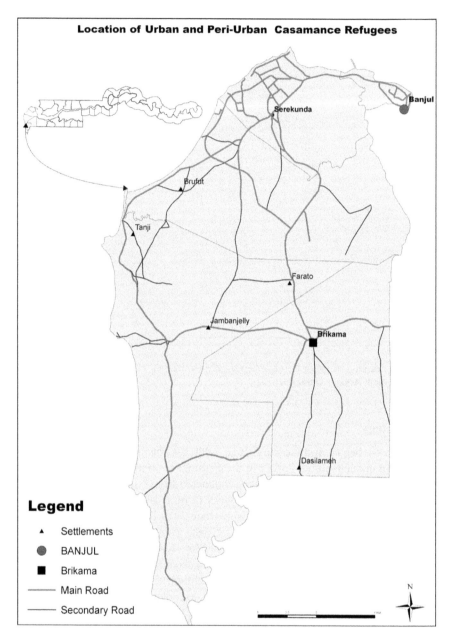

Figure 2.4 Location of urban and peri-urban Casamance refugees (source: G. Hopkins/ National Environment Agency Gambia).

on one town or area. This was more practical because areas tend to sprawl one into the other rather than having distinct beginnings and ends. Furthermore it was anticipated that focusing on one area would prove problematic because a researcher's presence would quickly become known by the Casamance community and they might have questioned why some and not others were 'selected' as respondents. Due to proximity of areas, some areas were unintentionally combined when members of one household joined the interview of another. In other cases, an area was documented to have a refugee presence but on attempting to contact those refugees they were found to have moved on.

Semi-structured interviews formed the basis of the methodology and comprised individual interviews, small groups of men and women, but mainly interviews at household level. Twenty-two interviews were held, with a total of approximately 60–70 respondents. There were no interviews with one gender only, as it was common during discussions for people to wander in and out of the house, an occurrence that was difficult to control politely. It was also common to find two different households present in the compound because word had travelled that a researcher was visiting. General interaction and participant observation supplemented by cultural detail from the translator complemented interviews. Discussions with staff members of international and national agencies and departments working with Casamance refugees formed a background to the research in terms of the history of aid provision and the present and projected presence of assistance. Women were encouraged to speak out and to put forward their experiences but in fact did not appear reluctant to talk once it was commented that they were 'keeping very quiet' – seemingly giving them permission to speak.[12] Men willingly allowed women to speak once the differences between 'men problems' and 'women problems' were highlighted and acknowledged.

All interviews and group discussions were recorded with the permission of respondents.[13] A male and female research assistant acted as translator where necessary and appropriate. Interviews were mainly in Jola or Mandinka using the translator, or occasionally in English or French or a combination. For the sake of anonymity, names of respondents have been altered.

'We are just managing': life in The Gambia

Fieldwork confirmed the majority of post-2006 urban Casamance refugees in The Gambia had almost exclusively relocated from rural settings in Gambian border areas. The urban population of post-2006 Casamance refugees therefore consists almost entirely of arrivals who are adapting to urban living.

This poses challenges beyond those such as orientation, loss of networks, possible language and ethnic barriers (UNHCR 2009), and entering an unfamiliar employment environment, all of which could be expected in other forms of migration, but which are emphasized in forced migration due to the unpreparedness of flight. But adapting to urban living poses additional challenges which may include a lack of, or limited, formal education, inappropriate skills for an urban setting, an unfamiliarity with urban living, and general psychological and physical trauma. This becomes increasingly important to understand as more and more refugees are located in urban areas rather than camps (Jacobsen 2006).

One initial question I asked refugees was why some are settled in urban areas and not in Gambian border villages, where donor assistance and programming has been, and is still, more focused and where, overall, it may be expected that they are more 'at home' in terms of environment, and more able to follow a farming lifestyle.

Challenges and frustrations as expressed by urban Casamance refugees focused largely on housing and on employment or livelihoods in the rural area, and therefore inevitably on food, and combined to create the major driving force to urban living. Secondary to these were anxieties over children's education and health. Relinquishing the debatable 'security' and support of the various assist-ance programmes in rural areas in exchange for the potential in urban areas, appeared to those refugees who relocated to be a gamble they were willing to take.

Rural refugees' initial motivation to move to urban centres varied and included finding family members who would be able to assist them with improv-ing their housing situation, seeking employment to support themselves and their family, and in a few cases seeking secondary education. Once in the urban area, wider concerns developed over children's education and over health and added frustrations to everyday life. Improving accommodation was the primary driving force for refugee families, and employment was the primary driving force for individual refugees. The daily challenges of life in the village as expressed by refugees (Hopkins 2011) – food, housing, livelihoods, health, education, more or less in that order of priority – both motivated refugees to move to urban centres and also followed them there, as fieldwork showed. However, fieldwork revealed that misunderstandings, lack of clarity, poor circulation of information, possible employment discrimination and lack of resources of national and international organizations all play their part in forming barriers to refugees achieving suc-cessful integration and regaining self-reliance in the urban setting.

Housing

It is typical for a refugee family arriving in the rural areas to stay with Gambian host families (Hopkins 2011). This may be short term or may be for several years. Family connections, however tenuous or distant, provide the possibility of moving on to secure accommodation, or at least accommodation that felt relatively more secure than living with a host family, where the rules and habits of another house-hold must be observed and where feelings of resentment could build on either side. These two brothers living on a relative's urban compound explain,

> It's not easy because every time you're in [the hosts] control. Also why we move from Siwol [village] – when we were in Siwol and the children touch their mangoes, they say "you see your children are spoiling the mangoes, so talk to your children and stop this". So you see, here [Brikama] is not our

own compound but it is like ours. So we can plant the mangoes and nobody will disturb us.

(Sene and Ebrima Jerju, Brikama)

If a family connection was discovered in an urban area the refugee may move in that direction. Initially the individual or family may live with the urban-based relatives, later finding their own accommodation. In other cases, a relative owned, or knew of, an empty compound in the town that could be occupied rent-free, sometimes short term, sometimes long term. However or whenever the rural to urban move was made, it was typical to be facilitated by an already urban-based relative.

A common strategy for obtaining free housing for Casamançais in urban Gambia is to act as unpaid watchmen for a part-built compound. As payment, the refugee family would have permission to stay, often long term, on the compound. The conditions of such living arrangements varied. In some cases the basic outer structure and roof would be built but walls would remain unrendered, in other cases only the outer compound wall and a watchman's hut would be built. In either case, refugee families would create makeshift additional structures to house the family and provide cooking and bathing areas.

Since land away from central urban areas is cheaper and building a house may occur over a span of years due to cost, there are many plots of land and part-built compounds in urban areas and on the outskirts or in peri-urban areas owned by Gambians, and these attract refugee families using the above watchman arrangement. Casamance refugees are therefore spread over a wide urban area in the Kanifing locality and the peri-urban area.

Refugees have the right to own a house and to buy land in The Gambia. Living on a part-built compound rent free can enable a few families to save a little money to buy land in peri-urban or less favourable urban areas to start building their own house. However, this is unusual and difficult to achieve due to available finances, and only one respondent family had collectively already bought land in Brikama with the intention to build gradually. For most, owning their own land or house in the urban area remains an aspiration in the face of employment challenges, education responsibilities and family needs. In the overwhelming majority of cases, such hardships prevent the most fundamental indicator of integration being achieved: owning one's own home (Hopkins 2010: 529; Murdie 2004). As Zetter and Pearl (1999: 2) argue,

> Housing is a key resource in the resettlement of asylum seekers and refugees themselves. The security, shelter and personal space which housing provides are vital elements in the process of regaining the dignity and independence often denied to them through persecution, incarceration and torture in their countries of origin.

Having a secure and stable home is a valuable asset in readjustment and recovery.

If you are staying on your own compound you will be free, no-one will disturb you.... But here, if the wife did not take care of the compound, when the owner of the compound gets in, what will he say? "Look you did not take care of my compound, will you please leave". So that idea will make us unhappy.

(Mustafa Jammeh, Lamin)

Security of tenure is not only related to owning one's home. It is evident where refugees act as watchmen and where they rent on a room and parlour basis and can relate to fear of eviction and to quality of housing. This man explains the anxiety around acting as watchman:

He's not renting it, he is living free. It's just given to him to stay here and look after it. When the owner of the compound is ready, he will just come and say, 'look, I am ready to build my compound.' So when the [owner] takes the compound he's in trouble.

(Sekou Badjie, Jabang)

Living with a host family and even living in rent-free accommodation on a part-built compound, security of tenure is always a concern for refugee families, as Mustafa and Sekou explained: at any time, the landlord may decide to complete the compound and ask the family to leave and find another home. Others who rent a room and parlour have similar concerns regarding security should they fall behind with rent or should the landlord tell them to leave. In addition, overcrowding is often an issue in a room and parlour, with a husband and wife plus three or four children sharing two rooms typically each barely three metres square. Such accommodation will often be situated on a compound with seven or eight similar homes, a small amount of shared outside space for domestic washing, a tiny cooking area and one shared water tap. Depending on location, these two rooms may cost 400–600 dalasi per month (£10–15).[14] In a precarious and uncertain housing situation, it is challenging for refugees to regain the dignity Zetter and Pearl (1999) advocate.

Furthermore, it is not simply a matter of finding a house or a room in the urban area and affording the rent; in The Gambia landlords rarely repair housing of this type and the tenant is often expected to do so. In this way, landlords are able to exploit vulnerable tenants who have little option but to do what the landlord demands. Although this is less relevant with part-built compounds, repairing a room and parlour can be a drain on limited resources and may mean rooms are not watertight for the rainy season. On several occasions evidence was shown of water ingress and structural damage from last year's rains, such as partially collapsed walls making some rooms uninhabitable especially during rainy season. Whilst this situation is not unique to refugees in urban areas and many Gambians face similar pressures, it does add to refugee vulnerability in a location where they lack an extended family network.

Even though the initial motivation may be to improve the family's housing situation, moving to the urban areas itself poses housing problems and in many urban cases situations of overcrowding will follow families from the rural host setting. Small or damaged accommodation, paying rent, paying for water, no space to grow food to make small money at market would appear a discouragement from making the rural to urban move. However, the potential for employment that exists in urban and peri-urban areas is, for some, worth pursuing in its own right, but for many is especially so when combined with independent living space which equates to *relative* security of housing, and potentially larger living space.

Employment

> When you come from a different place – he said an example is an old man, he comes to The Gambia and lives in Serrekunda, he can't do farming because there is no space. People have their feeling of barriers and that forces you to look into a job that means you don't earn [much money]. Me, I look for work as a watchman. That's why he decide the women go into soap making to make some income. That will relax your mind to forget your situation here and also bring you some earnings.
>
> (Alieu Jerju and family, Kunkujang)

One major barrier to livelihoods in urban areas or obtaining employment is that the majority of post-2006 Casamance refugees have a farming skill-set and farming expertise has limited use in urban areas. Those refugees who live on part-build compounds may grow small amounts of cassava and benefit from mango or papaya trees on the compound for their own use and for re-sale. Those who do not live on a compound have no opportunity to use their farming skills and must buy all their food. Refugees are therefore compelled by necessity to find alternatives to farming, and in the urban areas engage in a variety of income generating activities, some more successfully than others.

During interviews, a family member's absence from the house would often indicate they were at some type of work or actively looking for work. Their work is largely casual and unpredictable in frequency. Men may secure a small contract for making mud bricks or collecting and delivering water. Others would travel to the villages to collect and bring firewood. A small number owned or borrowed a donkey and cart and moved items for a fee. Some worked as night watchmen. Women would take in washing, some would sell firewood brought from rural areas by the men; others would buy vegetables at market and re-sell them in their local area. Some girls had employment as maids. It was difficult to gauge average monthly income as most work was not regular or predictable, apart from those who worked as watchmen and maids who, respectively, would earn 1,500 and 1,000 dalasi each month (£33 and £22). Men without work would often walk all day visiting compounds looking for work.

> Block building is everywhere. You go round in the morning to find work. So at the end of the day you could be lucky and go home with 200 or 300

[dalasi]. If you get that, in a week definitely … it's even easier than making charcoal in Bulock [village]. The way of managing things is easier here even though it's more expensive. It's just according to how you manage…. Definitely to stay here is easier than to stay in Foni [village area]. Here, you can manage and just have a small contract where you can get money. In Foni you can say let me just go to the bush, when you bring the firewood sometimes it will stay there, nobody will buy it. And the charcoal, you can [make it] but it will take time [to sell]. But here, no matter what it is, the day after you do the contract, when you are coming back they will pay you. Even if you have 50 butot [one penny] – for a man, 25 butot you can eat with it and keep the other 25 butot.

(Mustafa Jammeh, Lamin)

Even considering the pressures of urban living, respondents such as Mustafa felt the town was preferable to the village where there is an almost complete absence of opportunity for employment and where farming, collecting firewood or making charcoal can be financially unproductive (Hopkins 2011).

For men, their financial focus was rent and rice. For women, their responsibilities were broader and they saw the benefit of urban living as providing an opportunity for business, although that goal was often frustrated. Asked what would be the two most important things for them, the women replied:

Interpreter: They are debating. She says food number one. But she says money, 'cos if you don't have money where will you have money to buy fish? They decide money is the second thing, you can buy books for your children to go to school…. If they have the money they help each other to make a business. So you can go to the shopkeeper, buy oranges, bananas, come and sell it again to gain profit. That's why she said money. Something to start a small business.

(Mustafa Jammeh's wife, Lamin)

Another major barrier to obtaining employment in urban areas is a lack of, or limited, formal education amongst those refugees originating in rural areas. Most respondents over 35 years had no formal education at all. This was the same for both men and women, with men marginally more educated. Those in their twenties had some primary education, but as they had formerly been preparing for a farming life, it seemed continuing schooling had not been prioritized by their families. Lack of literacy skills results in refugees being forced to seek employment at the bottom end of the employment chain with little hope of advancement.

However, even for those refugee children now being raised in The Gambia and attending school, a completed formal education may not lead to jobs, as unemployment is high amongst the Gambian population itself. This relates to a third barrier to employment success for urban refugees: actual or perceived prejudice from employers toward refugees combined with misunderstandings or misinformation around refugees' legal entitlement to work in The Gambia.

Throughout interviews, understandings of entitlement to work or self-employment in The Gambia as a registered refugee seemed confused. Initially, I had been advised by those working with refugees that registered refugees had an entitlement to work which appeared to put them on a par with Gambian nationals. Refugee respondents insisted they did not have the right to work. After persistent investigation and questioning, I was told that registered refugees have the right to work freely in the informal sector, but not in the formal sector. Advertised positions in government and some organizations are open only to Gambian nationals, but this is not the case for many other jobs. However, all non-Gambians working in the formal sector are required to have a valid residence/work permit and alien card and this applies also to refugees. The work permit is a tax usually paid by the employer and in the case of ECOWAS nationals is 15,000 dalasi (£375) per year. Given the high unemployment in The Gambia, there is considerable competition for salaried jobs and an employer is reluctant to employ a non-Gambian as they bring a tax liability with them.

However, further confusion and perception of prejudice arises due to the general belief amongst refugee respondents that refugees are not permitted to work in The Gambia at all and that the annual tax payable by the employer is 40,000 dalasi (£900) – the tax level for non-ECOWAS nationals. Refugees reported that when they apply for a job the employer turns them down because of the required 40,000 dalasi. Either this myth has evolved due to misunderstandings spreading amongst the refugee community, or some employers have used this as a way to justify not employing refugees. The answer is not clear from interviews with refugees. UNHCR, GAFNA, GRC and GID each expressed with frustration that refugees were informed 'again and again' that registered refugees have the right to work (conversations UNHCR, GAFNA, GRC, March and June 2011). The stumbling block seems to be misperception of the employment tax level and this is perhaps an area in which UNHCR and GID could further sensitize employers and refugees alike as some refugees are unable to use existing skills to seek salaried positions such as carpentry, furniture making and upholstery, which are in demand in The Gambia.

The right to work without taxation in the formal sector may not currently be highly relevant in the case of many adult Casamance refugees given their farming background and often limited education, but is relevant for some and is important for refugee children currently in Gambian schools whose future is likely to be in The Gambia. The principle of the right to work without reservation or impediment for registered refugees is therefore important. With greater comprehension amongst refugees and the wider Gambian community of the rights of registered refugees to work in The Gambia, additionally waiving the employment tax for registered refugees could act to motivate refugees to add to their own education or retrain in appropriate skill areas, and to prioritize their children's education. Doing so will lead to higher levels of effective integration and to regaining self-reliance.

Not having employment sufficient to provide for one's family has an emotional and psychological impact on refugee men and women who are already vulnerable, and undermines the security sought from and offered by The Gambia as a host country and by UNHCR as an organization supporting and protecting refugees. Anxiety, frustration and despair, and damage to self-esteem, especially amongst men who cannot fulfil their obligations as breadwinner, are born out of a lack of employment, which also creates a very real physical impact of a nutritionally deficient diet which characterizes Casamance refugees in The Gambia, not only in urban areas. However, when discussing with respondents the withdrawal of food assistance by WFP, which was partly justified on the basis that Casamance refugees were in no worse position than Gambians (conversation WFP, April 2011), the particular situation of the urban refugee came to light as they disagreed with WFP's rationale. Referring to his situation living in a house borrowed from his brother when he arrived 'empty-handed', Sekou's mind is divided across the many things he must fulfil in starting over:

Can he [the refugee] compare his life to Gambian people's life? Does he feel they have equal life?

It's different. Because the … Gambian has his own compound with the family, no disturbance, the only thing he has to think of is to find work and get money. But for [the refugee] he is thinking how to get a compound, how to feed the family, how to clothe the family and how to pay the school. For him today the wife is just coming to cook Nankatan, rice and salt. No sauce. But as a Gambian you would not want to eat that type of food, and he would take out money and say cook another type of food. [The refugee] has not the power, cos power means economically. So coming here empty-handed … [Gambians] are already here piling things over years, so you can say they are better.… To start everything from scratch, that's why he can't have an equal standard. If you're a Gambian, if you don't have Jumbo [seasoning] you can go to a neighbour or a relative, they will help you with that.… A Gambian would have so many other people. He says he has no help at all. That's another example [of inequality]. The first is house, the second is the food they are eating, the third is he is a refugee.

(Sekou Badjie, Jabang)

It will not be the same as their Gambian neighbours. For their [Gambian] neighbours they will not even bother to go and hustle 'cos everything they will have. But for them [refugees] they will go and hustle to give to their children.

(Alieu Jerju and family, Kunkujang)

It is true however that some poorer Gambians are in a similar situation of poverty as are refugees, but the absence of social networks that Sekou points out above is crucial in assessing whether or not refugees are able to support themselves in terms of food security. Furthermore, it should not be supposed that this situation is therefore acceptable just because the situation of refugees is no worse than some poorer Gambians. Refugee self-reliance is an objective of UNHCR and in their most recent guidelines on urban refugees they draw

> a clear distinction between self-reliance and a refugee's ability to survive without assistance. Unassisted refugees cannot be regarded as self-reliant if they are living in abject poverty, or if they are obliged to survive by [illicit means] or degrading activities.
>
> (UNHCR 2009: 17)

Many urban Casamance refugees *are* living in abject poverty: unable to replace tattered clothes or buy soap for washing clothes, unable to provide enough rice to eat every day, unable to buy shoes for their children with the result that they are turned away from school, surely counts as degrading.

A will to work exists amongst Casamance refugees, but in a context such as The Gambia with existing poverty and widespread urban unemployment, the absence of good communication and a lack of awareness of available assistance compounds the struggle toward integration and self-reliance for refugees who are often in competition with poorer Gambians for work in the informal sector.

Education

Education is a further, yet more minor, motivation for the rural to urban move made by some families and some youth as individuals. Access to secondary education, which is unavailable in village areas or which requires long daily travel to larger villages, sometimes acts as a push to urban areas. This is especially relevant for Casamançais as some families prefer their children to be educated in French in preparation for a one-day return to Casamance. Choosing urban secondary education may involve a family move or sending the youth to live with a family member. In the latter case, on completion of their education, it is typical for the youth to remain in the urban area and send remittances to their rural-based family, returning to the village only during the rainy season to assist with farming if their work allows.

The drive to urban areas for education combines with housing and employment prospects as a motivational force for some families. But as stated previously, having completed education does not automatically equal gaining a job, due to high unemployment in The Gambia. This, in addition to the possible prejudice of employers or confusion over refugees' right to work, may act to demotivate some parents from prioritizing their children's education. However, most respondents expressed a desire for their children to be educated:

> His first decision was to have a job and help the brother to educate the children. But there was no job so it was difficult.
>
> (Alieu Jerju, Kunkujang)

Alieu's difficulty was repeated many times during other interviews. Additionally, successfully accessing education for refugee children in The Gambia was problematic for parents as the UNHCR/GAFNA method of funding and, again, poor understanding of the system of entitlement, often led to exclusion from school, or interrupted education.

In some respondent families lack of money together with poor or non-existent understanding of entitlement and despair with the 'system' would lead to choices being forced as to which child would go to school and which would take an apprenticeship or be sent to farm with relatives in the village. Girls seemed to be a little worse off in this respect; however, boys also suffered from this forced choice.

> This boy is working in the Kombos (urban area). He was at [secondary] school and his brother too, but the father didn't have enough money to pay the school fees to complete his education. So that's the time when the boy decided to come back home and help the father. He is seventeen now. [*The boy continued*] Sometimes I would be at school for three months then have to stop and go and get money to start. When I get money I start again. Stop and start. That's why I said let me stop and help the family. I said, this thing is becoming useless, let me work.
>
> (Family at Sutusinjang)

Education in The Gambia is free at government schools for Gambians and for children of registered refugees up to grade 12. However, the mention of the word 'free' is the point at which refugees and Gambians alike shake their heads in disagreement. Fees at government schools are paid by the Gambian government, or by UNHCR in the case of registered refugees, but uniforms, writing materials and textbooks have to be purchased. Without writing materials and textbooks, academic progress is uncertain, and without a uniform and shoes it is highly likely a child will be sent home. On this basis, many in The Gambia consider education not to be free and as some families – refugees and Gambians alike – cannot afford these costs some children are excluded from the education system or suffer interrupted education as illustrated by the youth in Sutusinjang. Additionally, refugees reported delays in the arrival of funding from UNHCR/ GAFNA, which resulted in refugee children being sent home because one or other of the two six-monthly tranches of educational funding had not been paid to the school for the refugee child.

Lacking a formal education themselves, and perhaps recognizing a changing world, most refugee parents, like Alieu, actively acknowledged the importance of sending their children to school. In addition, youth now have life aspirations that do not include farming. The world of technology – mobile phones, internet, satellite television – has drawn youth toward the potential of a lifestyle in urban centres that goes beyond simply education and subsequent employment opportunities.[15] However, the little money earned by urban refugee families is often depleted by unnecessarily paying children's school fees due to a widespread lack of awareness, or misunderstanding, of 'the rules' for accessing free education.

In its educational provision for refugees, UNHCR provides up to 1,500 and 5,000 dalasi respectively per child per year (£33 and £110) in primary and secondary schools to cover school fees, uniform and books. The child may go to any school in The Gambia, state run or private, and the fees are paid direct to the school once the child is registered at that school as a student. Whilst this amount does not cover the full cost of a private school, it fully covers government school fees and would cover between half and two thirds of private school fees in most cases, depending on the choice of school. This system potentially works well, especially in the case of Casamance refugees, as many refugee children will have started their education in French and are enabled to continue French-based education in The Gambia if they wish.

During fieldwork, it seemed that 'the rules' were misunderstood by refugees but also perhaps by some NGO staff members. In conversations, different NGO staff members and fieldworkers supplied incomplete information regarding the provision for education. Whether this was due to staff members' own misunderstanding of provision or because they were not being thorough in their explanation was unclear. What was clear, however, was that the large majority of refugee respondents lacked awareness of education provision for refugee children. Time and again during interviews I would ask if respondents knew their children could have free education. Greeted by blank faces, providing this information became an integral part of the interview process. As with employment, representatives of UNHCR and GAFNA stressed that refugees were informed of their entitlements but it was clear the necessary information was not effectively reaching individual refugees or it was becoming distorted.

Navigating policy and accessing assistance

Whilst integration of refugees is actively encouraged and state-led in The Gambia, as outlined earlier, and is supported by UNHCR as one of their stated durable solutions, there are aspects of living in The Gambia that challenge rebuilding lives and integration for Casamance urban refugees but that also challenge delivering refugee provision. Refugees in The Gambia have legal and civil rights which enable them to work, live and move freely, unlike in some other Southern receiving countries as stated earlier. However it is how, and if, refugees are able to exercise those rights which influences how, and if, they are able to achieve self-reliance and successful, or meaningful, integration. Exercising rights is dependent not only on what assistance is given, but how it is given or, in the case of information, how it is disseminated. If Gambian government and UNHCR strategies and interventions exist around, for example, employment and education but remain unknown or unclear to refugees and Gambian employers or are distorted, as described above, the potential positive effect of those strategies and interventions is likely to be highly limited.

Gambian government strategy of allowing registered refugees free movement encourages and facilitates integration at one level, but at another level can render urban refugees invisible to agencies that provide assistance and disseminate information, such as UNHCR, and to GID. In rural areas, refugees are more visible because village Alkalos and refugee leaders know those who live in their villages and communities are more close-knit. Thus information and material assistance is easily distributed, either by refugee leaders or the Alkalo. In urban areas, refugees are more scattered and communities more fragmented; keeping track of refugees becomes problematic for those engaged in providing assistance and information. The system of refugee leaders that functions well in rural areas through daily personal contact, does not function effectively in urban areas due to the distance between refugees and cost of transport and phone credit.[16]

The high mobility of refugees decreases in urban areas as refugees tend to become more settled once they have a home, whether owned, rented or provided free. However, when refugees move, they do so without informing GID or GAFNA – indeed, they do not need to do so. The consequence of their mobility in the more fragmented urban area – which is a necessary survival strategy in terms of housing and livelihoods – is that accessing assistance relies on word of mouth which is often unreliable or haphazard in the urban situation. Urban refugees therefore tend to 'disappear' in a way that they do not in the rural situation where GID, GRC and UNHCR staff members and others involved in refugee assistance can refer to a list of refugees and their locations.

Losing track of where refugees are effectively renders them invisible to donors and even to each other. This can affect employment and education opportunities and ultimately life chances, simply because information does not reach refugees or becomes distorted through a 'Chinese whispers' style of passing information, as may be the case in the example of employment rights as described earlier.

Urban refugees are organized to a limited degree, with refugee leaders in theory being a conduit for information between refugees, UNHCR, their partners and GID. In practice, this is less effective than it might be, as information in urban areas is spread largely by phone or by calling meetings, both of which incur financial cost. Attendance at such meetings is often poor, due to lack of funds for transport and the disincentive of travelling time, and telephone dissemination of information by refugee leaders may be curtailed by lack of funds for phone credit. These constraints therefore present a challenge for refugees and agencies alike and are one negative result of refugees being allowed free movement.

The logistics of providing assistance to urban Casamance refugees, and therefore the effectiveness of that assistance, meets with specific challenges that are rooted partly in refugees' mobility and their consequent invisibility, and partly in a lack of workable networks of refugee leaders in the urban areas, but also are located in unnecessarily complex systems of access to services. As UNHCR note in their policy regarding urban refugees:

> It is often assumed that persons of concern who reside in urban areas enjoy easy access to UNHCR. That is not necessarily the case. Refugees are often confined to slum areas, shanty towns or suburbs, a long and expensive journey away from the UNHCR office.
>
> (UNHCR 2009: 12)

This has particular impact when assistance occurs within a context of existing poverty.

An illustration of this can be given by relating a story told in connection to another area of concern where the prescribed system of access impacts on assistance – health care. First, some background to refugee health provision in The Gambia.

Most respondents were aware of the location of UNHCR field office and the GAFNA office, and refugees regularly presented their concerns at these two locations. The official focal centre for concerns, however, is the UNHCR Counselling Centre in Bakoteh where refugees were redirected. The Counselling Centre is also the first point of contact for health care. The system for accessing health care by registered refugees in The Gambia is for the patient to first go to the Counselling Centre. They will then be sent to the Centre's nearby health clinic. If the clinic cannot assist, the patient is referred to the hospital and at the hospital must present the letter of referral from the clinic and their refugee identity card in order to access free health care.

There are several issues with this system. First, there is confusion over whether or not free health care includes free medicine when required. On investigation it seemed, in practice, this decision is made on an ad hoc basis and that no rule is applied. Second, the Counselling Centre is closed at weekends, therefore effectively no free health care can be accessed from 1.30 p.m. on Friday until 8 a.m. on Monday. Third, travelling from home to the Counselling Centre and then to the hospital if referred can be an expensive and time-consuming journey for the patient. These issues often result in the patient going to the nearest (private) clinic: because it is quickest, because the patient believes s/he will have to pay for medicine even if seeing the doctor is free, or because the Counselling Centre is closed.

This woman lived with her family 15 minutes from the Counselling Centre by local transport. She was pregnant and ready to deliver but was having problems. Her husband was not there so she asked a friend to take her to the Counselling Centre clinic:

> ... by then she was bleeding. Those people [at the clinic] said I cannot take care of you, I will refer you to Sukuta, so go and tell them nobody is at the [clinic] here and I don't know what to do. So she went to Sukuta with her friend and by then the blood is continuous. [At Sukuta] they said, where is your husband? She said my husband is not here. [The nurse] said, if your husband is here he will just buy 75 dalasi of fuel and we will refer you to Banjul so you can be well treated. She said, for me my husband is not with me and this Fula woman is just helping me, so where can I get the money and me I am a stranger here? The nurse said well I cannot help you.... Later the doctor came and looked at her. That man sympathize with her and said, if people don't help you in the next minute you might die because you are bleeding and you need to fix [a drip] on you. So that's the time when the doctor just came and fixed [a drip] to her and ... a bottle of blood. (*Through*

tears) She said when the doctor checked her later, he said she should have delivered [by now].... She said, I am no more alive. I will die because of the way I am suffering today.... When she delivered, because of the bleeding the blood got inside the nose of the child and it could not breathe.... That makes the child to pass away....

Since this happened have you received any hospital treatment, a check-up?

Only once. She went there with the husband. To the clinic.
(*Husband*) They just gave her some Ibuprofen. I asked if they were going to send her to Banjul to clean her womb, but they didn't answer.
(*Husband*) What we ask of UNHCR is that they give us access to the hospital and not just to clinics where they cannot do things.

(Husband and Wife – Latrikunda German)

Had the wife had direct access to the hospital for free treatment rather than via the Counselling Centre and Clinic, the baby may have survived.

Conclusion

Challenges faced in providing assistance to urban refugees in The Gambia impact on employment and livelihood possibilities that have a consequent effect on health and nutrition and are linked to unclear or complex systems of access and poorly understood or inconsistently implemented policy on, for example, employment rights and health care, such as discussed above. This manifests itself in levels of assistance received (or not) and is a source of frustration and despair for refugees in urban areas. Furthermore, this acts to retain many urban refugees in situations of abject poverty, negating one of the tenets of refugee protection: the 'right to adequate food, shelter, health and education, as well as a livelihood' (UNHCR 2009: 4).

It might be argued that whilst there is no spatial segregation of refugees, there is some level of societal segregation which results from limitations of material and service provision. Whilst not part of an official intentional or protectionist policy, the impact of unclear or insufficiently communicated information regarding refugee assistance is nonetheless to disadvantage refugees in their transition to full integration and to regaining a level of self-reliance comparable to their lives in Casamance. This has increased impact in a context of existing poverty such as The Gambia where urban refugees find themselves struggling for work and food without the support of wide and established social and family networks as 'back-up', as enjoyed by poorer Gambians, and also without the support of host families as in the rural setting (see Hopkins 2011). The equality between Gambians and refugees cited by WFP as one reason to cease food assistance exists only partially in terms of food poverty, as consideration was not given to the impact of loss of networks as a support mechanism for food and general support, and this is not the same for refugees and Gambians.

Although refugee mobility contributes to absent or distorted information and poor understanding regarding entitlement, mobility is utilized by Casamance refugees as a strategy to secure improved housing and to gain some sort of adequate livelihood in an effort to regain self-reliance. However, other more easily overcome reasons impact on effectiveness of assistance. First, there is a lack of UNHCR and Government resources to provide a wide network of urban outreach workers and urban community centres through which refugees may gain access to accurate information and where they may confirm methods of access and entitlement as recommended by UNHCR (2009: 13). Second, in the case of health and education, for example, unavailability of medicine in clinics may serve to affirm the notion amongst refugees that medicine is not free or that it is given on an ad hoc basis. Similarly, the late arrival of funds for school fees may strengthen the uncertainty of educational entitlement and is particularly critical for the self-reliance of future generations if they are not to remain dependant on long-term assistance or descend further into poverty.

In moving to an urban area, refugees accept they must find alternatives to their traditional farming lifestyle. But with the exception of a few items grown on compounds if space allows, refugees rely on casual work which renders them vulnerable to the possibility of days or weeks without income. Although initiatives of UNHCR via implementing partners offer training to women in soap-making and tie-and-dye, this training is largely focused on the rural areas, with only one training centre in the urban area in Bakoteh near the UNHCR Counselling Centre. Even these training programmes at the Centre were unknown to many respondents, and training programmes focused on men were limited and again unknown to most respondents. Since refugees' farming expertise has limited use in urban areas, more attention to providing urban-relevant skills for urban refugees would assist them in establishing new livelihoods. For example, improved support in basic training in starting and running a small business, combined with the reintroduction of micro-credit facilities, would answer a need expressed by refugees.

The hardships and challenges encountered by Casamance refugees relocating to urban areas are outweighed by the combined potential of work and independent housing, but the mobility of urban refugees and their geographical fragmentation hinder access to assistance and information and pose challenges to organizations providing assistance. Becoming invisible by the very strategy employed by refugees to improve their standard of living in the receiving country ironically places them in danger of not being able to receive the benefit of assistance and interventions which exist to support them. This impacts on the level to which they are able to integrate 'successfully' in The Gambia and become contributing members of Gambian society.

Lack of established social and family networks results in refugees often subsisting below the level of poorer Gambians and is of concern in the short, medium and long term. This has direct relevance to UNHCR urban policy, to UNHCR objectives on refugee populations, and to global goals relating to poverty.

It has been the aim of this chapter to outline the specific issues faced by urban Casamance refugees in The Gambia. According to UNHCR, almost half the world's 10.5 million refugees now reside in cities and towns (UNHCR 2009: 2). Casamance refugees in The Gambia are an example of the changing profile of refugees in urban areas: no longer primarily men and youth relocating to urban areas for employment, whole families are moving to urban centres for a variety of reasons which, when combined, make the move worthwhile despite the hardships encountered. In the case of The Gambia, the system – preferred by the Government of The Gambia and refugees alike – of self-settlement rather than camps, and of integration rather than segregation, itself produces a significant reason for refugees to move: to improve overcrowded and tense housing situations with host families. Although not providing wholly secure housing, the urban move has improved quality of life for Casamance refugees and, when combined with livelihood potential, attracts and retains them in urban areas.

The situation of refugees in The Gambia is particular insofar as registered refugees are allowed free movement and are not restricted to specific areas and there are currently no refugee camps. This is positive in that refugees are not corralled into camps, meaning to a large degree they escape degrading head-counts. The Government of The Gambia's approach toward managing refugee populations means the framework for integration of refugees is already in place. With adequate support this ought to provide the platform from which refugees can become contributing members of Gambian society and enable them to become self-reliant to, or beyond, the level they previously held in Casamance. However, this framework, and the support given by UNHCR, is being undermined by poor dissemination of information and complex or unclear systems of access which act to form barriers for urban refugees, and by the absence of an effective network of urban community offices through which to reach scattered urban refugees. The invisibility resulting from refugee mobility poses new challenges for urban refugee provision – challenges which can largely be overcome by the establishment of informed, and informative, refugee community centres which operate closely with UNHCR and its implementing partners, as UNHCR (2009) already suggests, and with existing networks of refugee leaders. Whilst the system of free movement has the potential to facilitate integration and acts to 'de-victimize' refugees to a large extent, it presents international and national organizations involved in refugee assistance with challenges to overcome in terms of provision.

Notes

1 Those arriving prior to 2006 are considered by GID and UN agencies – contentiously by these earlier refugees themselves – as integrated. Their integration is debatable and a comparison of pre- and post-2006 urban refugees is the subject of a forthcoming article. Although issued with refugee cards by GID in 2006, most pre-2006 refugees claim not to have received food and other material assistance in recent years as have post-2006 arrivals.

2 See Hopkins 2011 on rural Casamance refugees in The Gambia.

3 The region was under colonization for around 200 years variously by the British, Portuguese and finally the French, gaining independence in 1960 as part of Senegal.

4 1991 Peace Treaty between MFDC and Senegalese Government signed in Guinea Bissau; 1992 Casamance Peace Commission held in Guinea Bissau; 1993 Peace Treaty signed in Ziguinchor; 1998 crisis meeting held in The Gambia between The Gambia, Senegal and Guinea Bissau; 1999 peace talks in Banjul resulting in the short-lived January 2000 Resolution; 2003 peace talks resulting in a peace deal and amnesty for MFDC; 2004, new MFDC leader rejects Senegal's amnesty; 2007, ECOWAS produces a conflict prevention framework (sources: Gambian newspapers 1982–2007).

5 Newspapers which were the most prominent at the time of publication were reviewed, although the *Point* and the *Observer* dominate. The *Observer* later became the *Daily Observer*.

6 Although a de-mining programme has cleared many mines, there continue to be occasional instances of death and injury by landmines in Casamance. For example, in September 2006, Jeanie Waddell-Fournier, an ICRC delegate, was killed; in May 2008, a man was killed and 20 passengers injured when a bus drove over a landmine (www.icrc.org/eng/resources/documents/feature/senegal-feature-231209.htm). Warning signs along the border indicate where walking or driving off a well-travelled path is not advised.

7 This report usefully provides a chronology of events of the Casamance conflict.

8 Most of the 15,000 Sierra Leonean refugee population in The Gambia originated from urban areas and settled in urban areas of The Gambia. Similarly, Liberian refugees preferred to situate themselves in the urban area.

9 An Alkalo is a village chief.

10 www.unhcr.org/524d832b9.html.

11 Cessation refers to an evaluated decision by UNHCR that the circumstances in the country of origin have changed. Articles 1 C (5) and (6) of the Refugee Convention specify that a refugee shall no longer be considered as such when 'the circumstances in connection with which he [or she] has been recognized as a refugee have ceased to exist'. Refugees from that country no longer enjoy refugee status after a specified date.

12 In previous rural interviews, the occasions when both men and women were likely to keep quiet were the times when an Alkalo was present: he was allowed to speak for others, or when refugees spoke, their information was measured and diplomatic. This effect was not encountered in the urban context.

13 Ethical considerations were guided by the advice laid down by The Oral History Society, www.oralhistory.org.uk/ethics/index.php.

14 An exchange rate of £1 = 45 dalasi was used.

15 For more on refugee youth and urban youth, see for example De Boeck and Honwana (2005); Whitehead *et al.* (2007); Jeffrey (2010); Thorsen and Hashim (2011); Newhouse (2012).

16 A comparison between urban and rural refugees is presented in a forthcoming publication.

References

CIA 2012. Available at: www.cia.gov/library/publications/the-world-factbook/geos/ga.html (accessed 12 February 2012).

Conway, C. 2004. 'Refugee livelihoods: A case study of The Gambia'. Report for UNHCR Evaluation and Policy Analysis Unit, Refugee Livelihood Studies series. Geneva: UNHCR. Available at: www.unhcr.org/41b417724.pdf (accessed 20 January 2013).

Crisp, J. 2003. 'UNHCR, refugee livelihoods and self-reliance: A brief history'. Policy Development and Evaluation Service Background Documents, 22 October 2003. Geneva: UNHCR. Available at: www.unhcr.org/3f978a894.html (accessed 20 January 2013).

Crisp, J. 2004. 'The local integration and local settlement of refugees: A conceptual and historical analysis'. *New Issues in Refugee Research*, Working Paper No. 102. Geneva: UNHCR Evaluation and Policy Analysis Unit.

Daily Observer 2010. 'Gambia: Casamance rebels strike in Foni.' 6 October. Available at: http://observer.gm/africa/gambia/article/casamance-rebels-strike-in-foni (accessed 20 January 2013).

De Boeck, F. and Honwana, A. 2005. 'Children and youth in Africa: Agency, identity, and place'. In Honwana, A. and De Boeck, F. (eds) (2005) *Makers and breakers: Children and youth in Postcolonial Africa*. Oxford: James Currey.

De Jong, F. and Gasser, G. 2005. 'Contested Casamance: Introduction'. *Canadian Journal of African Studies* 39(2): 213–229.

ECOWAS 1979. Protocol Relating to the Free Movement of Persons, Residence and Establishment. Available at: www.ehu.es/ceinik/tratados/11TRATADOSSOBREINTEGRACIO NYCOOPERACIONENAFRICA/111ECOWAS/IC1116.pdf (accessed 30 March 2012).

Evans, M. 2000. 'Briefing: Senegal: Wade and the Casamance dossier'. *African Affairs* 99(397): 649–658.

Evans, M. 2002. 'The Casamance conflict: Out of sight, out of mind'. *Humanitarian Exchange Magazine* 20: 5–7. Available at: www.odihpn.org/report.asp?id=2408 (accessed 3 February 2011).

Evans, M. 2004. 'Sénégal: Mouvement des forces démocratiques de la Casamance (MFDC)'. Africa Programme Armed Non-State Actors Project Briefing Paper No. 2. London: Chatham House.

Evans, M. 2007. ' "The suffering is too great": Urban Internally Displaced Persons in the Casamance conflict, Senegal'. *Journal of Refugee Studies* 20(1): 60–85.

Fábos, A. and Kibreab, G. 2007. 'Urban refugees: Introduction'. In Fábos, A. and Kibreab, G. (eds) *Refugees in urban settings of the Global South*. Special Issue of *Refuge: Canada's Periodical on Refugees* 24(1) 1–19.

Government of The Gambia 2008. *Refugee Act, 2008* [Gambia], 23 October 2008. Available at: www.unhcr.org/refworld/docid/4a71a8202.html (accessed 11 March 2012).

Harrell-Bond, B. 2010. 'Building the infrastructure for the observance of refugee rights in the Global South'. *Refuge* 25(2): 12–28.

Home Office UK Border Agency 2010. 'Senegal: Country of Origin Information (COI) report'. London: Home Office UK Border Agency.

Hopkins, G. 2010. 'A changing sense of Somaliness: Somali women in London and Toronto'. *Gender, Place and Culture* 17(4): 519–538.

Hopkins, G. 2011. 'Casamance refugees in The Gambia: Self-settlement and the challenges of integration'. UNHCR PDES New Issues in Refugee Research Series, Research Paper No. 220. Available at: www.unhcr.org/4e79ef7c9.html (accessed 20 January 2013).

Jacobsen, K. 2006. 'Refugees and asylum seekers in urban areas: A livelihoods perspective'. *Journal of Refugee Studies* 19: 273–275.

Jeffrey, C. 2010. 'Geographies of children and youth: Eroding maps of life'. *Progress in Human Geography* 34(4): 496–505.

Kibreab, G. 1996. 'Eritrean and Ethiopian urban refugees in Khartoum: What the eye refuses to see'. *African Studies Review* 39(3): 131–178.

Murdie, R.A. 2004. ' "House as home" as a measure of immigrant integration: Evidence from the housing experiences of new Canadians in Greater Toronto study'. Seventh National Metropolis Conference, Montreal, March 2004.

Newhouse, L. 2012. 'Urban attractions: Returnee youth, mobility and the search for a future in South Sudan's regional towns'. UNHCR Policy Development and Evaluation Service series, *New Issues in Refugee Research*, Research Paper No. 232. Available at: www.unhcr.org/cgi-bin/texis/vtx/home/opendocPDFViewer.html?docid=4f560a3d9&q uery=refugee%20youth (accessed 20 January 2013).

Observer, 9 August 2001, page 1. The Gambia.

Sommers, M. 1999. 'Urbanization and its discontents: Urban refugees in Tanzania'. *Forced Migration Review* 4: 22–24.

Sonko, B. 2004. 'The Casamance conflict: A forgotten civil war?' *CODESRIA Bulletin* 3/4: 30–33.

The Oral History Society. Available at: www.oralhistory.org.uk/ethics/index.php (accessed 18 August 2010).

Thorsen, D. and Hashim, I. 2011. *Child migration in Africa*. London and New York: Zed Books

UNHCR 2009. 'UNHCR Policy on Refugee Protection and Solutions in Urban Areas'. Available at: www.unhcr.org/4ab356ab6.html (accessed 20 January 2013).

UNHCR 2010. 'Assessment Mission'. Unpublished report. UNHCR The Gambia.

UNHCR 2011. 'Promoting Livelihoods and Self-Reliance: operational guidance on refugee protection and solutions in urban areas'. Geneva: UNHCR. Available at: www.unhcr.org/ cgi-bin/texis/vtx/home/opendocPDFViewer.html?docid=4eeb19f49&query=self-reliance (accessed 20 January 2013).

UNHCR–World Food Programme 2009. 'Joint Assessment Mission with The Department of State for the Interior and NGOs in The Gambia'. Unpublished report. WFP The Gambia, December.

Whitehead, A., Hashim, I. and Iversen, V. 2007. 'Child migration, child agency and inter-generational relations in Africa and South Asia'. Working Paper T24, Development Research Centre on Migration, Globalization and Poverty. Brighton: University of Sussex, UK.

Zetter, R. and Pearl, M. 1999. *Managing to survive: Asylum seekers, refugees and access to social housing.* Bristol: The Policy Press.

3   The politics of mistrust

Congolese refugees and the institutions providing refugee protection in Kampala, Uganda

Eveliina Lyytinen

Introduction

The aim and objectives of the chapter

The aim of this chapter is to interrogate the protection–trust nexus by examining mistrust and insecurity amongst refugees (i.e. social mistrust), and between refugees and institutions providing protection (i.e. institutional mistrust). The term 'refugee' is used here broadly to refer to people who have fled their country of origin, regardless of their official legal status. The term 'protection' is also used in a broad sense to include not only refugee assistance, physical and legal protection, and 'durable' solutions; in general the upholding of both general human rights and refugee-specific rights (Verdirame and Harrell-Bond 2005: 7), but also a sense of security embedded in trust that is often established in community settings.

The analysis has three objectives. First, to examine the extent to which Congolese refugees in Kampala are united. Second, to contextualize and deconstruct the nature of the relationships between Congolese refugee communities. Finally, to critically examine refugee communities' mistrust of each other and of the institutions providing them with protection.

Data regarding refugees was limited to Congolese refugee, asylum seeker, and undocumented men, women and youth. Analysis of refugee communities was narrowed to the specific community structures that refugees identified during data collection. The term 'community' is used in this chapter to refer to various non-territorially defined Congolese refugee groups characterized by some level of trust amongst their members. Thus, trust is established here as a medium of community, with inter-community relations of particular interest. This chapter focuses mostly on mistrust between the 'Congolese Refugee Community in Uganda' which was trying to unite all of the Congolese refugees in Kampala, and various smaller communities, some of which were cooperating with the protection institutions. The smaller communities include, for instance, the 'Banyamulenge community', different refugee community organizations (RCOs), Congolese churches, and various refugee support groups. In relation to the institutions providing protection for refugees this chapter is limited to an analysis of key institutions, namely the Department of Refugees of the Office of the Prime Minister (OPM); the Refugee Law Project (RLP), a semi-autonomous project within the Faculty of Law of Makerere University; and the United Nation High Commissioner for Refugees (UNHCR) and its only implementing partner in Kampala, InterAid Uganda (IAU).[1]

Rethinking the relationship between trust and community-based protection

Trust is required for establishing meaningful communities where refugees can feel protected and secure. These elements – trust and community – are vital in understanding refugee protection. Moreover, the political nature of protection must be acknowledged (Ferris 2011; Addison 2009; Huysmans *et al.* 2006) likewise the fact that 'not all community-driven and -determined action is positive or protective ...' (Ferris 2011: 199). These concepts of 'protection', 'community' and 'trust' cannot, therefore, be taken for granted.

Being a refugee is often accompanied by feelings of insecurity and mistrust of fellow nationals, host communities and also the institutions providing protection. The very definition of refugee refers to the well-founded fear of persecution. Protection and insecurity are interrelated as refugees who flee insecurity seek protection from their country of asylum. Yet, they might not feel completely safe in their country or city of exile, which blurs the distinction between protection and insecurity. Deconstructing what insecurity means and how it influences trust and community-building is, therefore, essential.

Criminology studies have suggested that people's experiences and fears of crime or, more broadly, insecurity, were mediated by the strength of their relationship with their local community and their structural position within that community. Understanding the nature of these relationships suggests that the question of trust is, therefore, significant in dealing with issues of crime, violence, and insecurity (Walklate 1998: 550). Giddens (1991) states that trust is evident in traditional societies through personal relations and feelings of belonging to various communities. Ontological security, which can be understood as 'the notion of safety, routine and trust in a stable environment' (Hawkins and Maurer 2011: 143) would thus provide a structure for one's life and can then lead to a regaining of trust in people (Padgett 2007).

Refugees often feel that their ontological security – their safety, routine and sense of trust – is, if not completely absent, at least under threat (T. Hynes 2003), because, in the context of forced displacement people normally lose trust and routine, at least temporarily (Kibreab 2004). Losing trust in others often leads to weak community ties. Yet there is a clear link between belonging to communities and feeling protected. For instance, in the 2009 UNHCR 'Policy on Refugee Protection and Solutions in Urban Areas', the importance of community structures and community support in refugee protection is emphasized.

UNHCR's approach in urban settings will be community-based.... The Office will strive to mobilize and capacitate the refugee population, so as to preserve and promote their dignity, self-esteem, productive and creative potential.... UNHCR will foster the development of harmonious relationships amongst the different refugee groups residing in the same city. Similarly, the Office will encourage refugees and their local hosts to interact in a positive manner.

(UNHCR 2009a: 7)

In addition, various UNHCR policies establish an explicit link between protection and community: 'Protection, which includes physical security and the restoration of human dignity, involves supporting communities to rebuild their social structures, realize their rights, and find durable solutions' (UNHCR 2008: 11). Furthermore, a study of four African cities by Landau and Duponchel (2011: 1) suggests a clear link between protection, particularly physical and community structures as the 'primary determinants of urban protection have less to do with direct assistance and policy frameworks than individuals' choices and positions in social and institutional networks'.

However, 'mutual support mechanisms and community-structures are probably less effective in urban than in rural areas' (Ferris 2011: 254). Most commonly, in the urban context, territorially-defined community structures are weakly developed, thus not providing a platform to reinforce community-based protection mechanisms. Urban communities are rather perceived as communities in process and cannot merely be fixed in space (Amin and Thrift 2002: 43). Latham et al. (2009: 149–150) have suggested that there are three principal ways in which the term 'community' has been used: community as place or neighbourhood; community as a set of shared values, practices, and ways-of-being-in-the-world; and community as shared interests. Community can also be defined as symbolic and imagined (O'Neill 2010: 12). Moreover, it is essential to understand community in relational terms: it involves (trust-based) connections between people and provides a means of resistance (Amin and Thrift 2002).

Forming communities requires trust. Yet, in various academic studies, trust routinely is taken for granted (Rubbers 2009) and seen more as an indicator than a process worth examining (P. Hynes 2009). In refugee studies, trust has been perceived as an ambiguous term (Voutira and Harrell-Bond 1995: 219) referring to 'habitus' or 'being-in-the-world' (Daniel and Knudsen 1995: 1). Trust is a crucial element in analysing relations between refugees and their hosts and between refugees and the institutions providing protection for refugees.

P. Hynes (2009: 100) has suggested four forms of trust that are particularly useful in conceptualizing the experience of forced migration: social trust (an individual being able to have confidence in another person); institutional trust (having confidence in political institutions such as police, parliament and courts); political trust (being satisfied with democracy); and restorative trust (the process by which an individual regains social, political and institutional trust).

This chapter discusses briefly social trust amongst refugees, as a background, but the main focus is on institutional trust between refugees and the institutions providing protection for refugees. In this chapter institutional trust is discussed from the viewpoint of refugees. It is, however, realized that social and institutional trust are sometimes interlinked and mutually reinforcing.

Previous studies have discussed social trust from at least two viewpoints. First, there have been debates about how much the culture and society of the refugee's country of origin – which often was affected by violence and conflict – contributes to mistrust and community formation in exile (T. Hynes 2003; Malkki 1995; Sommers 2001; Turner 2010; Russell 2011). Second, it has been suggested that perceptions vary on whether or not there is something inherent in the refugee experience that leads to and sustains this mistrust (Daniel and Knudsen 1995; T. Hynes 2003; Kibreab 2004).

In addition to social trust amongst the Congolese refugees and between refugee communities, this chapter focuses on institutional trust (P. Hynes 2009) which is understood to underpin personal trust (Rubbers 2009). A considerable volume of literature on trust between refugees and institutions focuses on the relationship between UNHCR and refugees. These analyses are indicative of widespread mistrust between refugees and humanitarian organizations and authorities in general. Verdirame and Harrell-Bond (2005) have provided an interesting analogy between UNHCR and a welfare state, which revealed the widespread mistrust between refugees and UNHCR. According to Verdirame and Harrell-Bond (2005: 291), as a result of the increasing care and maintenance aspect of UNHCR's work, they 'developed many problems–or pathologies–that are common to national welfare institutions: the negative stereotyping and disempowerment of the beneficiaries; ... the gradual dehumanization of helping; blaming the refugees for institutional failures ...'. This all contributed to the breakdown of trust between refugees and UNHCR.

Discussion of mistrust between refugees and the institutions providing protection for them has included not only analyses of various humanitarian practices, but also the ways of representing refugees, and acts of humanitarianism. Hyndman (2000) has suggested that because displaced people often have refused to be categorized or managed, the politics of representation have become, in some cases, more important than humanitarian operations on the ground. This has sometimes led to semio-violence, a 'representational practice that purports to speak for others but at the same time effaces their voices' (Hyndman 2000: xxii).

Data and methods

Data used in this chapter is part of a larger doctoral study[2] and was collected over eight months of fieldwork in Kampala between May 2010 and September 2011. Overall, the data comprised 73 semi-structured individual interviews and 13 focus group discussions with Congolese refugees, as well as semi-structured interviews with 18 refugee community leaders, 16 Ugandan authorities (at the national, city, division, parish and zone levels), and 22 officers of protection institutions (including UNHCR and NGOs working on refugee-related projects). Finally, observational data was collected during various community meetings and at the offices of the institutions of protection. Additionally, written documents produced by refugees and institutions were used. Most of the interviews were conducted in Kiswahili (and some in French) and translated into English by an interpreter. Many were also conducted in English without an interpreter. More than 90 per cent of the interviews were audio recorded with the informant's permission, and transcribed. Transcriptions were coded and analysed using NVivo 9 software for qualitative analysis.

The 73 Congolese refugees individually interviewed for this study had left Congo between 2000 and 2011. On average they had stayed in Uganda for three years, with a range from a couple of days up to 11 years. The majority of the refugees were from the eastern parts of the DRC: 48 per cent were from South Kivu and 45 per cent from North Kivu. Most of them had an urban background and were relatively well educated. Of the individual refugees interviewed, 67 per cent were male and 33 per cent female. Refugees were between 15 and 64 years old, the average age being 32 years. Sixty-three per cent of them had refugee status, 30 per cent were asylum seekers, and 7 per cent were not registered with the authorities. While the Government of Uganda encourages refugees to live in rural settlements, 77 per cent of the refugees interviewed had never been in the settlements. Most of the refugees who identified themselves as a member of a particular sub-ethnic group were Bashi (23 per cent), Banyamulenge (17 per cent), Barega (10 per cent), Nande (10 per cent) or Bembe (8 per cent).

When the issue of trust is investigated, it is important to recognize that data collected with refugees was subjective and based on people's interpretations. Thus, refugees' perceptions, rumours and stories were considered to be as valuable as their testimonies and experiences. This is because different forms of articulation were understood to form an integral part of the informants' lives (Coulter 2009; White 2000). The data used in this chapter focuses on Congolese refugees' discourses on mistrust and insecurity in the context of living in Kampala.

The context: Congolese refugees and their protection in Kampala

The number of refugees in Kampala expanded rapidly in the end of the 2000s. In July 2008 the number of recognized refugees and asylum seekers in Kampala was around 20,000; by 2011 the number had nearly doubled. In addition, a large number of unrecognized refugees were living in the city (Bernstein and Okello 2007). In July 2011, of the total 150,000 recognized refugees and asylum seekers in Uganda, more than 26 per cent, that is 39,921, were living in Kampala (UNHCR 2011).

Uganda is a party to the 1951 Convention relating to the Status of Refugees and its 1967 Protocol, and is also a signatory to the 1969 OAU Convention Governing the Specific Aspects of Refugee Problems in Africa. Uganda replaced its heavily criticized 1960 Control of Alien Refugees Act (CARA) by the 2006 Refugee Act, which is more in line with international laws governing refugee protection. There are, however, 'some deficits ... which potentially erode the progressive and protection orientation of the Act and threaten to lower its compliance with international protection standards considerably' (Refugee Law Project, undated: 3). Those deficits mostly relate to freedom of movement and residence, freedom of association and expression, and the right to work (Sharpe and Namusobya 2010). There were also delays in the actual implementation of the 2006 Refugee Act because even though it entered into force in 2008, the regulations necessary for it to become operational were only finalized in 2010.

Despite the overall positive changes in refugee law in Uganda, and the adoption of the 2009 UNHCR urban refugee policy by the UNHCR country office, both of which recognized cities as legitimate places for refugees to reside, 'what exists today – and for the foreseeable future – is a policy that focuses assistance and protection on refugees living in settlements, and not those refugees who chose, for various reasons, to live outside such restrictive spaces' (Bernstein and Okello 2007: 47). Therefore, refugees in Kampala were still expected to be self-sufficient since there exists hardly any social assistance and minimal protection. Also, even though freedom of movement is recognized in the 2006 Refugee Act, explicit permission was required to leave the settlement in order to live in a city, and this 'provision in effect curtails a refugee's right to choose his/her place of abode and his/her freedom of movement' (Refugee Law Project, undated: 19). Thus, it could be argued that Uganda is still heavily relying on its settlement policy (Hovil and Okello 2008), and the situation of urban refugees has not improved significantly despite the newly adopted legal frameworks and policies.

Congolese refugees are the focus on this study. In July 2011 80,221 registered Congolese refugees and asylum seekers were living in Uganda with 18,075 (22.5 per cent) living in Kampala (UNHCR 2011). The number was constantly increasing due to ongoing violence in the DRC, particularly in the eastern parts. Of the nearly 40,000 registered refugees and asylum seekers in Kampala, 45 per cent were Congolese, with 48 per cent women and 52 per cent men. In 2010 the governments of Uganda and the DRC signed a tripartite agreement with UNHCR to govern Congolese refugees' voluntary repatriation. However, since there was little or no security in eastern Congo, repatriation there remained highly unlikely.

Social mistrust and insecurity amongst the Congolese refugees

The DRC has experienced two decades of armed conflict (Prunier 2009, Stearns 2011) and consequent forced migration both inside and outside its borders (Tamm and Lauterbach 2010). The intertwined regional, national and local armed conflicts since the early 1990s have caused the deaths of nearly four million people and displaced around the same number (Clark 2008: 2). The refugees informing this chapter came mostly from the North and South Kivu provinces of the DRC. They arrived in Uganda between 2000 and 2011. There are, however, older vintages (Kunz 1973: 173) of Congolese living in Ugandan refugee settlements and in cities. Some of them have lived in Uganda since the 1960s (Lomo *et al.* 2001: 3) whereas a number of them fled the DRC during the First (1996–1997) and the Second (1998–2003) Congo Wars (Lammers 2006: 18). Eastern Congolese also fled to Uganda due to military operations since 2008 and in the post-election violence since 2011.

Refugees informing this study had been displaced due to generalized violence as the result of the fighting or because of more targeted forms of persecution; usually a combination of these. Refugees expressed fear of persecution spilling into Uganda and affecting their lives. Based on the data collected, it appeared that the persecution and violence experienced in the DRC was a central element of refugees' feeling of insecurity in Kampala. When discussing insecurity in Kampala, refugees often expressed fear of the Congolese state and the rebel groups and militias which either opposed or cooperated with the state. Refugees who had fled their homes because of personal, often ethnically driven, conflicts caused by their neighbours, friends, church members, business associates or relatives also seemed to be afraid of possible revenge for having witnessed atrocities in the DRC. Many refugees in Kampala feared that the people who had caused them to flee would come to Kampala on pretexts such as business or seeking refuge and retaliate against them as witnesses. The refugees feared being killed, abducted, raped, tortured, poisoned or subjected to witchcraft and forced repatriation, among other forms of retribution. Consequently, some of the refugees did not see that Uganda could provide a truly safe space of exile where their rights could be upheld because of its close proximity to the DRC and its involvement in the history of the conflict in the DRC.

Feelings of insecurity towards other Congolese were a limiting factor in some refugees' daily lives in Kampala. They adopted strategies for living with this, at times, overwhelming sense of fear by attempting to hide their Congolese identity. In practical terms, some avoided living in neighbourhoods where the majority of Congolese nationals lived. Others avoided places where Congolese normally congregated, such as the offices of the protection institutions and the refugee communities.

Given the mistrust amongst Congolese refugees, some refugees claimed they could only be open with a limited number of trusted friends and close family. It has been suggested in other studies that Congolese typically relied on limited social networks both as refugees (Amisi 2006; Clark 2006; Lammers 2006; Women's Refugee Commission 2011) and as citizens (Rubbers 2009). Yet these kinship and family ties could be broken due to displacement and refugees were obliged to form new types of trust-based communities.

When discussing the challenges of uniting the Congolese refugees and forming tight communities, refugees mostly referred to two dividing factors: the prevalent ethnic divisions and other associated spillover effects of the conflict in the DRC, and the disunity caused by the institutions providing protection for refugees.

While the conflict in the DRC should not be simplified into an ethnic conflict, refugees in this study residing in Kampala mostly referred to ethnic divisions (i.e. tribalism) when they explained the difficulties of forming strong communities. The majority of interviews conducted with Congolese refugees in Kampala indicated a point of origin in North and South Kivu provinces where the conflict was most severe. According to these refugees, primarily the refugee leaders, the ethnic divisions that were vivid in the DRC, notably in the eastern part, were also visible in Kampala. For instance, in South Kivu there were two major sub-ethnicities, Bashi and Barega, and their relationship was characterized by tensions. In Kampala these divisions, amongst others, were evident to the Congolese. A refugee community leader explained how he saw these divisions having an impact in his community work:

> They [Bashi and Barega] cannot be together. And they always have two [football] teams and they have to fight. Seriously, even in football.... Those two tribes are in a permanent fight. Even here, it is not really very visible, but if you can just see how people behave, the relationship, it seems to be the same but not really ...[3]

Another form of social mistrust within the Congolese refugee population was demonstrated between the 'Banyamulenge Tutsi'[4] and the non-Banyarwanda[5] Congolese, being one of the clearest divisions between the Kivutian refugees in Kampala. This deep-rooted division has its origins in the history of the DRC but refugees suggested that it was reinforced by the institutions providing protection for refugees. The 'Banyamulenge' refugees argued that they were systematically discriminated against because of 'face, race and language'[6] both by non-Banyarwanda Congolese refugees and by the protection institutions. This perceived discrimination, like some of the conflicts in the DRC, centred on the question of who was a 'pure' (Malkki 1995) Congolese. The 'Banyamulenge' self-identified as Congolese, but they claimed that other Congolese saw them as Rwandan because of their background. Physical appearance could often escalate mistrust amongst the Congolese. Refugees whose parents were of two different sub-ethnicities were often caught between these clashes. The 'Banyamulenge Tutsi' refugees also claimed that they were discriminated against by other Congolese because of their distinctive appearance.

Another factor of mistrust was the refugee's place of origin in the DRC. Refugees with no or minor security concerns related to the conflict in the DRC often chose their residence in Kampala based on where people from their area of residence in the Congo were living. Most refugees lived in the central Nsambya and Katwe parishes of Kampala and were mainly from eastern Congo. Many 'Banyamulenge Tutsi', however, lived in Nakulabye. This spatial clustering affected attendance at Congolese churches, RCOs and other community forums. Personal relationships characterized by trust, both at the individual or household level, were also frequently limited to people who came from the same area of the DRC.

Often these divisions did not arise from conscious decisions but rather resulted from a practical response to circumstances. And most refugees spoke and acted against these kinds of divisions. For instance, the initial idea of setting up the 'Congolese Refugee Community in Uganda' with an elected leadership was to attempt to provide a forum to represent all Congolese refugees in the city irrespective of ethnic, linguistic or other background. In practice, as explained in this chapter, this proved to be extremely difficult. Some Congolese pastors fought against conflicts in the Congolese refugee population while others intentionally avoided communities known to be characterized by division and rivalry. All in all, despite the prevalent mistrust and disunity amongst Congolese refugees in Kampala, the desire for trust-based community was clear, through the formation of different types of refugee communities. This desire for a community in the midst of disunity has also been recognized in other studies (Clark 2006; Lammers 2006; Russell 2011).

Institutional mistrust between individual Congolese refugees and the protection institutions

> There are two things: The insecurity which is caused by our neighbours and our neighbourhood, but also other insecurity cases are from the organizations of refugees.[7]

Mistrust between Congolese refugees and official providers of protection was present in Kampala. This institutional mistrust manifested not only at the individual level between officers of the protection institutions and refugees, but also in a more established manner between the institutions and various refugee communities.

Refugees' dissatisfaction with the organizations and authorities was expressed in their common view that Congolese refugees were not protected in Kampala like other refugees because there were so many of them. As a consequence of this perceived discrimination, the Congolese reinforced their identity as refugees deserving of protection and 'durable' solutions to their plight. A male refugee explained how he understood this situation:

> [I]n general in the Congolese community we have a problem, and those problems do not just concern me as an individual. And the problems we are having are the way that the Ugandan different services work. Congolese cases are rarely worked on. The system is a way of making us suffer, a system of imposing things on us. We do not have a right to talk about our problems and to be heard.[8]

Most criticism from individual refugees arose from what they perceived to be negative attitudes by the officers and management practices which refugees believed to be harmful. Perceptions were a primary cause of mistrust between refugees and the institutions. For example, refugees repeatedly argued that the officers did not believe them and consequently did not take their claims for protection seriously. Many refugees informing this study suggested that the issue of mistrust was based on the perception by the officers that all of them wanted resettlement and were willing to do anything for it.

Some of the officers, however, suggested that 'untruthfulness' by refugees was a real challenge and that a substantial percentage of refugees' stories were not credible.[9] This was, suggested certain officers, related to the prevalent paranoia amongst the Congolese refugees. The officers frequently referred to the mindset that the Congolese refugees had especially in relation to resettlement:

> You see refugees are human beings who are capable of doing anything. And we have seen and we have known that these refugees, when they come, some of them come with set minds and they come with expectations.... [T]hey come expecting to go to a third country.[10]

Verdirame and Harrell-Bond (2005) have suggested that blaming refugees has been used by humanitarian institutions to rationalize their institutional failures, which has caused refugees to lose their trust in them. Yet most Congolese refugees argued that the protection institutions in general discouraged them from speaking the truth. Not telling the truth in every situation was also explained by some of the refugees as part of Congolese culture:

> In the Congolese culture when you say directly the truth to someone, you are like, you have disrespected that person. That is the reason why you are seeing that people are affected by a lot of problems, but they do not know how to talk about it freely. There is a saying that it is not good to talk about everything that is the truth.[11]

Some refugees suggested that they were punished when expressing their concerns and true feelings. A male refugee[12] who had been working as a human rights activist in the DRC explained that he thought that UNHCR could not accept criticism. Moreover, he claimed a UNHCR officer who had treated him 'like a rebel' advised him not to criticize the organizations if he wanted to be supported. Some Congolese refugees said that the officers showed a patronizing attitude towards them that further discouraged the refugees from being open.[13] Thus, refugees in Kampala often claimed that the protection institutions tended to dehumanize them or categorize them as a difficult population. These attitudes generally attributed to humanitarian agencies by refugees have been established in other studies (Verdirame and Harrell-Bond 2005; Hyndman 2000).

In addition, refugees involved in this study claimed that telling the truth to IAU officials was not feasible in the early 2000s because the Congolese intelligence services used to come to the IAU to spy on refugees. The nationality and the ethnic background of the officer at times impacted on whether refugees were willing to tell the truth. This was because of memories of colonialism as well as the regional nature of the conflict in the DRC. Refugees were occasionally wary of telling the truth of why they fled the DRC to the Ugandan authorities because Uganda had been involved in the conflict that made them flee. Refugees, however, also realized that not telling the truth could negatively impact their prospects of being protected and resettled, and they advocated for a behavioural change within their communities.[14]

Refugees perceived that authorities and organizations held particular attitudes about their presence in the city. Many refugees, and also some of the more independent organizations working for refugees in Kampala, suggested that despite the flexible laws and policies that allowed refugees to settle in cities, most Ugandan authorities and UNHCR officers did not want increasing numbers of refugees in urban areas because of the intensified pressure on infrastructure and host–refugee relations. Thus, what refugees often claimed was that the officers told them to go back to the rural refugee settlements. This, again, widened the gap between refugees who wanted or felt compelled to stay in Kampala and the institutions. In previous studies, it has also been suggested that officers perceived the refusal to relocate to a settlement as proof that refugees were in no real need of protection (Sandvik 2012: 115).

Among the Congolese refugees, physical appearance was another issue that created mistrust in the protection institutions. Congolese were known for making every effort to look presentable; looks were essential to their culture, particularly in an urban context. However, according to the OPM officer, Congolese refugees were 'flash, really flash',[15] and this led officers to suspect that the refugees did not need protection. Refugees, however, did not understand this perception:

> For the purpose of the hygiene you do everything you can to remain clean before you go to the office, and when you reach there, they tell you that 'You look healthy. What kind of assistance are you looking again?' Do they want us to look shaggy so that they can know that we are suffering?[16]

One male refugee rejected the media-driven perception of a crying, starving and unclean 'refugee'. His views stirred significant discussion with other refugees in the community, who reiterated his point. This was triggered in response to a UNHCR poster for the World Refugee Day (WRD) celebrations, which was used for advertising and hung at the celebration site and IAU and OPM offices.

> In IAU there was the same poster with a photo of a crying woman. I was worried because it gave many ideas. Are refugees a community where everyone just cries? My point of view is that this photo does not reflect on who is a refugee. We should have a positive attitude towards a refugee. He is a human being like any other. Taking pictures of refugees who are crying for donations and for manipulation is not a good idea.[17]

In terms of various management practices, refugees raised accusations regarding their legal protection. Some accused authorities dealing with their cases of using their files and life stories to resettle or to grant refugee status to someone else. A lawyer from an NGO working with refugees suggested that this was a challenge; in particular during the refugee status determination (RSD) process when refugees came in large numbers.[18] In addition, the OPM practice of posting the names of those who have received refugee status on the open wall outside the OPM office was criticized by some refugees. According to them, this was a risky practice for those who had fled because of personal persecution. Yet they saw no chance of changing this system.

Another factor that contributed to mistrust between Congolese refugees and the officers was the lack of responses and information received from the officers: their promises to follow up were often considered empty words.[19] Some refugees also accused the institutions of verbally and physical abusing them. These accusations of misbehaviour had a long history and have been documented in previous studies (Verdirame and Harrell-Bond 2005; Refugee Law Project 2005). Most cases referred to refugees being beaten by guards or police called in by the officers. Some repeated accusations included vivid descriptions of how refugees, even minors, had been 'tortured'[20] in the offices of the protection institutions.

Furthermore, rumours circulated in Kampala of authorities and organizations being corrupt. Accusations were laid, first and foremost, against the police and medical officials, but also against UNHCR and its implementing partner, the IAU. In Kampala, instead of refugees having direct access to UNHCR staff in their main office in the upper-class residential area of the city, UNHCR officers met refugees at the IAU office and at the OPM extension. To get an appointment with the UNHCR officers, Congolese refugees felt obliged to sleep outside the IAU office because the appointments were given on Mondays on a first come, first served basis. This put refugees in severe danger. Refugees claimed that IAU staff benefitted financially from this situation since, as was common knowledge among refugees and officers, young male refugees slept outside the IAU and then sold the appointments they obtained to other refugees. According to refugees, IAU staff got a share of that money. The lack of assistance in Kampala was also explained by some refugees as due to corruption amongst officers. UNHCR was accused by some refugees of offering resettlement to those who were willing and able to pay for it, even Ugandans: 'UNHCR is corrupted. UNHCR likes people with money. In order to get resettlement you need to pay UNHCR'.[21] Given the above, Congolese refugees claimed that to obtain UNHCR protection had become difficult and almost impossible – like 'a dream', as one refugee put it.[22]

'Congolese Refugee Community in Uganda' versus the protection institutions: a community perspective on institutional mistrust

According to data collected from protection institutions, Congolese refugees in Kampala were usually viewed as disunited by institutions providing protection for them.[23] Yet, as a consequence of the increasing number of Congolese, their

networks and communities became tighter. The institutions saw these community networks as both enhancing information sharing but also allowing
rumours and gossip to flourish.[24] The officers claimed that 'fighting' amongst the
Congolese refugees created confusion in the refugee population. The officers
observed that the larger the refugee community, the more prone it was to be
characterized by confrontation,[25] and the subsequent establishment of smaller
communities was seen as an outcome of these internal conflicts.[26] These rivalries
also made cooperation between the refugee communities and the protection institutions difficult. Consequently, a number of organizations and authorities preferred cooperating with various smaller community structures because they
perceived that they were more united and less driven by internal conflicts. Officially, UNHCR in Kampala recognized and worked with both the nationality-
based 'Congolese Refugee Community in Uganda' and with smaller Congolese
communities, because they wanted to respect 'the different identity and also the
different needs.'[27]

According to officers of the protection institutions, the overarching challenge
of working in cooperation with the refugee communities was the high and sometimes unrealistic expectations of the refugees. Of the nearly 19,000 registered
Congolese refugees and asylum seekers in Kampala, competition existed not
only among individuals trying to access minimal protection, but also among
refugee communities trying to cooperate with the protection institutions in order
to be able to support their fellow refugees but also to gain benefits. The mistrust
and consequent division was most obvious among the 'Congolese Refugee Community in Uganda', who criticized these smaller refugee communities for being
manipulated and corrupted by the institutions providing protection.

The 'Congolese Refugee Community in Uganda', like the other nationality-
based official refugee communities in Kampala, was established in 2008 after an
election initiated by UNHCR. The aim was to create a strong, united community
for the Congolese refugees so that they would be more involved in providing
protection and support for their own members. However, relations between the
protection institutions and the 'Congolese Refugee Community in Uganda'
deteriorated. The community observed that the protection institutions, consisting
mostly of IAU, RLP, UNHCR and OPM, were dividing the Congolese refugee
population and compromising their leadership by supporting the smaller refugee
communities. According to their members, UNHCR and other institutions could
have used the community in order to 'understand the problems of refugees', but
instead they chose to open 'a fight against the community'.[28]

This conflict escalated to the point that the elected community representative
and a few other central members of the community claimed they were not being
allowed access to the IAU. Their refugee statuses were also under review by the
Ugandan authorities for alleged severe misbehaviour.[29] Key community
members and the male representative of the community denied these charges and
requested an independent investigation into the accusations.

The 'Congolese Refugee Community in Uganda' was also undergoing
internal strife. The case of the previous female representative of the community

demonstrates external and internal rivalries which were not exclusive to Kampala. In other cities, Congolese refugees also had major difficulties forming an official elected leadership because of internal conflicts which were often reinforced by the institutions providing protection for refugees (Amisi 2006). In Kampala in 2008, the female representative was elected together with the male representative. According to stories circulating in Kampala, she had close relationships with some of the NGOs working with refugees. As a result of cooperation with a particular NGO, she was seen by other refugees as supporting LGBTI (Lesbian, Gay, Bisexual, Transgender and Intersexual) refugees as this NGO was commonly known for its work with sexual minorities. According to some of the members of the 'Congolese Refugee Community in Uganda', this protection institution only assisted 'Banyamulenge Tutsi' and LGBTI refugees. Again, the 'Congolese Refugee Community in Uganda' argued that they were being accused by this institution of discriminating against these two minority groups.[30] Internal politics and mistrust between the elected male and female community leaders also contributed to disunity within the community. The existing male-dominated leadership argued that the community became increasingly fragmented after 2009 because of the female leader: 'Congolese used to be best organized refugee community in Kampala.... But since we got two representatives and the female one was used [by the institutions], our community got destroyed'.[31] Later on, when the situation got worse, the female representative had to leave the community and was resettled in a third country.

Moreover, at the community meetings, which took place at the OPM compound, some members of the 'Congolese Refugee Community in Uganda' would remind everyone about the presence of spies amongst them.[32] They believed that a particular NGO sent refugees affiliated with it to report on the meeting. The leaders of the community also accused one of the NGOs of conspiring against them in order to take over the small office building that the community had rented after being 'displaced'[33] from the UNHCR established António Guterres Community Centre, which was supposed to form a community space for the urban refugee population in Kampala in order to 'enhance protection of urban refugees' (UNHCR 2009b: 23). OPM, however, took over this community centre shortly after its opening in 2010 because their office building was under renovation. Consequently, the Congolese community decided to open a small office for their community activities, but in 2011 they were evicted from their office. According to them, an NGO working on refugee matters had paid the landlord to do this in order to cause disunity amongst them: 'They [the NGO] did everything possible to fight that office up to the extent that they bribed the owner to take over the office.'[34]

Members of the 'Congolese Refugee Community in Uganda' claimed that the protection institutions were continually trying to divide them in order to maintain the dependency of the Congolese refugee population. According to community members, other refugee nationalities also used to have problems with the institutions. However, they organized themselves into tight communities to look after their own people, and only the Congolese were not that organized. Consequently, according to the refugees, the officers of the protection institutions did not want to lose their jobs, so they kept Congolese dependent on the institutions by disuniting them.[35]

This rivalry arguably manifested during the process of holding the next refugee elections in the city. According to the IAU, elections were postponed because they would clash with general elections in Uganda in February 2011 and this might have put the refugee candidates in danger if accusations were raised of refugees' involvement in politics. However, members of the Congolese community believed the true reason behind postponing the elections was different:

> The real reason was that most of us were not willing to go through the elections, because one reason: IAU had paid people who were a bit closer to them, that if they get the representative out they will get to control more.[36]

Because of these challenges, the 'Congolese Refugee Community in Uganda' increasingly argued for total independence from the protection institutions. The overall reason had to do with the institutions' negative, misleading perceptions of Congolese refugees. Aid workers' negative stereotyping of refugees based on their nationality has been documented in other situations (Turner 2010). Some of the officers of the protection institutions seemed to have viewed Congolese refugees as mean, egoistic, and unable to unite.[37] Yet the 'Congolese Refugee Community in Uganda' said that the institutions reinforced – if not created – this perceived egoism by manipulating and corrupting refugees to work and spy for them.[38] Moreover, this perceived egoism not only led to the disunity of the Congolese refugee population, but also to conflict between the community and the institutions:

> [P]eople's egoism.... Because people just want their own interest, not the interest for everyone.... That is the reason why they betrayed the community. Even if they have seen something wrong about that particular officer, they cannot talk about it. That is the reason why you hear our colleagues to say that they do not talk freely, just because they want their own interest. And that is the reason why many officers have managed to control us, control the community, because of that kind of an attitude.[39]

The perceived egoism, sometimes also recognized by refugees, was not seen by them as something internal to the Congolese way-of-being but rather as a result of living in Kampala. The lack of formal assistance and minimal protection forced refugees to compete for scarce support. Therefore, they suggested that they were vulnerable to the manipulation of the protection institutions. Thus, they argued, if the refugees had more independence from the institutions, they would be more united:

> If you are rearing a chicken and your neighbour is doing the same, why you release your chicken to the compound? If you do not give food [assistance]

to the chicken [other fellow refugees], it goes and eats neighbours' [institutions] food. Congolese in Uganda are not enemies to each other, but can lie to get food.[40]

Refugees' quest for communal action and independence

Given the heavy constraints both in terms of funding and human resources, some of the UNHCR officers acknowledged that UNHCR in Kampala was not as active with community work as they aimed to be in the future. Yet they realized that the role of refugee communities in urban areas was crucial for refugee protection. The increasing focus on urban communities was seen as imperative, given the clear community-based approach in the 2009 UNHCR urban refugee policy. Implementing this policy was, however, perceived as difficult by the UNHCR officers because of extremely limited financial and human resources.[41]

Given the changes in laws and policies governing refugee protection in Uganda, and the fact that an increasing number of refugees stayed in Kampala, the IAU, together with UNHCR and OPM, renewed its urban refugee programme in 2009. The programme had been running since 1995, but the older, more limited, approach was no longer seen as appropriate given the situation. The new approach largely focused on facilitating refugee organization and 'empowering' refugee communities to get more involved in issues affecting their lives. Through this community-focused programme, the IAU aimed to 'ensure continued adherence to the highest standards of protection of refugees and asylum seekers' (InterAid Uganda 2009: 21).

Despite the apparent willingness of the protection institutions to work more in partnership with refugees, most Congolese refugees argued that they had no meaningful ways of cooperating with these institutions. Regarding this lack of participation in refugee protection at the official level, there seemed to be two conflicting discourses referred to by some of the Congolese refugees in Kampala. First, refugees often presented themselves both individually and collectively as vulnerable and dependent on the protection institutions. However, because these institutions were providing such minimal protection and few opportunities for refugee participation, refugees felt that they were neglected and mistreated. Consequently, the second discourse formed around the quest for independence from these institutions. These two discourses were often used fluidly in various situations depending on the desired outcome.

The narrative of victimhood shifted towards a discourse of independence due to repeated disappointments experienced by the refugees and the consequent mistrust of the protection institutions in Kampala. This quest for independence often took a communal form. Because of individually experienced disappointment and neglect, refugee men, women and youth looked for others who shared similar experiences. By forming communities, they aimed to reinforce their independence from the institutions, preferring to rely on other refugees. This desire for communal action was often first aimed at gaining the ability to be heard by institutions that would cooperate with them. However, if this aim was not fulfilled, communal action could also be used for gaining total independence from these institutions. Sometimes this led to conflicts between the protection institutions and the refugees, as was seen in the case of the 'Congolese Refugee Community in Uganda'.

The perception that refugees were seen by institutions as a burden that needed to be managed,[42] led some of the refugees to come to the conclusion that there were no real advantages to be gained by cooperating with the protection institutions. Some refugees even felt the institutions were working against them because everything that the institutions would do without meaningful refugee participation would be seen to be against refugees.[43] Trusting the institutions' willingness to give refugees meaningful power was also jeopardized, according to refugees, by competition for the 'refugee national cake'[44] between institutions as well as between refugees and these institutions. This competition over resources and influence further fragmented the relations between refugees and the institutions:

> I can say that this is not an asylum country. It is a country where we are being captured and we are now slaves because they are eating on our name of refugee. Now they are receiving money and funds from everywhere, now they are eating on our name, refugees, when we are still here dying.[45]

The lack of meaningful ways of cooperating with the protection institutions was manifested in the World Refugee Day (WRD) celebrations on 20 June 2011. The events of that day were analysed and debated in the meeting of the 'Congolese Refugee Community in Uganda' the following Saturday. The members of this community did not accept that the celebrations, organized by UNHCR, IAU and OPM,[46] were useful for their cause, represented them in a proper manner, or were appropriate when, at the same time, their real needs were not being addressed by these institutions. This arguably semio-violent (Hyndman 2000: xxii) way of representing the urban refugee population during the WRD left refugees feeling that their voices went unheard. Therefore, some of the refugees were glad to hear that few refugees attended the celebration; this was taken as a sign of increasing independence. At large, the 'Congolese Refugee Community in Uganda' decided to boycott the celebrations and instead spend the day in prayer which they deemed more appropriate.

The official WRD 2011 celebrations in Kampala began at the IAU premises where refugees were given free T-shirts. After assembling, they marched in a parade accompanied by a band, to the Old Kampala primary school football pitch where they heard and gave speeches.[47] They were also given 'one samosa and one soft drink'.[48] Several refugees expressed their dissatisfaction about the celebrations during the following community meeting. According to them, they needed protection, not celebrations. Also, during the celebrations one refugee was admitted to Mulago, the local hospital, where he later died. This made the refugees object that the institutions provided them with soft drinks instead of health care.[49] All in all, the refugees suggested that they had learnt from this event and that they were willing and ready to take more responsibility for their own people and to act independently from these protection institutions:

> I think the time for crying is over. It is time to resolve our own problems. Refugee community can solve their problems on their own. Sometimes crying does not help at all. We have the bad idea that organizations will help us, or carry our crosses. We must take the Jesus principle and carry our own cross. There are signs that we are now reaching the independence from those organizations. We work together towards that independence. That is the idea I conceived during the World Refugee Day celebrations. We have refugees with qualities to achieve what we need. We just lack the will or the persistence.[50]

Conclusions

In this study, it has emerged that many Congolese refugees felt a sense of mistrust in their relationships both with other Congolese refugees, as well as with the institutions claiming to provide protection for refugees. The intertwined and mutually reinforcing nature of this social and institutional mistrust contributed significantly to difficulties in community-building.

Mistrust amongst the Congolese refugees was to some extent embedded in the historical and current conflicts in the DRC, and this conclusion is supported by other studies (Amisi 2006; Lammers 2006; Rubbers 2009; Russell 2011). Some of the conflicts in the DRC were reflected in refugees' attempts to unite in their city of exile, causing the Congolese refugee population to remain fragmented. Therefore, even though refugees longed for a supportive community structure in their city of exile, the attempt to establish one strong Congolese community under elected leadership proved to be challenging. Consequently, there were a number of smaller communities, some of which cooperated with the protection institutions. This fragmentation, however, led to competition amongst the refugee communities.

Even though the institutions, which were working under increasing pressure as the number of refugees in Kampala was expanding, seemed to be willing to empower refugees by working in cooperation with various refugee communities, their relationship with the Congolese refugee population was characterized by a mutual mistrust. The Congolese refugee population was, according to the refugees, seen by the institutions as egoistic, untruthful and unable to be united. Given this perception, refugees saw the institutions as disuniting them in order to gain power over them; they were a burden which needed to be managed. Furthermore, refugees were provided specified roles in which they were able to participate in their own protection, by institutions. Refugees who refused to conform to this 'ideal' set forth by the institutions were viewed as 'troublemakers', destroying the ideals of the community. Consequently, the power imbalance in refugees' interactions with the institutions caused the refugees to articulate their disappointments through narratives of mistrust and insecurity. All these conflicts between refugees and institutions represented a form of resistance and non-acceptance of the situation.

Acknowledgements

I would like to thank the Congolese refugees who kindly agreed to participate in this research project, and the people working for the protection institutions who gave their time to be interviewed. I am also grateful to the Ugandan government for granting me permission to conduct this research. During my fieldwork I was affiliated with the Refugee Law Project and their staff provided me invaluable support. I would also like to thank Emeritus Professor Roger Zetter and Dr Patricia Daley for their supervision, and Dr Barbara Harrell-Bond for brainstorming the initial outline of this paper with me. I am grateful to the entire project for supporting my fieldwork financially. Many thanks also to the anonymous reviewers and the editors. Thanks also to the proofreaders (Tanya Kumar, Carly Leighton and Vicky Mason). Lastly, responsibility for any errors and interpretation is exclusively mine.

Notes

1 It was, however, realized that numerous other actors, such as the police, health officers, city level authorities, civil society actors and the local chairmen, amongst others, played a role in refugee protection in Kampala.
2 This chapter is part of a larger doctoral research project conducted at the School of Geography, University of Oxford, in affiliation with the Refugee Studies Centre, University of Oxford. The research project (2009–2013) was funded by Oskari Huttunen Foundation, Finnish Cultural Foundation, Emil Aaltonen Foundation and Alfred Kordelin Foundation.
3 Male refugee, interview, 15 July 2011.
4 'Banyamulenge' (singular Munyamulenge) refers to 'those living in the hills of Mulenge'. 'Banyamulenge Tutsis' are believed to have started to move from Rwanda to the DRC in the seventeenth century onwards (Autesserre 2010: 138). The 'Banyamulenge Tutsis' of Kampala referred to themselves using this term and presented themselves as a very tight community within the Congolese refugee population.
5 Banyarwanda refers to people with Rwandese ethnic origin.
6 Munyamulenge woman, focus group discussion, 21 June 2011.
7 Male refugee, focus group discussion, 25 June 2011.
8 Male refugee, interview, 28 June 2011.
9 UNHCR Officer, interview, 20 September 2011; OPM Officer, interview, 7 January 2011.
10 OPM Officer, interview, 7 January 2011.
11 Male refugee, focus group discussion, 2 July 2011.
12 Male refugee, focus group discussion, 2 July 2011.
13 Male refugee, focus group discussion, 2 July 2011.
14 Male refugee, 'Congolese Refugee Community in Uganda', meeting, general discussion, 20 August 2011.
15 OPM Officer, interview, 7 January 2011.
16 Female refugee, focus group discussion, 23 July 2011.
17 Male refugee, 'Congolese Refugee Community in Uganda', meeting, general discussion, 25 June 2011.
18 NGO, Legal Officer, interview, 16 August 2011.
19 Female refugee, focus group discussion, 23 July 2011.
20 Female refugee, focus group discussion, 23 July 2011.
21 Male refugee, interview, 7 September 2011.
22 Male refugee, interview, 28 June 2011.
23 OPM Officer, interview, 7 January 2011; UNHCR Officer, interview 20.9.2011; NGO, Legal Officer, interview, 16 August 2011.

24 OPM Officer, interview, 7 January 2011.
25 NGO, Legal officer, interview, 16 August 2011.
26 UNHCR Officer, interview, 20 September 2011.
27 UNHCR Officer, interview, 18 January 2011.
28 Male refugee, focus group discussion, 2 July 2011.
29 Letter from the Commissioner for Refugees, OPM, 'Review of your refugee status in Uganda', 21 December 2010.
30 Male refugee, 'Congolese Refugee Community in Uganda', meeting, general discussion, 27 August 2011.
31 Leadership of the 'Congolese Refugee Community in Uganda', unofficial discussion, 20 August 2011.
32 Male refugee, 'Congolese Refugee Community in Uganda', meeting, general discussion, 23 July 2011.
33 OPM Officer, interview, 7 January 2011.
34 Male refugee, 'Congolese Refugee Community in Uganda', meeting, general discussion, 20 August 2011.
35 Leadership of the 'Congolese Refugee Community in Uganda', unofficial discussion, 20 August 2011.
36 Male refugee, focus group discussion, 25 June 2011.
37 UNHCR Officer, interview, 21 September 2011; NGO, Legal officer, interview, 16 August 2011.
38 Male refugee, 'Congolese Refugee Community in Uganda', meeting, general discussion, 23 July 2011.
39 Male refugee, focus group discussion, 2 July 2011.
40 Male refugee, 'Congolese Refugee Community in Uganda', meeting, general discussion, 23 July 2011.
41 UNHCR Officer, interview, 18 January 2011.
42 Male refugee, 'Congolese Refugee Community in Uganda', meeting, general discussion, 9 July 2011.
43 Male refugee, 'Congolese Refugee Community in Uganda', meeting, general discussion, 9 July 2011.
44 Written statement by a male refugee, 7 December 2011.
45 Male refugee, focus group discussion, 9 September 2011.
46 RLP also organized separate celebrations from these other institutions.
47 Observation at IAU and at the football pitch, 20 June 2011.
48 Male refugee, 'Congolese Refugee Community in Uganda', meeting, general discussion, 25 June 2011.
49 Male refugee, 'Congolese Refugee Community in Uganda', meeting, general discussion, 25 June 2011.
50 Male refugee, 'Congolese Refugee Community in Uganda', meeting, general discussion, 25 June 2011.

References

Amin, A. and Thrift, N. 2002. *Cities: Reimagining the urban*, Cambridge: Polity Press.
Addison, S. 2009. *Protecting people in conflict and crisis: Responding to the challenges of a changing world.* Reflections on an international conference. Oxford: Refugee Studies Centre.
Amisi, B.B. 2006. 'An exploration of the livelihood strategies of Durban Congolese refugees'. *New Issues in Refugee Research*. Working paper 123. UNHCR, Geneva.
Autesserre, S. 2010. *The trouble with the Congo: Local violence and the failure of international peacebuilding.* Cambridge: Cambridge University Press.

Bernstein, J. and Okello, M.C. 2007. 'To be or not to be: Urban refugees in Kampala'. *Refuge* 24(1): 46–56.

Clark, C. 2006. 'Beyond borders: Political marginalisation of Congolese young people in Uganda'. D.Phil. Thesis, QEH, Department of International Development, University of Oxford, Oxford.

Clark, P. 2008. 'Ethnicity, leadership and conflict resolution in eastern Democratic Republic of Congo: The case of the Barza Inter-Communautaire'. *Journal of Eastern African Studies*, 2(1): 1–17.

Coulter, C. 2009. *Bush wives and girl soldiers: Women's lives through war and peace in Sierra Leone.* Cornell University Press, Ithaca, London.

Daniel, E.V. and Knudsen, J.C. 1995. 'Introduction', 1–12 in Daniel, E. and Knudsen, J.C. (eds) *Mistrusting refugees.* Berkeley, London: University California Press.

Ferris, E.G. 2011. *The politics of protection: The limits of humanitarian action.* Washington DC: Bookings Institution Press,.

Giddens, A. 1991. *Modernity and self-identity: Self and society in the late modern age.* Cambridge: Polity Press.

Hawkins, R.L. and Maurer, K. 2011. '"You fix my community, you have fixed my life": The disruption and rebuilding of ontological security in New Orleans'. *Disasters*, 35 (1): 143–159.

Hovil, L. and Okello, M.C. 2008. 'The right to freedom of movement for refugees in Uganda', 77–90 in Hollenbach, D. (ed.) *Refugee rights: Ethics, advocacy, and Africa.* Washington, DC: Georgetown University Press.

Huysmans, J., Dobson, A. and Prokhovnik, R. (eds) 2006. 'Preface'. *The politics of protection: Sites of insecurity and political agency.* London, New York: Routledge.

Hyndman, J. 2000. *Managing displacement: Refugees and the politics of humanitarianism.* Borderlines Vol. 16, University of Minnesota Press, Minneapolis, London.

Hynes, P. 2009. 'Contemporary compulsory dispersal and the absence of space for the restoration of trust'. *Journal of Refugee Studies*, 22 (1): 97–121.

Hynes, T. 2003. 'The issue of "trust" or "mistrust" in research with refugees: Choices, caveats and considerations for researchers'. *New Issues in Refugee Research.* Working paper 98. UNHCR, Geneva.

InterAid Uganda 2009. *InterAid Uganda annual review: Working with the disadvantaged persons.* January–September 2009 Report. IAU, Kampala.

Kibreab, G. 2004. 'Pulling the wool over the eyes of the strangers: Refugee deceit and trickery in institutionalized settings'. *Journal of Refugee Studies*, 17 (1): 1–26.

Kunz, E.F. 1973. 'The refugee in flight: Kinetic models and forms of displacement'. *International Migration Review*, 7 (2): 125–146.

Lammers, E. 2006. *War, refuge and self: Soldiers, students and artists in Kampala, Uganda.* PhD thesis. Thela Thesis, Amsterdam.

Landau, L.B. and Duponchel, M. 2011. 'Laws, policies, or social position? Capabilities and the determinants of effective protection in four African cities'. *Journal of Refugee Studies*, 24 (1): 1–22.

Latham, A., McCormack, D., McNamara, K. and McNeill, D. 2009. *Key concepts in urban geography.* London: Sage.

Lomo, Z., Naggaga, A. and Hovil, L. 2001. *The phenomenon of forced migration in Uganda: An overview of policy and practice in an historical context.* Refugee Law Project, Working Paper No. 1.

Malkki, L.H. 1995. *Purity and exile: Violence, memory, and national cosmology among Hutu refugees in Tanzania.* Chicago, London: The University of Chicago Press.

O'Neill, M. 2010. *Asylum, migration and community*. Bristol: Polity Press.

Padgett, D.K. 2007. 'There's no place like (a) home: Ontological security among persons with serious mental illness in the United States'. *Social Science and Medicine*, 64: 1925–1936.

Prunier, G. 2009. *Africa's world war: Congo, the Rwandan genocide, and the making of a continental catastrophe*. Oxford: Oxford University Press.

Refugee Law Project. *Critique of the Refugee Act 2006*. Refugee Law Project, Kampala, undated.

Refugee Law Project 2005. *A drop in the ocean: Assistance and protection for forced migrants in Kampala*. Working paper 16. Refugee Law Project, Kampala.

Rubbers, B. 2009. ' "We, the Congolese, we cannot trust each other". Trust, norms and relations among traders in Katanga, Democratic Republic of Congo'. *The British Journal of Sociology*, 60 (3): 623–642.

Russell, A. 2011. 'Home, music and memory for the Congolese in Kampala'. *Journal of Eastern African Studies*, 5 (2): 294–312.

Sandvik, K.B. 2012. 'Negotiating the humanitarian past: history, memory, and unstable cityscapes in Kampala, Uganda'. *Refugee Studies Quarterly*, 31 (1): 108–122.

Sharpe, M. and Namusobya, S. 2010. 'Refugee status determination and the rights of recognised refugees under Uganda's new Refugees Act'. Paper Prepared for Refugee Studies Centre Workshop: Refugee Status Determination and Rights in sub-Saharan Africa. Kampala, 16–17 November 2010.

Sommers, M. 2001. Fear in Bongoland: Burundi refugees in urban Tanzania. *Refugee and Forced Migration Studies*, Vol. 8, New York, Oxford: Berghahn Books.

Stearns, J.K. 2011. *Dancing in the glory of monsters: The collapse of the Congo and the Great War of Africa*. New York: Public Affairs.

Tamm, H. and Lauterbach, C. 2010. *Dynamics of conflict and forced migration in the Democratic Republic of Congo*. Expert workshop: report, 30 November–1 December 2010, Refugee Studies Centre, Oxford.

Turner, S. 2010. *Politics of innocence: Hutu identity, conflict and camp life*. Studies in Forced Migration, Vol. 30. New York, Oxford: Berghahn Books.

UNHCR 2008. *A community-based approach in UNHCR operations*. UNHCR, Geneva.

UNHCR 2009a. *UNHCR policy on refugee protection and solutions in urban areas*. UNHCR, Geneva.

UNHCR 2009b. *Refugee education in urban settings: Case studies from Nairobi–Kampala–Amman–Damascus*. Division for Programme Support and Management (DPSM), UNHCR Geneva.

UNHCR 2011. Uganda Statistics as of 1 July 2011. UNHCR, Kampala.

Verdirame, G. and Harrell-Bond, B. 2005. Rights in exile: Janus-faced humanitarianism. *Studies in Forced Migration*, Vol. 17. New York, Oxford: Berghahn Books.

Voutira, E. and Harrell-Bond, B.E. 1995. 'In search of the locus of trust: The social world of the refugee camp', in Daniel, E. and Knudsen, J.C. (eds) *Mistrusting refugees*. pp. 207–224. Berkeley, London: University California Press,.

Walklate, S. 1998. 'Crime and community: Fear or trust?' *British Journal of Sociology*, 49 (4): 550–569.

White, L. 2000. *Speaking with vampires: Rumor and history in colonial Africa*. California: University of California Press.

Women's Refugee Commission 2011. *The living ain't easy: Urban refugees in Kampala*. New York: Women's Refugee Commission.

4 Increasing urban refugee protection in Nairobi

Political will or additional resources?

Elizabeth H. Campbell

In recent years the protection for refugees in Nairobi has improved. Today many refugee children can access free primary education, most refugees can access primary health care at the same costs as Kenyan nationals, and rampant police abuse and arbitrary arrest and detention has been addressed more systematically. The UN Refugee Agency (UNHCR) and its partners, Kenyan civil society organizations, and Kenyan government officials have achieved some of these favourable results through deploying new community-based outreach methodologies and not by substantially increasing financial resources. Stronger and more consistent commitment and will is needed to capitalize on the gains made and to further increase the protection space in Nairobi. Financial resources are of course necessary, but the case of Nairobi demonstrates that in places where there is also a large camp-based refugee population, UNHCR can help foster protection gains in urban settings without substantial additional financial resources and without a change in the national refugee policy and approach, which largely supports encampment.

Introduction

Since the early 1990s Kenya has hosted hundreds of thousands of refugees, mostly in two camp complexes: Dadaab in northeast Kenya, and Kakuma in the northwest. Tens of thousands have self settled in Kenyan cities, especially Nairobi, where they maintain themselves with more or less success and with limited access to assistance from various agencies. The Government of Kenya continues to practise a policy of refugee encampment, but since 2005 some aspects of protection for refugees in Nairobi and other urban areas have improved: for example, many refugees now have access to primary education and health care. This paper examines how these changes came about in Nairobi and argues that these positive steps were largely a result of United Nations High Commissioner for Refugees (UNHCR) officials and other relevant stakeholders adopting a new attitude, perspective and methodology and not due to an infusion of new financial resources or a change in national refugee policy. This is an important finding that challenges a dominant view within UNHCR and the larger humanitarian community that the lack of urban protection, especially in sub-Saharan Africa, is largely due to a lack of financial resources and/or the host government's encampment policy. To be sure, each situation is context specific, and there is not a single model that can be applied successfully across all countries. Nevertheless, the case of Nairobi demonstrates that with the right leadership, will, and commitment, it is possible to improve some aspects of urban protection even under an encampment policy and in light of an ongoing camp-based emergency.[1] Using Nairobi as a case study, the purpose of this article is to inform current and ongoing policy discussions about the prospects of further implementing UNHCR's new urban refugee policy. The article challenges the argument made by many in UNHCR that urban protection efforts are not possible without additional financial resources, by illustrating the concrete steps that UNHCR Nairobi has taken since 2005 to help facilitate marked protection improvements in some areas.

Financial resources are of course needed, but urban protection efforts in Nairobi are much more cost effective than the camp-based care and maintenance programmes. Protecting refugees in Nairobi is about securing their access to national and local services and enhancing their basic human rights, including the right to work or to earn a living (even if only in the informal economy), the right to secure adequate housing, freedom of movement, and freedom from arbitrary arrest and detention. In most instances, it is not about the provision of material assistance. This approach is fundamentally different from the care and maintenance approach found throughout much of sub-Saharan Africa. Increasing protection for urban refugees requires a cultural shift within much of UNHCR and in many of its more traditional partners, including international non-governmental organizations (NGOs). The most effective programming requires investments in community services, full implementation of UNHCR's Age, Gender and Diversity Mainstreaming Initiative (AGDM – see UNHCR 2010a), partnerships with refugees and local civil society and governmental organizations, and the leveraging of resources and expertise from other UN agencies, development actors, and the private sector, particularly those who specialize in micro-finance and lending (Campbell 2010; Campbell *et al.* 2011 Pavanello *et al.* 2010; UNHCR 2009). In sub-Saharan Africa especially, urban refugee protection efforts have lagged behind camp-based protection, not only due to host government policy but also due to the dominant culture within UNHCR and other service providers who are more accustomed to and comfortable with a camp-based model. Moving out of this mould requires leadership, creativity, commitment, and will – not only financial resources. Addressing urban needs requires moving away from an expensive model of assistance, to one where UNHCR and its partners and government officials play the role of facilitating access to services.

UNHCR's 2009 Policy on Refugee Protection and Solutions in Urban Areas is an excellent guide for UNHCR, its partners, host governments and civil society in further enhancing and supporting current protection efforts in cities, including Nairobi. The new policy clearly states that UNHCR is responsible for providing protection to all refugees, irrespective of their location and status (or lack thereof) in national legislation (UNHCR 2009). Through revising its urban refugee policy and thereby replacing the 1997 policy that was based on the assumption that such refugees were more the exception than the norm and that protection and assistance should be prioritized in camps, UNHCR now considers urban areas as a legitimate place for refugees to enjoy their rights. Camp based assistance can no longer be considered or justified as a higher priority, and, as the case of Nairobi demonstrates, the lack of effort to improve urban protection cannot simply be explained away as a lack of resources.

Methodology

Field research informing this article was conducted in Kenya in October and November 2010. The author interviewed refugees in Dadaab Refugee Camp Nairobi, and in towns and suburbs neighbouring Nairobi, including Dagoretti, Eastleigh, Githurai, Kayole, Kasarani, Kawangware, Kitengela, Kangemi, Muthaiga, Ruiri, Umoja and Westlands. The author asked open-ended questions of refugee community leaders (both individuals and groups), and other refugee men, women, and youth, from the Somali, Sudanese, Democratic Republic of Congo, Rwandan, Ethiopian and Eritrean communities. The author met extensively with UNHCR in Nairobi, Dadaab and Geneva as well as with international, national and local non-governmental organizations, some of whom are UNHCR's implementing partners. Interviews with the Department of Refugee Affairs took place in Nairobi and Dadaab. Additional meetings were held with the Kenyan Ministry of Immigration Services and the State Ministry for Provincial Administration and Internal Security.

The author also met with other United Nations organizations, including the UN Organization for the Coordination of Humanitarian Assistance (OCHA), the UN Human Settlements Programme (UN-Habitat), the UN Development Programme (UNDP) and the UN Childern's Fund (UNICEF), and with various embassies, including those of the United States, Canada, Switzerland, Denmark, Norway and Sweden.

The author further examined secondary literature, including Kenyan law, academic articles, and recent reports by nongovernmental organizations (NGOs).

The legal status of urban refugees in Nairobi

The Refugees Act of 2006, which came into effect in May 2007, provided the first legal framework in Kenya for the protection of refugees (Refugees Act 2006). The Act helped to establish the Department of Refugee Affairs (DRA) and the first ever Refugee Commissioner. Based in Nairobi, the DRA is currently building its capacity to take on greater responsibility for refugees and asylum seekers. It has a small presence in both Dadaab and Kakuma refugee camps and has also opened a new office in Malindi with plans to expand its presence to other cities. The Government of Kenya, in close cooperation with UNHCR, civil society, and other stakeholders, is actively working on the development of the national refugee policy that will help clarify refugee rights and responsibilities in

Kenya, including residency rights. The DRA continues to take on more respon-sibilities and to develop its capacity and standards, including reception and reg-istration, but for now UNHCR remains the key agency providing protection and assistance to refugees in Kenya.

Since the early 1990s Kenyan authorities have practised an encampment policy that requires refugees to reside in one of the two established camps, Kakuma or Dadaab. Both camps are in remote and underdeveloped areas of the country where there is little to no agriculture, jobs or business opportunities. Refugees' movement is restricted. They are technically not allowed to leave the camps without written permission from the Government of Kenya, though thou-sands move between the camps and Nairobi regularly. Most of the refugees in the camps are dependent upon international organizations to provide food, health care and other basic necessities. As of mid-2012 the law criminalizes refugees who live outside of designated areas without authority; however, no areas have yet been officially demarcated for refugees – creating a situation of limbo. The encampment policy is thus not written into law, because areas for refugee resid-ence have never been designated (Interview, Kituo Cha Sheria 2010). In 2012 the Government was still working to establish its national refugee policy. Encampment in Dadaab or Kakuma has been the working policy of the govern-ment for at least the last 20 years (Human Rights Watch 2009; Interview, UNHCR 2010a). UNHCR has worked out some officially agreed criteria with the Government for refugees to leave the camps temporarily for education, health related reasons and protection concerns.

Despite the encampment policy, tens of thousands of refugees reside perman-ently outside of the camps, especially in Nairobi, the country's capital and largest city. At the end of 2011 Kenya was hosting over 600,000 refugees and asylum seekers, the majority from Somalia (UNHCR 2011a). Somalis are recog-nized on a prima facie basis. Once registered, the majority are given a Mandate Refugee Certificate. UNHCR estimates that there are upwards of 80,000 refugees residing in Nairobi (Interview, UNHCR 2010a), of whom about 52,000 have been issued a Mandate Refugee Certificate by UNHCR or have been registered and are awaiting status determination (UNHCR 2011b). Some NGOs and others argue there are as many as 100,000 refugees living in Nairobi (Pavanello *et al.* 2010).

After the large-scale influx of refugees into Kenya in the early 1990s, UNHCR supported an encampment policy (Verdirame and Harrell-Bond 2005). The core of its work (and budget) has been camp management, in harmony with the Government of Kenya's own policy. Until 2005 refugees who approached UNHCR's office in Nairobi were told that they had to report to and reside in one of the two camps. Upon commencing refugee status determination for non-prima facie refugees, UNHCR Branch Office Nairobi has historically functioned like a transit centre, urging all new arrivals to move to one of the camps. Documenta-tion to remain in Nairobi was rarely issued. Assistance was only provided to a very few of the most vulnerable cases, and always on a short term if not ad hoc basis until the person was able to report to the camp. Identity documentation, legal aid, access to schools and health care, as well as psychosocial support were basically unavailable to the majority of Nairobi's refugees. Only those refugees who could afford to pay for private services or who received remittances or help from others were able to benefit. Refugees who chose to remain in Nairobi did so at their own risk – largely without documentation or even the most basic protection or interventions from UNHCR or the Government (Human Rights Watch 2002).

Some of the main reasons refugees give for preferring to remain in Nairobi include (1) lack of access to services in camps; (2) increased educational and employment opportunities in Nairobi; (3) follow up on third country resettlement opportunities; (4) physical insecurity in camps; and (5) family reunification (Refugee interviews 2010; Pavanello *et al.* 2010). Despite years of encouraging refugees to reside in camps, UNHCR and the Government were unable to fully implement the encampment policy and instead faced a rather sizeable – and growing – urban population.

UNHCR and the government's new approach to urban refugees

Since the policy had been to encourage refugees to move to the camps, UNHCR knew very little about the situation of refugees in Nairobi, making it impossible to devise a protection strategy to address their concerns. Due to new staff, leadership, and the will to change the existing approach, beginning in 2005 UNHCR launched a new initiative to respond to urban refugee needs. Through the so-called 2005 Nairobi Initiative, UNHCR and its partners began to re-examine the protection and assistance provided to refugees living in Nairobi (UNHCR 2007). This was a real turning point for UNHCR that previously had a negative and distrustful relationship with both NGOs and refugees (Interview, Mapendo International 2010; Human Rights Watch 2002; Interview, Refugee Consortium of Kenya 2010). This poor relationship was only compounded by the infamous corruption scandal that broke out in 2000–2001 in which UNHCR staff, among others, were implicated in selling third country resettlement slots to refugees (see Human Rights Watch 2002; UNHCR 2008).

As a first step to reorient its approach, UNHCR conducted a survey of NGOs working with refugees in Nairobi and then held a series of meetings and consultations. Several NGOs joined UNHCR to carry out a participatory assessment with refugee communities to learn about their protection risks, coping mechanisms and community structures (Interview, Nairobi NGO Network 2010). This participatory assessment, which entailed meetings with refugee men, women, boys, girls, as well as other groups of various nationalities represented in the city, marked the beginning of stronger relationships among UNHCR, NGOs, local officials and institutions, and the refugee communities. UNHCR shifted its programme from one where refugees were required to approach the office with protection and assistance concerns to one where UNHCR proactively reached out to the refugee community and relevant local institutions and stakeholders

(Interview, Refugee Outreach Workers 2010). This was possible in part due to the increased number of NGO partners that had begun working on behalf of urban refugees, such as Kituo Cha Sharia, Mapendo International (now known as RefugePoint), and Hashima Kenya. At the same time the Government of Kenya, through the passage of the Refugees Act of 2006 and the establishment of the Department of Refugee Affairs, also began to be more involved in refugee affairs, prioritizing among its activities the documentation of refugees residing in Nairobi. These initiatives took place within the framework of the current encampment policy and with no clear guidance from national government officials on whether or not refugees have a legal right to reside in areas outside of the camps. To date there is no official Government policy that provides clarity on whether or not Nairobi is an officially designated area of residence for refugees. Verdirame and Harrell-Bond (2005) refer to this ambiguity as 'benign neglect.'

Despite this ambiguity, both the newly established Department of Refugee Affairs as well as UNHCR Branch Office Nairobi continue to register, document and provide some aspects of protection to many refugees in Nairobi. Since 2007 more refugees have been registered, documented and provided access to health care and education than in the last 15 years combined (UNHCR 2010b). These achievements occurred within a vague legal framework and without clear policy guidance from the government. They were largely a result of UNHCR leadership, will and initiative manifested in the new community-based outreach methodologies that were instituted. With UNHCR's release of its 2009 urban refugee policy, initiatives already underway in Nairobi were further solidified and supported. The policy in and of itself did not initially usher in new practices or programming, but it provided support to Nairobi's urban programme and gave advocates for greater urban protection a strong foundation on which to base their arguments and recommendations. The Government of Kenya, however, currently views the policy as a UNHCR policy and not something which the Government has adopted or agreed to implement (Interview, DRA Commissioner 2010). There is no doubt that much more needs to be done to solidify further protection gains, but the small advances that have already been made should not be overlooked.

Current urban refugee programme in Nairobi

Compared to the mid to late 1990s, today there is a formal established and expanding urban refugee programme in Nairobi. Both UNHCR and the Department of Refugee Affairs have joined NGOs and other community partners in acknowledging the presence of refugees in Nairobi, providing some with identity documents and working to ensure their access to schools and health clinics.[2] For the first time since the early 1990s, legal aid is also available. Direct assistance and other programmes like micro-credit remain limited.[3] At the same time refugees living in camps – the overwhelming majority – continue to be the priority in terms of staffing and resources in Kenya. Within UNHCR many officials

believe that refugees residing in camps should be prioritized over urban populations (Interview, UNHCR 2010b). Despite the 2009 urban refugee policy, assistance and protection to urban populations tends to be seen as a luxury or a
privilege that the agency cannot afford. Many officials continue to argue that
urban refugees are better off economically and are not in need of 'assistance.' In
2010 the urban refugee programme, serving over 50,000 refugees, had a budget
of less than US$3 million out of a total of $90 million for the entire country
(Interview, UNHCR 2010b). Attempts to garner small amounts of additional
resources for urban refugees were met with reluctance within UNHCR, due
partly to the emergency unfolding in Dadaab camp but especially since registration and refugee status determination are functions that should soon – in theory
– be fully absorbed by the government as it continues to expand its capacity.

Likewise, the Government of Kenya has made positive steps forward in some
areas of protection of refugees in Nairobi, yet still regularly reiterates its
encampment policy for all refugees. The increased protection of refugees stems
from the creation of the Department of Refugee Affairs, efforts to register urban
refugees, and the provision of documentation to some.[4] This has however coincided with growing security concerns by the Government of Kenya and the
internal security forces in particular, who believe that national security is best
achieved by refugee encampment (Interview, DRA Commissioner 2010). In
other words, while the Department of Refugee Affairs is taking steps to ensure
greater protection for urban refugees, other ministries within the government are
more reluctant to grant refugees freedom of movement and legal residency rights
in urban centres (Human Rights Watch 2010). This has especially been the case
since 'Operation Linda Nchi,' (Protect the Country) launched in October 2011,
the Government's response to eliminate the threat of Al-Shabaab in Somalia.
The military operation is part of a broader political strategy that includes creating 'safe zones' inside Somalia to which the growing Somali refugee population
could return (International Crisis Group 2012).

Despite many positive steps forward in the urban programme with measur-
able, successful outcomes, both UNHCR (from the headquarters to the field
level) and the Government remain somewhat conflicted about how the urban
programme should fit into the overall refugee protection efforts in Kenya. With
very few resources, UNHCR Branch Office Nairobi and its partners as well as
the Department of Refugee Affairs have been able to achieve important results
in a short period of time. The change in attitude and in approach to refugees in
Nairobi has resulted in greater protection in some areas. Of course many gaps
remain. Police abuse and arbitrary arrest continue. Most refugees cannot access
work permits in the formal economy. There is extreme discrimination and xenophobia directed toward Somali refugees in particular, and until recently, few
micro-credit or lending programmes have been developed to promote greater
self-reliance (see Danish Refugee Council and UNHCR 2012).

Refugee registration and documentation

Despite the encampment policy, UNHCR has long registered refugees in Nairobi. Throughout most of the 1990s and into the early 2000s, refugees were given a UNHCR Mandate Refugee Certificate stating that the person was a recognized refugee and should report to the camp.[5] This document reflected the clear policy of UNHCR at the time that registered refugees were to report to the camps. Around 2003–2004, as UNHCR became more open to providing some forms of protection to refugees in Nairobi, though not material assistance, the text of the document was changed to read that the person was recognized as a refugee by UNHCR but was not entitled to any assistance in Nairobi. Unfortunately, this wording proved to be a barrier for many urban refugees who complained that the 'no assistance' clause made it difficult for them to obtain assistance from other NGOs and religious institutions. The clause was even used by banks and schools as a reason to deny access to services. Soon thereafter UNHCR again changed the document to read that the person was a recognized refugee in Kenya under the 1951 Refugee Convention and the 1969 OAU Convention. There was no reference to a place of residency or to the denial of material assistance. Most of the refugees in Nairobi now have this document, and it has helped many access services and assistance that may be available to them. The wording also illustrates UNHCR's evolution on the thinking of urban refugee protection, which occurred independently of any shift in Kenya's national refugee policy or infusion of additional resources.

Historically, many UNHCR and Government officials argued that the provision of protection services in Nairobi would create a pull factor for refugees to move from the camp to the city. Yet very few refugees seeking registration in Nairobi come from the camps, only about 7 per cent (Interview, UNHCR 2010b). The majority come directly from Somalia, though there are also some groups of Congolese (DRC) and the Oromo Ethiopians. Since 2007, given the escalation of violence and persistence of drought in Somalia, there has been a steady and well documented influx of refugees to Dadaab refugee camp as well as to Nairobi (Interview, UNHCR 2010b). The majority have, however, sought asylum in Dadaab, given its proximity to the border.

For refugees recognized on a prima facie basis, mandate refugee certificates are issued within six months of registration. Historically, there has often been a large backlog of asylum cases in UNHCR Branch Office Nairobi. In 2010, the backlog was approximately 10,000 individuals and was due to (1) new influxes of refugees from Somalia since 2007; (2) UNHCR's proactive efforts to register refugees who may have been residing in Nairobi without documentation for some period of time and are only now coming forward; and (3) chronic understaffing and under-resourcing of the urban programme, in part due to the belief that the Government should be taking over greater responsibilities in this area now that there is national legislation on refugee issues (Interview, UNHCR 2010c).

Documentation is the foundation of protection for refugees residing in Nairobi – and in most other cities. Documents help protect refugees from police abuse and maltreatment, secure their release from detention, protect against *refoulement*, provide them with access to schools and health clinics, and help them find employment. In almost all of the participatory assessments and urban refugee surveys that have been conducted in Nairobi, refugees rank the need for identity documents highest on their list of demands (Campbell 2006; Human Rights Watch 2002; Pavanello *et al.* 2010). The need to provide urban refugees with timely documentation is thus critical. With a modest financial increase and stronger or more consistent commitment and will, UNHCR is well poised to provide more timely registration and documentation and vastly improve the protection environment for many refugees in Nairobi. Instead, in tight budget environments such protection efforts are often viewed as 'non-life saving services' and attention and resources are instead directed to the camps (Interview, UNHCR 2010b).[6]

Efforts beginning in 2010 by the Government of Kenya to register refugees are a welcome step forward but remain inadequate and incomplete. UNHCR is in the process of handing over the registration and refugee status determination functions, though a specific timeline and deadlines have not yet been set. It is desirable for the Government to take on greater registration, status determination, and documentation responsibilities, but it is likely that UNHCR will continue to be the main provider of these services in the short term. Even when the Government is better equipped to take on greater responsibilities, the partnership and shared responsibilities between the two entities will likely persist for many years to come.

Community outreach and partnerships

UNHCR once had a strained relationship with refugees and civil society in Nairobi. Distrust and bad feelings were common and communication and cooperation were limited. UNHCR was viewed largely as a gatekeeper and not as a partner. In 2005 when UNHCR initiated a new approach to its urban programme, it started with the premise that refugees do reside in Nairobi and will likely continue to do so irrespective of the encampment policy. UNHCR also understood that it could only increase its presence and protection of urban refugees through the building of partnerships and formal communication networks with refugee communities. Further, UNHCR sought to change the dynamic of serving only those refugees who were able to approach their offices to one where the agency would proactively engage in community outreach. UNHCR's 2005 Age, Gender and Diversity Mainstreaming (AGDM) participatory assessment also helped to apply a collaborative inter-agency approach to improve coordination, protection, and provision of services to refugees in Nairobi (see Pittaway 2010). Issues identified in the assessment were later incorporated into urban refugee programming. Today this is the standard practice. In 2011 refugees identified their priorities as documentation, followed by livelihood support (Danish Refugee Council and UNHCR 2012).

When UNHCR began to implement its new approach, it started with very little knowledge about the urban refugee population. Basic questions like how many refugees resided in Nairobi and from which nationality were unanswerable. The agency began to re-examine its efforts and its ad hoc and very reactionary programme, largely designed to meet the needs of those refugees who were willing and able to queue outside of their offices, by engaging in a series of meetings with refugee partners and refugee communities, the so-called Nairobi Initiative. UNHCR was for the first time since the early 1990s able to establish contacts with refugee communities at large and begin to better understand community structures in Nairobi. The agency did this in part by working with an international implementing partner, GTZ, to establish a community of almost 20 refugee outreach workers. GTZ currently convenes monthly meetings to train refugees on a variety of issues, including how to respond to refugee arrest and detention, how to refer sexual and gender based violence (SGBV) survivors to appropriate services, and how to identify extremely vulnerable individuals who may need additional assistance. Community outreach workers also serve as points of communication between UNHCR, its NGO partners, local officials and the wider refugee community. Outreach workers are well known to local police stations and are often on the front lines of negotiating refugees' release from detention.

UNHCR also expanded its formal partnerships to more NGOs, including especially national legal aid organizations. Two organizations in particular, Kituo Cha Sheria and the Refugee Consortium of Kenya, both long-standing Kenyan human rights organizations, have been providing invaluable legal aid services to refugees, among other things. Kituo Cha Sheria provides training to local officials on refugee and human rights law. It works with community mobilizers to increase information sharing. The organization was a key player in promoting the passage of the Refugees Act of 2006 and is now working closely with the Government on implementing regulations. It also advocates for work permits in the formal economy. The organization also spends a lot of time counselling refugees on their rights and how best to respond to police abuse. The Refugee Consortium of Kenya, working on behalf of refugees in the region since 1998, represents refugees in status determination procedures, and refers SGBV survivors to appropriate services. It has well established relationships with officials in the Department of Refugee Affairs as well as other governmental ministries and offices. It too has been an advocate for policy change, working to help pass the Refugees Act of 2006 and implement regulations. These local Kenyan organizations are great examples of the powerful role civil society plays in helping to protect refugees in Kenya and especially in Nairobi.[7]

UNHCR and its partners' training of Kenyan lawyers, magistrates and judges has been an important means to improving refugee protection in Nairobi. These officials have been trained on the provisions in the Refugees Act of 2006 as well as on identifying and responding to vulnerable refugee groups, such as SGBV survivors and unaccompanied refugee minors. Some have gone on to become trainers themselves, holding meetings and workshops for their colleagues, including trainings on the AGDM Initiative. One magistrate who is herself a trainer helped to organize a 'court users committee', which included members from the judiciary, police, the Criminal Investigation Department (CID), anti-terrorism and anti-narcotics units, district commissioners, staff from the Children's Department and the After Care Department, community members, staff from the Department of Prisons and NGOs (Interview, Kibera Courts Magistrate 2010). The committee was trained on refugee law. UNHCR's outreach to and partnership with the Kenya Magistrates and Judges Association has ensured fairly large scale training of magistrates and judges over the last several years. This valuable (and non-traditional) partnership has greatly increased refugee protection, particularly when it comes to the arrest and detention of persons of concern based on charges related to their illegal entry or presence within the country.

UNHCR has also worked to establish relationships with local authorities, including district commissioners responsible for areas in which there is a large refugee presence. These are new partnerships that did not previously exist. As a result of this relationship, many of the district commissioners are for the first time more aware of refugees' rights and the particular challenges they face. In one instance a district commissioner invited UNHCR to participate on the district board as a key stakeholder and also encouraged the agency to support efforts to enrol refugee children in schools (Interview, District Commissioner 2010). This example highlights the crucial role that local officials can play in helping to protect refugees in Nairobi. Unlike in refugee camp situations where UNHCR more frequently liaises with the central government, in urban areas local governance is often the key to increasing the protection space.

UNHCR and its partners have also reached out directly to police stations as well as the Department of Prisons that oversees the prison for those refugees who have been convicted and sentenced for illegal entry or lack of documentation. Due to the frequent communication and coordination with these entities, UNHCR is now better placed to call the Officer in Charge or Law Courts and ask if there are any refugees present in their facilities that should be released. Often Kenyan officials call UNHCR to alert them to the presence of refugees at police stations or in the prisons. One officer at a police station actually called for the agency to hold more training so that more of his officers could benefit from the knowledge (Interview, Police Commissioner 2010).

This proactive engagement is a direct result of UNHCR and its partners' outreach efforts, partnership building exercises, and trainings. As a result police abuse and court convictions for illegal presence in Kenya have declined, even though police abuse and arbitrary arrest remain endemic.[8] Most refugees will agree that communication between UNHCR and refugees has improved. In a meeting with refugees from the Ethiopian Oromo community, one noted, 'Our situation remains difficult – very hard, but after 15 years living here in the city, we are better off with the new legal aid organizations and with the formation of the refugee outreach workers than we were without them' (Interview in Eastleigh, 2010). A Sudanese family living in Githurai commented, 'We moved from the camp to Nairobi once we knew that we could attend schools for free in Nairobi. They are better in the city than in the camp, and here we feel like we might have a chance' (Interview in Githurai, 2010).

Self-reliance and work opportunities

Most refugees living in Nairobi have been able to provide for themselves and their families with no support from UNHCR, NGOs or the government. There is no widespread food distribution, provision of shelter, or refugee clinics and hospitals. As in all societies there is a clear class aspect to the refugee population in Nairobi. Some refugees are engaged in lucrative commercial activities and have well entrenched private enterprises in Nairobi (Herz 2008). The large majority of refugees living in Nairobi, however, work in the informal economy like most Kenyans (Campbell 2005). For 20 Kenyan Shillings (Ksh.) (US$0.25) refugees can secure a permit from the City Council to sell their foodstuffs or wares on the street corner. They survive on a day-to-day basis, for example selling tea and snacks on the street, in hopes of earning enough money to buy food for their families. Many of these families live in acute poverty. While refugees are legally able to work in the informal sector, many report abuse from the City Council authorities who frequent refugee neighbourhoods and harass and intimidate sellers (Campbell *et al.* 2006).

For those who came to Kenya with professional backgrounds or who have earned a degree while residing in Kenya, gaining employment in the formal sector is very challenging. UNHCR has had some success in obtaining Class A work permits for professionals, but it is very costly for refugees (60,000 Ksh./ US$750) (Interview, UNHCR 2010c). Many refugees with skills and university degrees are unable to practise their professions in Kenya. The provision of official work permits will not unlock the issue of a lack of refugee livelihoods or work opportunities, but it will definitely help improve the lives of those who are able to make important contributions in the professional fields.

Many of the refugee communities in Nairobi, but especially the Somalis and Ethiopians, have lived in Nairobi for almost 20 years. There is extensive migration within the region and elsewhere, a long-standing third country resettlement programme, and well established business networks throughout the Horn, East, and Southern Africa – not to mention the Middle East, the Gulf, Europe and the Americas. These trade networks and migration and resettlement opportunities have allowed for significant resources to be sent back to families and individuals residing in Nairobi. Remittances are thus an important source of income for many refugees living in the city (Campbell 2006; Horst 2007).

Given the entrepreneurial nature of many refugees, it is important that UNHCR has now developed a livelihood strategy, something that was unthinkable a decade ago and that has occurred without any commitment by the Government to integrate refugees (Danish Refugee Council and UNHCR 2012). The development of such a strategy signals strong intent by UNHCR to try to further improve the protection environment for urban refugees.

Access to health care

Until the 2005 Nairobi Initiative, UNHCR ran a parallel health care system for refugees. The agency had its own nurse at the Branch Office that would see refugees daily and then refer them to its implementing partner that ran a health clinic only for refugees. Due in large part to UNHCR's community outreach and partnership with the City Council's Health Department, most refugees now have access to primary health care, something they did not have in the 1990s. By showing one's Mandate Refugee Certificate and paying the fee required for all patients of 20 Ksh., refugees are largely able to access city run health clinics and receive treatment, including access to maternity wards.

In areas where there is a large concentration of refugees, such as Eastleigh, UNHCR provides support to the clinics to ensure refugee access and treatment. This effort started in Eastleigh and has now expanded to six other locations. When refugees are in need of secondary treatment, they can visit one of these clinics and seek a referral from the nurse who will send them to a public hospital. Urban refugees must 'compete' with camp-based refugees to obtain referrals from UNHCR or its implementing partner. According to UNHCR Kenya's Health Coordinator, camp referrals are prioritized over urban cases, and there is currently no real budget dedicated to urban refugee health issues (Interview, UNHCR Health Coordinator 2010). Refugees with chronic illnesses such as cancer and diabetes are largely not treated due to a lack of funding. The lack of access to secondary treatment, especially for those with chronic illnesses, is a growing gap – and one faced by most Kenyan nationals as well.

The success of the urban health programme, however, is that UNHCR and its partner GTZ have been able successfully to advocate for refugee access within Kenya's national health care system, and UNHCR no longer operates a parallel structure for urban refugees. Through the provision of some staff and resources, Kenyan authorities have been receptive to UNHCR's efforts to ensure that refugees have access to public health clinics.

Access to education

In 2003 primary education was made free for all children in Kenya. Through successful outreach campaigns and relationship building efforts with the Office of City Education in Nairobi (made possible by the 2005 Nairobi Initiative), UNHCR and its partner GTZ were able to capitalize on this change in law by ensuring refugee children had access to primary school. UNHCR and GTZ began this initiative by reaching out to head teachers in schools located in refugee neighbourhoods to educate them about the right of refugees to attend school. Due to UNHCR's unprecedented engagement with them, the Office of City Education was also able to provide information about refugees directly to its schoolteachers.

As of mid-2012 UNHCR has formal partnerships with 60 schools, 40 of which are in Nairobi. These partnerships consist of teacher trainings about refugee rights and refugee experiences and the provision of very basic assistance such as a few desks and some school uniforms. In Mwiki Primary School in Githurai, a community just outside of Nairobi, there are 2,000 students enrolled, of whom 328 are refugees from Sudan, Ethiopia, Somalia, Rwanda and Congo. The head teacher has been extremely amenable to enrolling refugee children in her school and has worked closely with the community to educate members about the particular needs of refugees (Interview, Ochola 2010). As a result, since the head teacher, teachers, and fellow students have been well-sensitized to refugee needs, the community was able to quickly mobilize and protect some families from arbitrary arrest and detention. In urban settings these types of local partnerships are critical for ensuring refugee protection.

Long-term solutions

Repatriation in the near term, especially for Somalis, is highly unlikely. Kenya continues to witness large scale arrivals into the country and has not facilitated any voluntary repatriation of Somalis in recent years (Interview, UNHCR 2010b). In fact the last recorded voluntary repatriation of a Somali refugee was in 2006 when two persons were assisted to return home (UNHCR Branch Office Nairobi 2010). Likewise, there is no evidence that the Ethiopian community, the second largest refugee population, is planning to return anytime soon. UNHCR has only assisted with four voluntary repatriations of Ethiopian refugees since 2006 (UNHCR Branch Office Nairobi 2010).

Third country resettlement continues out of Kenya, albeit at a relatively small scale compared to the number of refugees hosted by the country. In 2009 about 7,500 refugees were resettled (UNHCR 2010c), the majority to the US. This was 2,500 less than originally anticipated. Currently the number of refugees resettled out of Kenya in one year equals the number of refugees entering Kenya in one month. While resettlement is offering durable solutions to some individuals, it is not a solution for the refugee situation in Kenya.

Like all protracted refugee situations, the one in Nairobi belies easy solutions. Many of the refugees have lived in Nairobi for 20 years. They work, send their children to school, speak English and Kiswahili, and know only Kenya as their home. At the same time most of these refugees have never been afforded the opportunity to acquire Kenyan citizenship and the full legal protections and rights that come with it.[9] Currently, the Government of Kenya is not willing to entertain the possibility of official legal integration of refugees in Nairobi or in any part of Kenya for that matter. At the same time, since the formation of the DRA, the Government has taken important steps to address some aspects of the protection needs of refugees in Nairobi. Access to primary education and health clinics has increased the quality of life for some urban refugees. Efforts to mitigate police abuse, arrest and detention have also been positive, though these efforts are easily disrupted when security incidents occur and the Kenyan police are ordered to engage in large-scale sweeps of areas with refugee residents. The intentional issuance of documentation for refugees choosing to reside in Nairobi has further been beneficial. Of course much more needs to be done in all of these areas. More freedom of movement, the right to reside officially in urban areas, the right to work in the formal economy, the systematic provision of documentation, unfettered access to schools, clinics and hospitals, and more resources directed at micro-loans and work opportunities in urban settings will help refugees who have no prospect of an immediate durable solution to live with greater protection and dignity.

Conclusion

In 2012 urban refugees in Nairobi enjoy more access to protection in some areas than they did in the 1990s and early 2000s. The creation of the DRA, UNHCR's 2005 Nairobi Initiative, newly established legal aid organizations, more positive local relationships with the courts, police and prisons, the formation of refugee outreach workers, and new partnerships with schools and health clinics have all led to an improvement in some aspects of the protection environment for many refugees residing in Nairobi. These positive steps forward are largely a result of increased will and commitment within some parts of UNHCR and the adoption of a new methodology and way of working with urban refugees. To date, few additional financial resources have been directed to UNHCR's urban programme. While the operation needs to be supported with adequate resources to further capitalize on the gains made, serving urban populations is more cost effective than care and maintenance operations in Kenya's camps. In 2010 there were roughly the same numbers of refugees residing in Kakuma camp and in Nairobi. The UNHCR operating budget in Kakuma was $12.6 million, not including the World Food Programme's costly food distribution operation. In Nairobi, with only a slightly smaller population, the UNHCR's overall operating budget was less than $3 million. As long as the approach in Nairobi continues to focus on securing greater access to services for Nairobi's refugees and not on the provision of widespread material assistance, UNHCR and its partners will continue to expand the protection space in a cost-effective manner.

This case study demonstrates that in places where there is an encampment policy and where UNHCR has historically only focused on camp-based populations, the agency needs to develop a new vision, stronger leadership, and a meaningful commitment to serve urban refugees, and in ways that are quite different from those used with camp-based populations. It is possible to advance the implementation of UNHCR's urban refugee policy if the will and leadership is there to do it. The case of Nairobi shows that negotiating protection in cities is largely about new forms of communication, building networks, developing new partnerships, and advocating for access to existing services. To be sure, this approach is labour intensive, but it does not require the same resources as sustaining a care and maintenance programme. The experience in Nairobi should help inspire UNHCR to 'do more with less' and to implement the tools already at its disposal, such as AGDM as well as other 'best practices' and 'lessons learned' from cities across the world. It is no longer convincing for UNHCR to argue that a lack of resources alone is what is keeping the organization from implementing its urban refugee policy.

Notes

1 By 2012 the population of Dadaab Refugee Camp had risen to half a million, while the population of Kakuma stood at 100,000.
2 In 2010 the DRA engaged in a refugee registration exercise in Nairobi for urban refugees, whereby the Government of Kenya recognized the presence of refugees in Nairobi and legitimized their residency by providing them with documentation to remain in the city. At the same time the more powerful Internal Security and Provincial Administration Department often undercuts the DRA's urban refugee protection efforts by insisting that all refugees return to the camps for security purposes. It is therefore difficult to argue that the government has a single policy and approach that is consistently applied by all officials. Even within the government, Kenya's refugee policy is often ambiguous and contradictory.
3 For a more detailed understanding of current livelihoods in Nairobi, see Danish Refugee Council and UNHCR 2012.
4 There are signs that the Government will begin issuing identification cards to all refugees in Nairobi more systematically, but at the time of this writing, that had not yet occurred.
5 Copies of this mandate refugee certificate are on file with the author and were acquired during previous research in 1998 and 2003.
6 To be clear, the author does not believe that camp-based and urban-based refugees should be pitted against one another. Both are in need of protection and are deserving of UNHCR's expertise and resources. The ultimate goal should be to serve both populations, albeit differently.
7 UNHCR also has operational and implementing partnerships with other organizations in the urban context such as HIAS, GIZ, IRC and Heshima Kenya, all of whom play an important role in enhancing refugees' protection and ability to access services in an urban context.
8 Police and other officials frequently refer to refugees as 'ATMs', i.e. cash dispensers, as they are routinely forced to pay bribes to avoid arbitrary arrest and detention (Interview, Kituo Cha Sharia 2010; Pavanello 2010).
9 Recent changes in the Kenyan Citizenship and Immigration Act of 2011 may offer possibilities for some refugees to acquire citizenship, but this has not yet been tested.

References

Campbell, E.H. 2005. 'Formalizing the informal economy'. Global Commission on International Migration: Global Migration Perspectives, no. 47. Available at: www.iom.int/jahia/webdav/site/myjahiasite/shared/shared/mainsite/policy_and_research/gcim/gmp/gmp47.pdf (accessed 20 January 2013).

Campbell, E.H. 2006. 'Urban refugees in Nairobi: Problems of protection, mechanisms of survival, and possibilities for integration'. *Journal of Refugee Studies* 19(3): 396–413.

Campbell, E.H. 2010. 'Age, gender, diversity mainstreaming initiative key to urban protection'. *Refugees International*. Available at: www.refintl.org/policy/field-report/age-gender-diversity-mainstreaming-initiative-key-to-urban-protection (accessed 20 January 2013).

Campbell, E.H., Kakusu, J. and Musyemi, I. 2006. 'Congolese refugee livelihoods in Nairobi and the prospects of legal, local integration'. *Refugee Survey Quarterly*, 25: 93–108.

Campbell, E.H., Crisp, J. and Kiragu, E. 2011. 'Navigating Nairobi: A review of the implementation of UNHCR's Urban Refugee Policy in Kenya's capital city'. Available at: www.unhcr.org/4d5511209.pdf (accessed 20 January 2013).

Danish Refugee Council and UNHCR 2012. 'Living on the edge: A livelihood status report of urban refugees living in Nairobi, Kenya'. Available at: www.drc.dk/fileadmin/uploads/pdf/IA_PDF/Horn_of_Africa_and_Yemen/livelihood_status_report_on_urban_refugee_in_nairobi.pdf (accessed 20 January 2013).

Herz, M. 2008. 'Somali refugees in Eastleigh, Nairobi'. Available at: http://roundtable.kein.org/files/roundtable/Somali%20Refugees%20in%20Eastleigh.pdf (accessed 20 January 2013).

Horst, C. 2007. *Transnational nomads: How Somalis cope with refugee life in the Dadaab camps of Kenya*. New York/Oxford: Berghahn Books.

Human Rights Watch 2002. 'Hidden in plain view: Refugees living without protection in Nairobi and Kampala'. Available at: www.hrw.org/reports/2002/kenyugan/ (accessed 20 January 2013).

Human Rights Watch 2009. 'From horror to hopelessness: Kenya's forgotten Somali refugee crisis'. Available at: www.hrw.org/sites/default/files/reports/kenya0309web_1.pdf (accessed 20 January 2013).

Human Rights Watch 2010. 'Welcome to Kenya: Police abuse of Somali refugees'. Available at: www.hrw.org/news/2010/06/09/kenya-police-abuse-somali-refugees (accessed 20 January 2013).

International Crisis Group February 2012. 'The Kenyan military intervention in Somalia'. *Africa Report* No 184. Available at: www.crisisgroup.org/en/regions/africa/horn-of-africa/kenya/184-the-kenyan-military-intervention-in-somalia.aspx (accessed 20 January 2013).

Pavanello, S., Elhawary, S. and Pantuliano, S. 2010. 'Hidden and exposed: Urban refugees in Nairobi, Kenya'. Available at: www.odi.org.uk/resources/details.asp?id=4786&title=urban-refugees-nairobi-kenya (accessed 20 January 2013).

Pittaway, E. 2010. 'Making mainstreaming a reality: Gender and the UNHCR policy on refugee protection and solutions in urban areas. A refugee perspective'. Centre for Refugee Research, University of New South Wales. Available at: www.unhcr.org/4b0bb83f9.pdf (accessed 20 January 2013).

The Refugees Act, 2006 30 December 2006. Kenya Gazette Supplement No. 97 (Acts No. 13). Available at: www.unhcr.org/refworld/docid/467654c52.html (accessed 20 January 2013).

UNHCR 2005. 'Urban refugees in Nairobi participatory assessment'. Document on file with author.

UNHCR 2007. 'The Nairobi Refugee Program 2005–2007: Working with partner agencies and refugee communities to strengthen urban refugee protection'. Available at: www.unhcr.org/refworld/docid/48abd53c3.html (accessed 20 January 2013).

UNHCR 2008. *Remarks by Sean Henderson, Senior Resettlement Officer (global issues), at the 2nd Meeting of the Expert Group on Resettlement Fraud (10–11 September 2008, Amman, Jordan)*, 11 September 2008. Available at: www.unhcr.org/refworld/docid/492a9c7a2.html (accessed 20 January 2013).

UNHCR 2009. 'UNHCR policy on refugee protection and solutions in urban areas'. Available at: www.unhcr.org/refworld/docid4ab8e7f72.html (accessed 20 January 2013).

UNHCR 2010a. 'Changing the way UNHCR does business? An evaluation of the age, gender and diversity mainstreaming strategy'. June 2010, PDES/2010/08. Available at: www.unhcr.org/refworld/docid/4c21ac3a2.html (accessed 20 January 2013).

UNHCR 2010b. 'Registration data'. Document on file with author.

UNHCR 2010c. 'Resettlement statistics'. Document on file with author.

UNHCR 2011a. 'East & Horn of Africa update: Somali displacement crisis at a glance'. 24 November. Document on file with author.

UNHCR 2011b. Email from UNHCR Branch Office Nairobi, 22 August. On file with author.

UNHCR Branch Office Nairobi 2010. 'Fact sheet on durable solutions'. Document on file with the author.

Verdirame, G. and Harrell-Bond, B. 2005. *Rights in exile: Janus-faced humanitarianism*, New York/Oxford: Berghahn Books.

Interviews

District Commissioner 2010. Interview with Nairobi official responsible for Eastleigh.

Department of Refugee Affairs Commissioner 2010, Nairobi.

Ethiopian Oromo Refugee Community Meeting 2010, Eastleigh.

Government of Kenya 2010. Interview with staff in Department of Refugee Affairs, Nairobi.

Kibera Courts Magistrate 2010. Interview, Nairobi.

Kituo Cha Sheria 2010. Interview with staff, Nairobi.

Mapendo International (now known as RefugePoint) 2010. Interview with staff, Nairobi.

Nairobi NGO Network 2010. Interview with staff, Nairobi.

Ochola, Joyce 2010. Interview with Head Teacher, Mwiki Primary School, Githurai.

Refugee Consortium of Kenya 2010. Interview with staff, Nairobi.

Refugee Interviews 2010. Interviews with refugees from Ethiopia, Somalia, Sudan and the DRC, Nairobi and surrounding areas.

Refugee Outreach Workers 2010. Interview with employees, Nairobi.

Sudanese Family 2010. Interview in Githurai, November.

Police Commissioner 2010. Interview with Pangani Police Station Commissioner, Nairobi.

UNHCR 2010a. Interview with UNHCR Protection Staff, Branch Office Nairobi.

UNHCR 2010b. Interview with UNHCR Protection Staff, Branch Office Nairobi.

UNHCR 2010c. Interview with UNHCR Eligibility Staff, Branch Office Nairobi.

UNHCR Health Coordinator 2010. Interview, Branch Office Nairobi.

5 Practices of reception and integration of urban refugees

The case of Ravenna, Italy

Barbara Sorgoni

This chapter draws from the findings of research conducted from 2008 to 2010 by a group composed of five post-graduate students in political sciences whom I supervised.[1] The aim of the research was to produce an ethnographic study of the local institutions in the city of Ravenna, Italy, in charge of the "protection and social integration" of refugees and asylum seekers, and to analyse the way the national asylum system functions from an anthropological perspective.[2] In particular, the objective was to understand how national and transnational policies were implemented in everyday practices at the local urban level within a specific refugees' reception project, how bureaucratic procedures for refugee status determination affected the lives and needs of asylum seekers, and how local administrative practices addressed issues such as housing, employment and health. An ethnographic approach was adopted to examine everyday life inside local offices and organizations in charge of what was referred to as the 'reception and integration" process for asylum seekers and refugees – including voluntary organizations. The researchers participated in their activities and attended their meetings. We also conducted selected in-depth interviews with social workers, policymakers, lawyers and medical doctors, as well as with refugees and asylum seekers hosted by the project.

At the time of our research, a dual channel was operating in Italy, whereby asylum seekers were mainly sent to large Asylum Seekers' Reception Centres named CARA (Centri di Accoglienza per Richiedenti Asilo), while a small number of both asylum seekers and refugees entered local projects like the one described here, under the national Protection System (SPRAR). While I will return to the difference between the Centres and the national System in the next section, it is important to bear in mind that, in this chapter, I do not address asylum seekers and refugees separately; despite their being two distinct juridical categories, they were treated as similar social entities by the project staff, who offered them the same services and reception patterns. This was not merely a local characteristic. In fact, in Italy there was no regular programme specifically directed to refugees' second-level integration, and national protection policies were conceived simply as 'first arrival emergency aid' and initial adjustment period for both asylum seekers and 'new' refugees (SPRAR 2010). No institutional programme existed specifically dedicated to refugees' integration as a long-term wider process of social inclusion and participation (Strang and Ager 2010).

Furthermore, in the course of our research we soon realized that the very concept of integration employed in the project documents was quite a narrow one, reduced merely to employment (temporary and low-waged) and accommodation of any kind, irrespective of refugees' wishes, needs, abilities or previous experiences and occupations. In project documents, managers' public speeches or interviews and discussions with case workers, issues related to refugees' social inclusion – such as social participation, access to and use of institutional services, citizenship, family reunification, right to vote, or higher education – were never taken into consideration. Such a narrow understanding of integration is also reflected at the national level where the issue is left to local experiences and specific projects rather than addressed as a structured state policy. A recent report from the national Protection System (SPRAR 2010) shows that only 42 per cent of refugees assisted by the System left the programme having 'achieved integration'.[3] In the report, integration is reduced to two basic indicators, employment and lodging (no matter of what kind), with no reference to socio-institutional domains. Moreover, more than half of those leaving the programme for reason of 'achieved integration' appeared to have found *either* a job *or* accommodation.

Partly as a consequence of this, we realized that what was locally referred to as 'support to social integration' in the project documents – merely assistance in finding employment and housing – remained an ideal stage, never actually implemented at the local level. Those who left the project – whether refugees or rejected claimants – usually had no accommodation or job awaiting them when assistance ended, also because they had received no support from the project in this respect. Indeed, most if not all those we met during the research moved to different towns or even countries, often following informal family or ethnic networks, to compensate for the inadequate character of local institutional assistance. On the other side, the social workers in charge presented themselves as being overwhelmed by the excessive workload related to the first reception stage (asylum seekers and 'new' refugees' initial adjustment period), having no time left for the project's second stage, i.e. refugees' social integration. Thus, an 'emergency rhetoric' pervaded the whole project, as case workers reported great frustration at hardly succeeding in meeting a few basic, provisional and initial needs of the hosted refugees, vis-à-vis an increasing number of both asylum seekers and refugees, and a parallel decrease in both human and material resources.

In this chapter I focus mainly on bureaucratic and administrative procedures and practices related to the initial adjustment period of both asylum seekers and refugees and implemented by the project analysed, showing in the final part how some tentative steps toward a narrowly defined integration concept were rather left to the care and good will of unpaid volunteers.

In the following section I provide some figures and basic information regarding the estimated number of asylum seekers and refugees as well as past policy trends and current legislative and administrative patterns of the asylum procedures at national and local levels. I then describe in more depth the work of legal and social institutions in Ravenna, the type of assistance provided at the reception stage and the related problems, focusing in particular on crucial steps of the refugee status determination procedure, namely legal advice in view of the first personal interview, everyday assistance provided during the determination process, and the work of local associations in supporting newcomers' temporary adjustment. The final section is a critical description of the services provided by public and volunteer sectors in relation to language courses, job-training and employment assistance, their shortcomings, and the often unresolved needs of asylum seekers and refugees as expressed by social workers, project officials and refugees.

Two caveats still need to be mentioned. The first is that data and information (let alone research) on the actual quality of life, forms and degree of social integration of refugees after the assistance ended are not available, either at the local or the national levels.[4] The second is that this mainly descriptive chapter deals with just a fragment of a much wider and varied picture: only about a third of all asylum seekers and refugees in Italy enter 'privileged' projects like the one analysed here; moreover, these same projects differ from one another in a substantial way (IntegraRef 2008).

Numbers, trends and administrative framework: an overview

In Italy, what is referred to as the 'refugee crisis' has been a relatively recent issue that entered public and political debates in the early 1990s when the war in the former Yugoslavia forced thousands of people seeking protection to cross into Italy either by the land border or the Adriatic Sea. Initially these people were taken care of by local NGOs and voluntary civil or religious associations, but as their numbers increased, a national network was created in order to offer a more homogeneous aid pattern, and the first Identification Centres were set up. Following the outbreak of the war in Kosovo in 1999, the EU Commission also started to finance local projects to support asylum seekers' social and educational integration, thus encouraging the creation of a more structured network at the national level. At the same time, both UNHCR and the Italian Home Office began to play a role in the asylum process, initially simply monitoring local policies, from 2001 directly coordinating them through the National Asylum Programme (PNA) and finally, the Protection System for Asylum Seekers and Refugees (SPRAR).

Founded in 2002, SPRAR is directly funded by the Italian Home Office. This system operates under the supervision of a main body or *Central Service* (*Servizio Centrale*) which coordinates activities, policies and practices relating to asylum seekers and refugees implemented at the provincial and municipal level. In the same year, the Home Office entrusted the National Association of Italian Municipalities (ANCI) to direct and manage the Central Service. SPRAR was described as a multilevel local and national governance system, a web of projects of 'assistance, protection and socio-economic integration promoted by local authorities through the activation of territorial networks engaging non-governmental organizations, agencies and institutions with experience and competence in social and productive matters' (IntegraRef 2008).

Over the last 10 years, the number of asylum seekers and refugees in Italy had steadily increased, and, despite the fact that in 2009 the actual number of those seeking international protection had dramatically decreased by 41 per cent over the previous year,[5] Italy was in sixth place among EU countries, with 55,000 refugees and about 4,300 asylum seekers. Compared to other EU member states, this number was still modest in proportion to that of citizens: in 2009 there were 290 refugees per one million Italian residents, while the average in the entire EU was 520 refugees per one million people (UNHCR 2010). In 2009, those seeking protection in Italy were 79.5 per cent men, coming mainly from Nigeria, Somalia, Bangladesh, Pakistan, Eritrea, Ghana, Ivory Coast, Afghanistan, Turkey and Serbia. In the same year, those who were granted refugee status came primarily from Eritrea, Somalia and Afghanistan. The status was denied to most claimants coming from Nigeria, Ghana and Pakistan (SPRAR 2010).

At the legislative level, despite Italy having only piecemeal legislation on asylum issues and still lacking a comprehensive national law, it is one of the few European countries to guarantee a right to asylum in the national Constitution. According to Article 10(3) of the Italian Constitution of 1948,

An alien who is denied the effective exercise of the democratic liberties guaranteed by the Italian Constitution in his or her own country has the right of asylum in the territory of the Italian Republic in accordance with the conditions established by law.

But as Lambert *et al.* (2008) have explained, constitutional asylum was applied only sporadically, international protection being primarily granted under the 1951 Refugee Convention and increasingly the new EU asylum legislation, in line with the current proliferation of supra-national actors.[6] Two main steps towards a common European asylum system – i.e. Council Qualification Directive 2004/83/EC and Council Procedure Directive 2005/85/EC – were adopted in Italy in 2007 and 2008 respectively, albeit partly altered by the subsequent Legislative Decrees. Thus a tension existed between a European homogeneous procedure and states' policies and practices; as some noted, a common EU system 'is still a myth rather than a reality' (ECRE 2010).[7]

Upon their arrival in Italy, asylum applicants are required to submit their claim directly at the border or at any Police Station. Usually they are then sent to a Reception Centre for Asylum Seekers or CARA. Created in 2008 after the reception of EU Procedure Directive to replace former Identification Centres (CID), CARA are large centres located far from urban centres; since they provide little or no social, economic, cultural or linguistic service, they are described as long-stay parking-places for human beings. A smaller number enters the SPRAR programme while waiting for their status determination. Thus a complicated 'dual channel' operates, whereby about 70 per cent of applicants are sent to a strict-control type of centre while only an ever decreasing number of asylum seekers have access to a SPRAR project of 'protection and integration', together with some refugees. The criterion behind the decision as to who should be sent where is confused and arbitrary (Schiavone, in SPRAR 2010: 57).

In 2010, first instance authorities who granted refugee status or a subsidiary protection – or denied any protection – numbered 10 in all of Italy. In the last resort, and under specific circumstances, they could issue a one-year humanitarian leave to remain. Rejected applicants have 30 days to appeal, reduced to 15 days for those inside CARA. If the response remains negative after a second appeal, they have to leave the country within a few days. Refugees or persons with a subsidiary protection can remain in a SPRAR project for about six months during which they are hosted by the project, receive initial economic support and free primary legal and medical assistance through registering with the National Health Service, and can attend Italian language and other courses for education or professional retraining. Asylum seekers admitted to SPRAR can remain in the project for the whole duration of the refugee status determination.

Because of the dual channel described above, the long waiting list to access SPRAR which in 2010 had 1,000 people on the waiting-list, the proliferation of different types of reception centres and informal support organizations running parallel to the SPRAR system, and finally the fact that the living conditions of applicants who never even entered the programme remained unrecorded, the monitoring of the actual integration processes of refugees is, to say the least, quite difficult. Furthermore, research focusing on asylum seekers within a SPRAR project like the one described here covers a small portion of a wider, complex picture. Thus these projects cannot be considered typical of the Italian case, especially in view of the fact that the country's asylum system is characterized by strong internal differences, with administrative, legal, social and relational practices varying substantially from city to city, and from project to project even within the SPRAR system.

The local project

Local administrations' participation in the national SPRAR system works on a voluntary basis. Those provinces and cities choosing to be part of the programme can apply for financial support in order to set up 'Projects for the reception and integration of refugees and asylum seekers'. They implement these projects in close collaboration with local institutions such as NGOs, voluntary associations, and civil society or religious organizations. In 2009/2010, 138 projects were funded and coordinated by SPRAR, spread across 69 provinces. Local administrations implementing SPRAR programmes numbered 123, with 110 municipalities (SPRAR 2010). Within this framework, the Emilia-Romagna region where Ravenna is located, in 2004 promoted the creation of a regional network for the integration of refugees and asylum seekers, thus starting the project Emilia-Romagna Land of

Asylum. Yearly, the project monitored the number of asylum seekers in the area, provided courses to train legal advisors and social workers, promoted the exchange of good practices and the identification of common issues among the partners, and organized events to inform the civil society on asylum issues (Fiorini and Palamidesi 2010).

Ravenna is a small city on Italy's northeast coast. It became part of the national protection system in 2001, implementing a project for the social integration of refugees and asylum seekers in the urban area. The project was part of a wider network of services coordinated by the municipality which also included economic migrants and vulnerable persons, mainly victims of trafficking. This network included a bureau for migrants that provided information on available local services, legal, administrative and procedural assistance, an anti-discrimination office, a job information service and a multicultural centre.[8] All migrants, including forced migrants, could access those services, but asylum seekers who entered the SPRAR project were entitled to additional support (namely lodging, some economic support and free legal assistance).

During the period of our research (2008–2010), and until the end of 2010, the project was implemented by the Social Security Agency; it hosted some 66 SPRAR beneficiaries coming mostly from Cameroon, Eritrea, Nigeria, Togo, Afghanistan and Iraq (mainly Kurds), with a significant number of Kurds from Turkey.[9] Structured as a *Consortium* for social services, the Social Security Office was also responsible for providing poor people, disabled persons, those affected by any form of addiction and ex-convicts with welfare support. The SPRAR project staff was composed of two social workers, a career advisor and a legal expert. The project was also supported by the local branch of the Catholic NGO Caritas which ran a dormitory and a canteen, by a women's support association providing psychological consultancy to women and a psychiatrist for men, a lawyer to represent asylum seekers at reviews and appeals, and various volunteers providing Italian language courses.

The first reception stage

As Elizabeth Colson writes, upon their arrival in new countries refugees and asylum seekers are defined as needing assistance from 'agencies that have an obligation both to educate them on how to behave in the new setting and what they can in turn expect' (2004: 112). This initial stage is characterized by a period of limbo and juridical uncertainty during which prolonged waiting for their claims to be adjudicated is a common experience for asylum seekers throughout Europe.[10] As a Kurdish refugee in Ravenna said, 'while we wait for the status determination and leave to remain, our life is left without dignity, it is a long waiting made of worries and fears ... mostly the fear to be sent back' (Berkendel 2010). The painful and often traumatic experience of prolonged waiting is directly linked to the determination of status, thus highlighting once again the symbolic salience of the labelling process. It has been noted that the actual fragmentation of categories intended to manage mixed migration flows, far from enhancing protection, is rather leading to an increase in vulnerability; and also that labelling is a useful tool to remind people that they are under the authority of those who have the power to define them.[11] The construction of a category of persons united not by a common culture, language or history, but because they share the same experience of having to learn how to 'become refugees', brings to mind the question Barbara Harrell-Bond (1999, 2002) posed in relation to refugee camps in the south: can the relationship between refugees and their 'helpers' become itself a source of psychological stress? She also argued that debilitating stress is likely to occur not only in camps but in any situation where refugees are dependent on others for their survival, especially during the liminal period of transition.

At the initial stage, while still in a juridical limbo waiting for their status to be determined, asylum seekers all over Europe encounter different institutional actors in various contexts who have – or appear to have – the power substantially to define and determine their future condition. It is before them that asylum seekers are expected to recall and recount their experience of persecution, often more than once. Refugees who were met in Ravenna first related their statements of why they had been forced to flee when claiming asylum at a police station. Here, they were requested to fill in a form with general information such as their name and that of their parents, profession, country of origin, ethnic and religious affiliation, other countries crossed, and the route to Italy. They also had to write in any preferred language the history of their flight, their motives for their flight, and their fear of persecution. In this context, official interpreters were sporadically available only for English or French, while for other languages police officers made use of informal contacts already established with individual foreigners or migrants' associations based in the area. As one police officer in charge explained, an official interpreter was not needed given the nature of the meeting which was described as 'not so relevant' since 'the real thing happens in front of the *Territorial Commission*', i.e. the first instance authority. Yet, as the lawyer who assisted claimants in appeal explained, when information provided in review or appeal partly diverged from the first account handed in at the police station, this often negatively affected the final decision, thus showing that the first narrative was, on the contrary, *quite relevant*.[12] During that first meeting, police officers could also forward official requests to the local prefecture for the claimant to enter a project within the SPRAR national system.

Those who did enter the SPRAR system in Ravenna soon met with the legal expert, who was employed only eight hours per week. This legal expert was expected to cover all initial administrative procedures: his role was to inform asylum seekers of their rights and on the procedures, to help them write down a detailed testimony of persecution, and particularly to prepare the personal interview in front of the first instance authority. This was a key moment in the asylum process given its centrality for the status determination, yet in Italy (and differently from other EU countries) only lawyers were allowed to attend the interview, while paralegals – who in most cases were the only one to provide legal assistance up to that moment – were excluded.[13]

The literature shows that the preparation of the written autobiographical statement is particularly important because it is used as the main evidence that the applicant truly deserves international protection: i.e. to ascertain the real identity of the person and the credibility of the persecution experience recounted in the testimony.[14] But as Liisa Malkki has noted during her fieldwork in Hutu refugee camps in Tanzania, administrators, as well as social workers and legal experts, found it necessary to 'cut through "the stories" to get to "the bare facts"'.[15] In a similar way in many European countries, along with a growing obsession with so-called *bogus* refugees, the simple narrative which until the 1990s had been 'the cornerstone of proof, no longer sufficed' (Fassin and D'Halluin 2007: 310).

In order to help asylum seekers effectively to write up a credible testimony, the legal advisor in Ravenna explicitly compared his role to that of an art director who literally helps an actor to stage the required performance. In his own words:

> I work on the discourse construction, almost on a script, on the cut, in view of the interview [...] when their story is very weak I suggest they stress the dreadful condition of their country, thus aiming at least to get subsidiary protection.[16]

As has been already mentioned, this was also a particularly painful experience during which the person was repeatedly urged to recall and retell in detail extremely traumatic events (Beneduce 2008).

The legal advisor clearly explained to the asylum seekers that the content of the story – and also its fine details and the way it was told – all played an important role for the determination of their status. For this reason, an additional meeting was explicitly dedicated to prepare the interview with the authority of first instance. In this case, the legal expert informed the asylum seekers about questions frequently asked and the type of experience they could expect. Those who wished to do so were also offered the possibility 'to rehearse' the interview in front of him shortly before the real one actually took place. He described this meeting as a simulation, a sort of training where he might 'play the bad cop' to prepare them in case the actual interview got 'nasty'.

What I have described here suggested that the legal advisor, far from being himself trapped in the rhetoric of truth and credibility, was clearly aware of the manipulative potential of performed narratives, and of his strategic position in the procedure. He reminded us of the type of social worker described by Kobelinsky (2008) who regarded refugee status as a bureaucratic construct and adopted a distant attitude while at the same time providing technical know-how. It also suggested that the relationship actively invited the asylum seekers to play a role (namely that of the victim) in order to become more credible. Finally, when the legal advisor noted that such rehearsal-meetings were increasingly being turned down, he also implicitly recognized the high degree of trauma and suffering the interview elicited – so much so that asylum seekers preferred to avoid experiencing it twice even when the first one was 'only staged'. The legal advisor confessed he had at times wondered whether his role was really that of a facilitator, or rather one who made the asylum seekers feel uncomfortable. Apparently the latter situation occurs more often when the legal advisor is also politically active. As Fassin and D'Halluin wrote, the practice of certification is crucial 'in terms of ethic and political issues and, therefore, of emotional involvement', so that 'doubt assails even the certificates' authors who are precisely those who defend the idea of political asylum … activists must know how to be convincing even if as experts they hesitate' (2007: 318–319).

Bureaucratic encounters

Along with preliminary legal support, the initial reception stage in the SPRAR system also usually includes other aspects such as individual meetings with new-comers, introduction to the hosting structure and its regulation, signing of a hospitality contract, and economic support. In the project we studied, these activities were handled by two social workers. They had had long experience with different aspects of social assistance before they had been assigned full time to the asylum seekers and refugees office. In that context they were expected to be responsible for a wide variety of issues relating to the socio-economic integration of the beneficiaries for the entire period of their stay in the project, which was about six months. From the perspective of refugees and asylum seekers, this meant that they were heavily – if not entirely – dependant on their 'helpers' for their survival and security, and were thus placed in a disempowering asymmetrical relationship. The process of objectification implied in the development of standardized protocols, while increasing staff authority, also translated into a progressive erasure of refugees' experience as political subjects (Fassin and D'Halluin 2007). In the words of a refugee woman who vividly expressed what many others have experienced: 'why do I always have to ask, why do I always have to beg? … here, in Italy, I have no pride left' (Ya Basta! 2010).

SPRAR's 'Operational Manual for the start-up and management of reception and integration services' acknowledged the fundamental role played by social workers for a successful reception and integration of any refugee, describing their many daily duties and thus defining the relationship simultaneously as characterized by 'reciprocal trust' and 'reciprocity', but also as a 'professional and not personal' interaction.[17] This document made evident – and betrayed – the structural ambiguity inherent in the role of social workers dealing with such a particular type of beneficiary. The social workers had to deal with the most intimate aspects of refugees' lives, taking care of their profound needs and having access to their autobiographical files, suggesting solutions, communicating decisions, which drastically impacted on their lives, or even making decisions on their behalf. Yet these social workers were expected to do so while remaining detached and professional. In sum, they had to learn how to manage a relationship where profession, politics, ethics, affects and emotions all intersected with one another – and to do so from a non-personal standpoint. Such a unique relationship between helper and beneficiary rather had to be defined as

artificial, in that the bureaucratic encounter was from the start structured as a fic-
tional reciprocal trust that neither social workers nor refugees could escape. The
refugees were condemned to endless explaining and recalling their narrative in
order to access their rights. The social workers were expected to perform their
duty pretending they believed what they were being told. As one social worker
in frustration said, 'How can we verify what they tell? They can even change
their name. We do not even know their *real* name!'[18] The way this relationship
of assistance was structured further encouraged the overlapping of juridical dis-
tinctions with moral judgments (Fassin 2010).

Trust and mistrust are major themes within refugee studies, and they are at
work in the relationship between asylum seekers and hosting state agencies from
the first personal interview to the later stages of the integration process, in the
global North as well as the South (Daniel and Knudsen 1995). Because refugees'
experiences with governance in the new country are often through relationships
comparable to those set up for legal minors (Colson 2004); and because of the
consequent crucial role of social workers in providing vital services, the social
workers – just like NGOs and agency employees elsewhere – were both caretak-
ers and gatekeepers, and had and were perceived to have more power and decision capability than some of them were ready to admit. This further complicated
the ambiguous nature of trust, reinscribing it within a hierarchical relationship.
When pressing needs on the part of asylum seekers got no other answer than 'we
are very sorry but we cannot help you', or 'come back tomorrow' – which could
be repeated time and again – asylum seekers interviewed were likely to explain
this attitude as a deliberate decision with different causes. To some of them, this
was a sign that the social workers were not really interested in what they did;
they did not really care. Others held social workers responsible for granting
access to services, including key information, provided at their own arbitrary
discretion. Frequently, a complicated ethnic or gender hierarchical order was
imagined, whereby social workers 'first help Arabs to find a job and leave us
Africans (or Kurds) aside', while men lamented that women were allowed to
stay in the lodgings longer, and women complained that men were given better
job opportunities. This may not have been the case; at least in the case of gender
discrimination things were rather more complicated, as I shall argue in the final
section. However, ethnic and gender stereotypes were quite common among
staff members. In a study of civil servants working in public bureaus for
migrants and refugees in Sweden, Mark Graham (2002) noted that clients were
expected to perform in accordance with stereotypes of appropriate 'refugee
behaviour' – i.e. passive and grateful – and were also nationally ranked through
the creation of categories of 'difficult ethnic groups'. While according to his
research, the 'most difficult' refugees in Stockholm were Iranians, in Ravenna
Eritreans probably came first.

In any event, when criteria behind specific decisions relating to the allocation
of resources or rights (including the circulation of information and guidance)
were not clear or clearly explained, resentment and suspicion arose among
asylum seekers and refugees, and arbitrariness was interpreted as a deliberate

display of carelessness or power. Arbitrariness in the actual delivery of services
– which instead were described by social workers and asylum seekers as
objective rights provided by public institutions expected to act fairly – was a
widespread experience in the whole region, according to a recent report on
refugees' integration in the four cities of Bologna, Modena, Parma and Ravenna
(Ya Basta! 2010). While conducting our research in Ravenna, we realized that
among asylum seekers and refuges selected to enter the SPRAR system, a
smaller portion was further allowed to take part in a 'project within the project'.
This 'smaller project' – as a caseworker himself called it – allowed the deserving few to attend a job-training course in the morning and an Italian language
course in the afternoon, five days a week for a few months, in a *Cooperative*
where they performed electric-component assembly. I will come back to this
'smaller project' in the following section where I focus on what was termed *secondary reception*, i.e. the actual integration stage which followed first aid and
reception. But it is important to stress that all the asylum seekers and refugees
we interviewed on this issue, both those selected and those excluded, admitted
that they could not explain why some were selected to be 'in' and others 'out' of
the 'smaller project'. Some of them pointed to the gender dimension, and indeed,
during the two years of our research, we noted that men selected for this sub-
project significantly outnumbered women. Notably, even though we repeatedly
asked the project manager, the career advisor, and the two social workers about
the criteria they had adopted for the selection, their explanations were opaque
and changed over time. Confronted with criticism or complaints by those who
were excluded, project staff often referred to the limitations of their own competence and authority on the matter: it was not up to them to decide.

As already argued, the asymmetric relation of dependence was also reinforced
by the fact that social workers were expected to know and take care of most
intimate and minor details of refugees' daily life. The initial hospitality contract
that asylum seekers and refugees were requested to sign upon admission to the
SPRAR project listed a series of rules and duties prescribing the correct
behaviour for single adults when residing inside the project's lodgings. This
arrangement showed many analogies with a familial setting, the relations
between helpers and beneficiaries resembling that of parents and children. For
instance, project staff were entitled to check periodically – sometimes even
without notice – that the premises were being kept clean to an 'appropriate'
standard, while inmates were not allowed to leave for more than 24 hours
without previous notice. Failure to accomplish the assigned duties or respect the
contract might result in different sanctions, the most serious being expulsion
from the project and the whole protection system. Lesser sanctions included temporary reduction or withdrawal of pocket-money.

Social workers justified the need for standardized rules and sanctions as
follows: 'it is expedient that they understand as soon as possible that *here with us*
there are clear rules to respect, and if they do not conform they will be the first to
suffer'. Here, cultural difference was what opposed them to 'us' (Italians/citizens),
and 'their' supposed future integration was predicated upon the inevitable adoption

of one-fits-all regulations. Moreover, since a smooth coexistence was directly linked to the guests' willingness to conform to given standards and needs determined by others, alternative views were discouraged and silenced (Sorgoni 2011b).

The provision of pocket-money to beneficiaries in the project was not compulsory, and single SPRAR projects were free to choose whether to adopt it and also how much and how often to provide this economic support. In Ravenna pocket-money was distributed weekly by one of the two social workers.[19] Thus, on Friday morning adults lined up in front of the office to receive money in their hands directly from the hands of their helper. According to *both* the social workers and the asylum seekers and refugees, this was a particularly disconcerting and humiliating encounter. As a refugee explained, social workers 'represent the Italian State, they are not your father'. Other projects in the region adopted less personalized strategies, for instance issuing a special bank card or a note to pay that beneficiaries could cash in at the project office like all other employees.[20] But the ambiguity implicit in a 'gift' that only one part of the relationship had the power to control, diminish or withdraw, made the money look more like a reward than a right, strongly reinforcing and reproducing the patronizing and paternalistic aspects of this asymmetrical relationship to the point that some refugees preferred to quit the SPRAR and any form of public assistance: 'When we refused to submit to their rules they withheld our pocket money or even fined us ... this is how they punished us' (Ya Basta! 2010: 39).

Steps towards integration: language courses and the role of volunteers

As mentioned earlier, not all asylum seekers or refugees residing in Ravenna entered the SPRAR project; many never did, some had to wait. These were the less monitored persons who tended to move across the country or to different destinations. In Ravenna there were two dormitories for both citizens and foreigners. One, The Good Samaritan, had been operated directly by the Catholic parish S. Rocco since the late 1980s and had a canteen. The other was managed by the municipality and opened in 2003. Both dormitories were run by volunteers and were conceived for temporary short stays. But recent trends suggested that more persons, especially asylum seekers, increasingly tended to prolong their stays. According to some volunteers – and despite initial mistrust on their part ('because you never know if you are meeting someone who is trying to pull a fast one, or a person really in need') – a longer stay encouraged asylum seekers to talk about their persecution experiences which, many volunteers believed, was 'their deepest need'. The above statement combines two assumptions often repeated by both public sector staff and volunteers: namely, that having lost all connections to family and friends, asylum seekers and refugees expressed an urge to talk about themselves with a friendly listener, and that voluntary organizations offered a more generous environment, while social workers represented a formal and authoritative type of approach. The latter assumption in particular is widespread and, in some cases, created competition between the two sectors (Graham 2002). Volunteers' work also implied a specific idea of charitable gift which affected their relationship with asylum seekers and refugees in different ways, as revealed by the case of Italian language courses described below.

Depending on different local contexts, SPRAR projects could encompass Italian language courses, psychological and psychiatric assistance, education or professional retraining, job placement or lodging. Ager and Strang (2008) suggest that the concept of integration is used with widely differing meanings across Europe. In our field experience, integration was locally conceived of as a narrow concept, namely as economic integration; yet a picture of its extent as well as the types of jobs and accommodation available to refugees in Ravenna was extremely opaque. The work of volunteers in supporting refugees' integration in the town was mostly relevant in the case of Italian language courses which were considered as a critical induction to finding employment. The courses were managed by the Social Security Office and implemented by two types of unpaid volunteers: retired teachers and graduate students. Since 2008 the project could also employ a graduate student to coordinate all teachers' activities, hired through a grant from a local Foundation.

As elsewhere in Europe (Wren 2007), so also in Ravenna the voluntary sector carried on its shoulders the weight and responsibility of filling the gaps in public projects, and volunteers in the local project always stressed this point, along with their indispensable role. Retired teachers also expressed preference for the freedom granted by their new position, which allowed them to be less authoritative with their students (as opposed to their previous role in state schools). They stressed that they were offering their time without expecting material reward, a position that in their opinion was highly appreciated among asylum seekers, who turned to them more as friends than as teachers. In their view, it was the perception of their altruism that enhanced their relationships. Some of them had been doing this work for three years, from one to three afternoons per week. It was mainly this group that emphasized the value of their generous gift, all the more precious because, as some stated, it was offered to people 'who have nothing, are left with nothing'.

On a different level were graduate volunteers who taught courses usually for a three-month period with the clear aim to gain credits or some practical experience for their final dissertation. Most were interested in migration issues for personal reasons and some were politically engaged. Yet neither graduate students nor retired teachers had any specific education in migration or asylum issues, nor were they trained to teach Italian to foreigners. This was justified by project staff and management as due to lack of funding, but it gave rise to a strong sense of frustration among both groups of volunteers who had to cope with difficulties they had not initially foreseen. They felt abandoned, inadequate and confused, as the following episode shows.

Under pressure from asylum seekers and refugees, some volunteers proposed to organize a meeting with the migration municipal official in order to present him with the difficult economic situation most beneficiaries faced. They thought the meeting was to be attended by three volunteers, their coordinator and one refugee. Social workers supported this idea but implied that it might be better if just the volunteers attended. This example shows that volunteers often willingly fill gaps in public services, in this case caring about refugees' economic condition and not just their linguistic education as their role required. In any event, it was left to a young temporary volunteer coordinator to organize the meeting and to decide whether or not to invite *one* refugee. She found this situation difficult and unfair and had many doubts about why she was allowed to choose just one refugee, how to select him or her, and how she would explain her final decision to all other beneficiaries. This episode and the actual meeting that eventually did take place also shows the progressive marginalization of refugees themselves whose needs and demands were defined and expressed by others. In the end, one refugee did attend the meeting but had no chance to speak. On the contrary, attending volunteers used the meeting to voice their own problems and frustration.[21]

Suitable jobs

Aside from language courses, asylum seekers and refugees selected for the 'smaller project' also attended a job retraining course in electrical component assemblage in a *Cooperative*, a course initially designed for mentally disabled persons. There are other cases where patients with mental difficulties perform assembly tasks or where refugees are placed in the same 'box' as disabled people.[22] While the issue deserves deeper analysis, I want to stress that in Ravenna the tool the project staff used to evaluate asylum seekers' skills attending language and job-retraining courses was a local version of ICF test (International Classification of Functioning, Disability and Health test) of the World Health Organization. This test is used to evaluate adult mental and physical disability.[23] Conceiving asylum seekers as disabled people had at least two immediate and related consequences: on the one hand it allowed social workers to explain any social, legal or economic contextual difficulty refugees face as their own fault, thus blaming them (Waldron 1987) for their failure in 'integrating'; on the other it circularly justified and reproduced refugees' dependency on charitable assistance.

Some asylum seekers were further proposed to take part in specific job-training arrangements known as *Borse lavoro*, literally 'job-grants'. This was an agreement between the Social Security Office and local employers whereby the local employer was encouraged to temporarily hire a refugee full-time, eight hours a day for five days a week, without having to pay a proper salary but rather a grant of about €400 per month. After this period, proper employment within the same company could follow. This peculiar type of arrangement was highly criticized by asylum seekers and refugees. Some described it as 'labour exploitation camouflaged as assistance', others compared it to a form of slavery since they were expected to work full-time virtually for free knowing that promised employment was only 'a dream'. But many held the actual economic crisis responsible for present unemployment, naming various cases of persons they knew who, in the past, had been successful in being hired after the short period of such training.

It was true that the region of Emilia-Romagna had at the time one of the highest rates of unemployment, with a sharp increase in short and temporary contracts with little or no protection for the employee. But insecure conditions among asylum seekers and refugees also had other causes, the primary one being a structural character of the country. In Italy migrants in general, including refugees, are prominently channelled towards unskilled, lower tasks and low paid jobs in marginal sectors with low labour protections, notwithstanding their former qualifications, education or specialities.[24] Within SPRAR projects, this translated into a higher number of refugee women being employed because, no matter what their previous education or occupation was, they were hired primarily in the private cleaning sector. Of 26,000 asylum seekers and refugees who entered SPRAR, 74 per cent were men and 26 per cent were women. Of that total number, only about 40 per cent left the programme with a job, 32 per cent of whom were mainly women hired in the cleaning sector (SPAR 2010). Based on our research in Ravenna, it seemed possible to affirm that job-retraining was offered mainly to men (although many of them considered the type of training available unsuitable for the actual market needs), on the sometimes explicit assumption that women would informally find a job as cleaners or carers for elderly persons within local families.

The above could be described as a form of institutionalized social exclusion (Hynes 2009), which served to keep refugees 'in their correct social place'. Social workers reacted with irritation or a patronizing attitude to asylum seekers' personal wishes or life projects that were labelled unrealistic. For instance, three refugees who expressed their wish to enrol at a local university branch received no support or guidance, but were asked with scepticism how they imagined they could support themselves and pay university fees if they had no job. And while no information was provided in relation to exemption from fees, in one case a social worker questioned the degree the asylum seeker presented, affirming it could not be considered 'a proper BA'. As Ong noted (2003: 135), obstacles to a promotion of refugees' employment possibilities are also their exclusion from information circuits. In the words of a refugee, 'it's difficult to understand how to live here ... it's like a secret' (Ya Basta! 2010: 28).

It was against this structural situation that the present global economic crisis could be read, but the global crisis was also used circularly as justification for the institutionalized marginalization of migrants within the Italian labour market. At the end of 2009, refugees attending the language course asked the teachers' coordinator to organize a meeting with the career advisor responsible for the organization of job-grants arrangements and retraining courses. They wished to ask him why none of them had been offered either form of assistance during an entire year. The career advisor explained that, in the past, the local market had offered many employment opportunities for low and unqualified jobs 'which are the types of jobs that better suit you all', but the present economic crisis had dramatically reduced those opportunities. He thus concluded, 'under present critical circumstances, it is all the more difficult to convince employers to hire most fragile and needy persons like you to perform even low-profile jobs'.

This chapter deals with a fragment of a much wider and varied picture, given that only a third of all asylum seekers and refugees in Italy were admitted to SPRAR projects like the one described here, and that these projects differed markedly from one another. Yet, some features that emerge from our research were consistent with the findings from other reports at the national level, while others bore strong continuity with policies and practices adopted across the EU and also in refugee camps in the Global South. At the national level, two aspects seem to affect the life and integration perspectives of urban refugees more negatively. First is the arbitrary nature of service provision, including the lack of circulation of crucial information. The second is the institutional marginalization of all migrants within the Italian economic market, including refugees. Parallel to these constraints, public and volunteer sectors providing integration services in Italy are also strongly influenced by that humanitarian ethos widely analysed in refugee studies, which constructs help as a charitable and asymmetrical transaction, and refugees as passive victims who can only aspire to a marginal and insecure position within Italian society. What is in question, once again, is not the need for help but the type of help needed.

In the period of our research there were no specific and formal refugees' associations to support newcomers in Ravenna. Although our study did not explicitly address this issue, during our field experience we became aware of informal global networks based on family, ethnic or national ties that were clearly at work and could sometimes effectively channel relevant information, especially in relation to legal procedures, housing or job opportunities. This information often entailed a comparison among various other European countries about which asylum seekers and refugees either had previous personal experience or had received news from friends and family members. The majority of the refugees we encountered during our research expressed a strong wish to move to other northern European countries on the grounds, whether real or not, that 'there' it would be easier to find better and more gratifying jobs. Or, as one of them said, because '*there* they construct you as a person'.

Notes

1 Sorgoni 2011a; the researchers were Giulia Gianfagna, Sara Pozzi, Elena Starna, Rossella Urru and Lorenzo Vianelli. I am particularly grateful to ANCI and Fondazione Flaminia in Ravenna for the financial support provided. I wish to thank Daniela Di Capua and Maria Silvia Olivieri for their encouragement and advice, and Alessandro Fiorini for commenting on the legal aspects addressed here. This chapter was submitted in 2011.
2 An early critical assessment of the specificity of the anthropological perspective in refugee studies is Harrell-Bond and Voutira 1992.
3 In fact the figures quoted above show that, at a national level, about 58 per cent of refugees left the programme having found *neither* a job *nor* accommodation.

132   *B. Sorgoni*

4 I refer here to the period from 2001, when Italy set up a national protection programme and later adhered to a common EU asylum legislation. For some initial considerations see Catarci 2012. On the previous period see Korać 2001, 2003; Però 1999; Puggioni 2005; Schuster 2000.
5 The 2009 dramatic decrease was linked to the 2009 Italy–Libya Treaty and repatriation policies that took place directly in the Mediterranean sea without applicants being able to claim protection, and to border control at sea (see Fischer-Lescano *et al.* 2009; Marchetti 2010; and Amnesty International 2010).
6 As Barbara Harrell-Bond argued some time ago (1986), the rise of an international regime of humanitarian intervention replaced the role previously exercised by the State or individual themselves, also imposing a specific *ethos* of humanitarian work. Within this framework, Constitutional asylum is commonly understood merely as a right to stay in the country while waiting for the claim to be processed according to more recent EU procedures.
7 For a historical sketch of the initial process of harmonization of EU asylum policy see – among many others – Sigona 2005, who also discusses concepts of refugee integration.
8 On the functioning of this 'integrated reception system' see Bernabini and Sutter 2010.
9 By 2011 the project was being operated by the municipality; it should have hosted about 45 refugees or asylum seekers per year, but the actual number exceeded that.
10 See Brekke 2004 among others; see also the contributions in Sorgoni 2011d for a comparison across some European countries on institutional uncertainties and waiting strategies.
11 See respectively Zetter 2007, Marchetti 2007; and Colson 2004.
12 This issue was addressed by Starna, in Sorgoni 2011a: 113–140.
13 As the ECRE/ELENA 2010 Report shows, proper legal aid in Italy was provided at later stages, mostly at revision or appeal.
14 See Sweeney 2009. For an anthropological critical discussion of credibility and asylum see Blommaert 2001; Bohmer and Shuman 2008; Cabot 2011; Good 2007; Kobelinsky 2010; Maryns 2006; Sorgoni 2011c.
15 'The "real refugee" was imagined as a particular kind of person: a victim whose judgment and reason had been compromised by his or her experiences. This was a tragic, and sometimes repulsive, figure who could be deciphered and healed only by professionals, and who was opaque even (or perhaps especially) to himself or herself', Malkki 1996: 384–385.
16 On narrative credibility and the role played by legal advisors, see Pozzi in Sorgoni 2011a: 35–60.
17 *Manuale operativo per l'attivazione e la gestione di servizi di accoglienza e integrazione per richiedenti e titolari di protezione internazionale*, n.d: 5–6.
18 On the concept of artificial trust in the analysed project, see Urru in Sorgoni 2011a: 61–86.
19 Interestingly, both Italian official manuals and case workers in conversation commonly use the English expression 'pocket money'. In Ravenna this weekly sum amounted to about €20, which asylum seekers and refugees could use for personal necessities excluding food, which was differently provided.
20 This is the solution adopted in the nearby city of Forlì where I have since expanded the research. In Ravenna, since 2011 the project has been operated directly by the municipality which is exploring different avenues, including using a bank.
21 On this issue see Vianelli (in Sorgoni 2011a: 87–112), who quotes Harrell-Bond's suggestion (1999), that sometimes helpers need refugees more than refugees need them.
22 See respectively McCourt Perring 1994, and Steen 1993, for similar cases in the United Kingdom and Denmark.

23 This issue is explored by Gianfagna, in Sorgoni 2011a: 141–162. In 2011, also on the basis of a preliminary report on our ethnographic study, the municipality cancelled both that particular assemblage training and the use of the ICF test.
24 Caritas Migrantes 2010. Indeed, in Italy it is also particularly difficult to get any foreign educational degree or professional certificate officially recognized; without a recognized degree or certificate, employment in the whole public sector is severely limited.

References

Ager, A. and Strang, A. 2008. 'Understanding integration: A conceptual framework'. *Journal of Refugee Studies* 21(2): 166–191.

Amnesty International 2010. *Seeking safety, finding fear: Refugees, asylum seekers and migrants in Libya and Malta*, December 2010 Report.

Beneduce, R. 2008. 'Undocumented bodies, burned identities: Refugees, *sans papiers, harraga* – when things fall apart'. *Social Science Information* 47: 505–527.

Berkendel, A. 2010. 'Per fare un mondo ci vuole di tutto'. In Gozzi, G. and Sorgoni, B. (eds) *I confini dei diritti*. Bologna: Il Mulino.

Bernabini, F. and Sutter, R. 2010. 'Ravenna terra d'asilo. Le politiche dell'ente locale per i rifugiati e richiedenti asilo'. In Gozzi, G. and Sorgoni, B. (eds) *I confini dei diritti*. Bologna: Il Mulino.

Blommaert, J. 2001. 'Investigating narrative inequality: African asylum seekers' stories in Belgium'. *Discourse and Society* 12(4): 413–449.

Bohmer, C. and Shuman, A. 2008. *Rejecting refugees: Political asylum in the 21st century*. London and New York: Routledge.

Brekke, J.-P. 2004. *While we are waiting: Uncertainty and empowerment among asylum seekers in Sweden*. Report. Oslo: Institutt for Samfunnsforskning.

Cabot, H. 2011. 'Rendere un rifugiato riconoscibile: performance, narrazione e intestualizzazione in una Ong Ateniese'. *Lares* LXXVII (1): 113–134.

Caritas Migrantes 2010. *Immigrazione. Dossier Statistico 2009*. Roma: Idos.

Catarci, M. 2012. 'Conceptions and strategies on integration across refugee services in Italy'. *Journal of Educational, Cultural and Psychological Studies*, 5: 51–83.

Colson, E. 2004. 'Displacement'. In Nugent, D. and Vincent, J. (eds) *A companion to the anthropology of politics*. Malden-Oxford-Victoria: Blackwell Publishing.

Daniel, V. and Knudsen, J.C. (eds) 1995. *Mistrusting refugees*. Berkeley–Los Angeles–London: University of California Press.

Ecre 2010. *Comments from the European Council on Refugees and Exiles on the European Commission Proposal to recast the Asylum Procedures Directives*. May 2010. Available at: www.ecre.org/resources/Policy_papers (accessed 20 January 2013).

Ecre/Elena 2010. *Survey on Legal Aid for Asylum Seekers in Europe*. October 2010. Available at: www.ecre.org/resources/Policy_papers (accessed 20 January 2013).

Fassin, D. and D'Halluin, E. 2007. 'Critical evidence: The politics of trauma in French asylum policies'. *Ethos* 35(3): 300–329.

Fassin, D. 2010. 'Introduction. Frontières extérieures, frontières intérieures'. In Fassin, D. (ed.) *Les nouvelles frontières de la société française*. Paris: Éditions La Découverte.

Fiorini, A. and Palamidesi, G. 2010. 'Il Progetto 'Emilia-Romagna Terra d'Asilo'. In Gozzi, G. and Sorgoni, B. (eds) *I confini dei diritti*. Bologna: Il Mulino.

Fischer-Lescano, A., Tillman, T. and Tohidipur, T. 2009. 'Border controls at sea: Requirements under international human rights and refugee law'. *International Journal of Refugee Law* 21(2): 256–296.

Good, A. 2007. *Anthropology and expertise in the asylum courts*. Abingdon and NY: Routledge-Cavendish.

Graham, M. 2002. 'Emotional bureaucracies: Emotions, civil servants, and immigrants in the Swedish welfare state'. *Ethos* 30(3): 199–226.

Harrell-Bond, B.E. 1986. *Imposing aid: Emergency assistance to refugees*. Oxford/New York/Nairobi: Oxford University Press.

Harrell-Bond, B.E. 1999. 'The experience of refugees as recipients of aid'. In Ager, A. (ed.) *Refugees: Perspectives on the experience of forced migration*. New York: Continuum.

Harrell-Bond, B.E. 2002. 'Can humanitarian work with refugees be humane?' *Human Rights Quarterly* 24: 51–85.

Harrell-Bond, B.E. and Voutira, E. 1992. 'Anthropology and the study of refugees'. *Anthropology Today* 8(4): 6–10.

Hynes, P. 2009. 'Contemporary compulsory dispersal and the absence of space for the restoration of trust'. *Journal of Refugee Studies* 22(1): 97–121.

IntegraRef 2008. *Local communities and refugees: Fostering social integration*, Final Report. Available at: www.evasp.eu/attachments/039_035_IntegraRef%20finalReport.pdf (accessed 20 January 2013).

Kobelinsky, C. 2008. 'The moral judgment of asylum seekers in French reception centres'. *Anthropology News* 5 and 11 May.

Kobelinsky, C. 2010. *L'accueil des demandeurs d'asile*. Paris: Éditions du Cygne.

Korać, M. 2001. 'Cross-ethnic networks, self-reception system and functional integration of refugees from the Former Yugoslavia in Rome'. *Journal of International Migration and Integration* 2: 1–26.

Korać, M. 2003. 'The lack of integration policy and the experiences of settlement: A case study of refugees in Rome'. *Journal of Refugee Studies* 16: 398–421

Lambert, H., Messineo, F., Tiedemann, P. 2008. 'Comparative perspectives of constitutional asylum in France, Italy and Germany: Requiescat in pace?' *Refugee Survey Quarterly* 27(3): 16–32.

McCourt Perring, C. 1994. 'Community care as de-institutionalization? Continuity and change in the transition from hospital to community-based care'. In Wright, S. (ed.) *Anthropology of organizations*. London and New York: Routledge.

Malkki, L. 1996. 'Speechless emissaries: Refugees, humanitarianism, and dehistoricization'. *Cultural Anthropology* 11(3): 377–404.

Marchetti, C. 2007. 'Blurring boundaries: "Refugee" definitions in policies, law and social discourse in Italy'. *Mediterranean Journal of Human Rights* 11(2): 71–94.

Marchetti, C. 2010. 'The expanded border: Policies and practices of preventive refoulement in Italy'. In Pecoud, A. and Geiger M. (eds) *The politics of international migration management*. Houndmills and Basingstoke: Palgrave Macmillan.

Maryns, K. 2006. *The asylum speaker: Language in the Belgian asylum procedure*. Manchester, Northampton: St. Jerome Publishing.

Ong, A. 2003. *Buddha is hiding: Refugees, citizenship, the New America*. Berkeley and London: University of California Press.

Però, D. 1999. 'Next to the dog pound: Institutional discourses and practices about Rom refugees in left-wing Bologna'. *Modern Italy* 4 (2): 207–224.

Puggioni, R. 2005. 'Refugees, self-help strategies, and institutional invisibility: Evaluating Kurdish experience in Rome'. *Journal of Refugee Studies* 18 (3): 319–339.

Schuster, L. 2000. 'A comparative analysis of the asylum policy of seven European governments'. *Journal of Refugee Studies.* 13: 118–132.

Sigona, N. 2005. 'Refugee integration(s): Policy and practice in the European Union'. *Refugee Survey Quarterly* 24 (4): 115–122.

Sorgoni, B. (ed.) 2011a. *Etnografia dell'accoglienza. Rifugiati e richiedenti asilo a Ravenna.* Roma: CISU.

Sorgoni, B. 2011b. 'Alcune note etnografiche su politiche istituzionali e pratiche sociali nella gestione dei richiedenti asilo'. In Faldini, L. and Pili, E. (eds) *Saperi antropologici, media e società civile nell'Italia contemporanea.* Roma: CISU.

Sorgoni, B. 2011c. 'Storie dati e prove. Il ruolo della credibilità nelle narrazioni di richiesta di asilo'. *Parole Chiave* 46: 113–131.

Sorgoni, B. (ed.) 2011d. 'Chiedere asilo in Europa. Confini, margini e soggettività'. *Lares*, LXXVII (1), special issue.

SPRAR 2010. *Rapporto annuale del Sistema di Protezione per Richiedenti Asilo e Rifugiati. Anno 2009–2010.* Roma.

Steen, A.B. 1993. 'Refugee resettlement: Denmark and Britain compared'. *Refugee Participation Network* 14: 8–11.

Strang, A. and Ager, A. 2010. 'Refugee integration: Emerging trends and remaining agenda'. *Journal of Refugee Studies* 23(4): 589–607.

Sweeney, J.A. 2009. 'Credibility, proof and refugee law'. *International Journal of Refugee Law* 21(4): 700–726.

UNHCR 2010. *2009 Global Trends.* Geneva.

Waldron, S.R. 1987. 'Blaming the refugees'. *Refugee Issues* 3(3): 1–19.

Wren, K. 2007. 'Supporting asylum seekers and refugees in Glasgow: The role of multi-agency networks'. *Journal of Refugee Studies* 20(3): 391–413.

Ya Basta! Bologna (ed.) 2010. *Vite da Rifugiati. Condizionale sociale, integrazione e prospettive dei rifugiati a Bologna e in Emilia-Romagna.* Bologna: VolaBo.

Zetter, R. 2007. 'More labels, fewer refugees: Remaking the refugee label in an era of globalization'. *Journal of Refugee Studies* 20(2): 172–192.

Part II

A country's population on the move

Burma's refugees in Asia

Conflict and persecution of ethnic minorities and political opponents continues to drive people from Burma to seek refuge in South Asia and Southeast Asia before they go on to safe third countries, such as Japan, for resettlement. Japan is an interesting case study, as it is a fairly new resettlement destination and one that is under-studied. Regional neighbours of Burma meanwhile struggle with the large refugee numbers from Burma seeking protection, shelter, education, health services and a livelihood. Chapters explore what services are available and what life is like for these refugees in India, Thailand and Malaysia.

6 Surviving the city

Refugees from Burma in New Delhi

Linda Bartolomei

Preamble

This chapter is primarily informed by material shared by refugees from Burma during community consultations held during 2009 and 2010. However, as the researcher has had the opportunity to return to New Delhi on a number of occasions during 2011, 2012 and in early 2013, it is also informed by more recent observations. The most important of these relate to the not insignificant progress made by the United Nations High Commissioner for Refugees (UNHCR), and current partners, and indeed the Government of India to improve the protection environment for the refugees. However, in spite of the commitment and efforts of UNHCR and partners, the problems faced by the refugee community often seem overwhelming. The refugees, while appreciating many of the positive changes, remain frustrated, afraid and often times desperate for safety and security. Sadly the refugees from Burma living in New Delhi continue to face an uncertain future. As a small number of political leaders return home in the hope of a place in the new government, those who remain suffer both from the loss of important community leadership and increasingly the withdrawal of international funding for many of the small community projects which have helped to mitigate the risks and challenges, particularly for women and girls. This chapter seeks to honour the insightful but often-desperate voices of the refugees while at the same time recognizing the genuine commitment of many UNHCR and partner staff who too continue to struggle in the face of ever-diminishing funds.

Introduction

In 2011, UNHCR reported that there were approximately 8,500 recognized refugees and asylum seekers from Burma living in New Delhi (UNHCR 2011). The majority of these were of Chin ethnicity, with smaller numbers from the Kachin, Burman and Kuki ethnic groups (UNHCR 2011). They have fled widespread human rights abuses including systematic rape, beatings, forced labour and forced relocations. Some, including political leaders and activists, have been in Delhi for over 20 years. Others have arrived more recently,[1] often driven by grinding poverty and fear of forced recruitment as soldiers or porters (Refugees

International 2009). It is important to note that most Chin refugees from Burma who have fled to India in search of safety live not in New Delhi but as undocumented refugees along the border with Burma in the Indian State of Mizoram. The Indian Government does not recognize this population as refugees and UNHCR does not have access to Mizoram. While no accurate numbers are available, the refugee community organizations in Mizoram put the figure at over 75,000 (Alexander 2009; Pittaway and Bartolomei 2008). Those who wish to seek formal recognition as refugees and the protection of UNHCR need to make the arduous journey by bus, taking several days to travel to New Delhi. They arrive hoping to find work and a safer place to live, and in some cases, to be resettled[2] to third countries.

For most, however, life in New Delhi brings a new set of risks and problems. Historically, refugees from Burma have had an ambiguous status in India. While the Indian authorities tolerated the presence of refugees from Burma (Ananthachari 2001), providing those who are recognized by UNHCR in New Delhi with residence permits and turning a blind eye to their working in the informal sector, most formal rights were denied for the refugees who temporarily called the urban centre of New Delhi 'home'. While this situation changed in late 2012 with an announcement by the Indian Government that all refugees registered with UNHCR would be allowed to apply for long-stay visas, which would allow them to work in the private sector, this new policy remains in the early stages of implementation. Most refugees from Burma are forced to work in the overcrowded and poorly paid positions in the informal sector in order to survive. Due to UNHCR's restricted budget, only the most vulnerable refugees receive some limited financial support. Similarly small numbers are able to access income generation activities conducted by one of UNHCR's implementing partners. This leads to interrelated issues of poor accommodation, lack of adequate food, limited access to education and health care, labour exploitation and abuse by landlords as well as psychosocial issues such as isolation, depression and an increase in domestic violence. When approaching Indian authorities for assistance or to report concerns, refugees often experience prejudice. Within the wider Indian community, they face discrimination, harassment and ridicule in their everyday lives. Living in the urban slums, refugee women and girls, and some boys and men, face extreme risks of rape and sexual violence (Buscher 2011; Pittaway 2009; Refugees International 2009; Saxena 2007).

For these people, faced with a life in limbo, the UNHCR (2009) Urban Refugee Policy – which for the first time, acknowledged both the legitimacy of refugees residing in urban areas and the range of rights to which they were entitled (UNHCR 2009: 3) – was clearly not only welcome but long overdue. In November 2010, following the release of the Urban Refugee Policy, I travelled to New Delhi to assess what impact this new policy framework might have had on refugee protection. I conducted a follow-up visit in November 2011 and have continued to monitor the situation during 2012/2013. As discussed below, one of the greatest barriers to the enjoyment of rights is the high level of discrimination faced by refugees living in New Delhi. The widespread and underlying poverty in India and the general lack of awareness and understanding of refugee issues, also further compound the risks, discrimination and exclusion faced by the refugees. While the policy provides the framework of possibility, there remain multiple political and social barriers to its delivering the range of rights to which refugees are entitled. However, as will be argued below, it nonetheless acts as a supportive and mobilizing framework to foster a new direction in refugee protection that continues to be pursued by UNHCR in New Delhi.

UNHCR's Urban Policy

While the UNHCR (2009) Urban Policy does not provide detailed operational guidelines, it does set out a clear and concise framework to guide the efforts of UNHCR in urban areas. Paragraph 17 clearly sets out the key rights to which refugees in urban areas are entitled:

> These rights include, but are not limited to, the right to life; the right not to be subjected to cruel or degrading treatment or punishment; the right not to be tortured or arbitrarily detained; the right to family unity; the right to adequate food, shelter, health and education, as well as livelihoods opportunities.
>
> (UNHCR 2009: 4)

The policy stresses the importance of active engagement by UNHCR with host country governments, city authorities and most importantly with refugee communities, emphasizing the importance of community outreach, community mobilization and regular communication (UNHCR 2009: 7–8). However, in an otherwise generally rights-based and comprehensive policy, its lack of adequate focus on the particular risks faced by women and girls in urban areas is disappointing. Like many recent policy initiatives from UNHCR, the rights of women and girls have been subsumed within the Age, Gender and Diversity Mainstreaming (AGDM) framework (UNHCR 2010b). As a result, a well-intentioned attempt at inclusivity often results in a loss of a critical focus on the particular needs of different groups and the loss of a gender analysis. This policy includes only two paragraphs under the AGDM heading, including only one fairly cursory reference to women. With the exception of paragraph 18 (p. 14), which refers to risks of survival sex, there is no significant focus on the risks of sexual and gender based violence faced by refugee women and girls in urban areas. In recognition of this significant omission, gender specialist, Eileen Pittaway, was invited to submit an addendum to the policy in the form of a paper entitled *Making Mainstreaming a Reality: Gender and the Urban Refugee Policy – A Refugee Perspective*, at the High Commissioners Dialogue on Urban Refugees held in Geneva in December 2009 (Pittaway *et al.* 2009). Read in concert with the Urban Policy, this paper provides a more thorough overview of the particular risks women faced in urban areas. It was clear from the research in New Delhi between 2005 and 2011 that unless the particular risks facing women and girls were acknowledged and addressed, not only women but whole communities would continue to suffer. Since 2009 and across multiple consultations with the refugee communities in Delhi, men from the communities joined with women to call for an end to this endemic violence (CRR 2007, 2010, 2011, 2012).

The research methodology

This chapter is informed by research undertaken with more than 300 refugees from Burma living in New Delhi, covering a period of over seven years from September 2005 to December 2012. During this period, researchers[3] from the Centre for Refugee Research worked directly with the refugee communities from Burma in New Delhi and with UNHCR to understand and document the problems faced by the refugee communities. We have also assisted, and continue to assist, both the communities and UNHCR to develop and implement participatory and rights-based responses to many of the protection problems identified.

Two members from the Centre for Refugee Research team first travelled to New Delhi in September 2005 at the invitation of UNHCR and undertook a two-week community consultation, risk-assessment and community-leadership training programme with refugee women from Burma (Pittaway and Bartolomei 2005). Following this initial visit and at the direct invitation of the refugee women's groups, we returned each year from 2007 until 2012[4] to conduct trainings in advocacy, livelihoods, human rights, counselling and community based responses to violence. Beginning in 2009, refugee men from Burma have also been included in the trainings.

A major component of each of the trainings was the preparation of a detailed situational report and analysis of the current living circumstances of the refugees. These reports were produced with the formal written consent of the refugee community and were sent back as drafts for comment and approval prior to be being shared with UNHCR and other stakeholders for advocacy. They provided detailed analysis not only of the range of human rights abuses facing the communities but also of their impact and, most importantly, included the refugee community's ideas for solutions.

In this methodology entitled *Reciprocal Research* (Pittaway and Bartolomei 2009; Hugman *et al.* 2011), rather than being viewed as the subjects of the research, the refugee women and men are active partners in the research process. The research methodology employs a range of interactive and participatory techniques including the use of the human rights framework, story circles and storyboards. Storyboarding involves the participants drawing a series of six pictures in which they explore the major problems they face as refugees and asylum seekers in finding protection in New Delhi. They then explore the impact, evaluate the current responses, and share their ideas for solutions. Each of these techniques is used as a methodological tool to develop a rich situational analysis of the protection environment from the perspectives of the refugee participants. In addition, the story circles and storyboards provide an opportunity for the researchers to test their understanding of the information shared by the refugees and to probe for additional information and a deeper understanding of the individual and community impact of the problems they face. Rather than conducting single interviews or short focus groups, the research data is collected over a period of between three and five days through an extended process of community consultation. This process enables CRR researchers to build trust with the refugee participants, to negotiate consent once the aims of the process have become clear to the participants and to agree a means of giving control of all the research data to the participants. Under this agreement reached, none of the research data collected is published without the prior written approval of the refugees' chosen representatives.

Refugee law and protection in India

UNHCR does not have a formal agreement (Accord de siège) with the Government of India but instead operates under the agreement of the United Nations Development Programme (UNDP). UNHCR was first granted permission to work in India in 1969 in order to provide vocational training to Tibetan refugees and was later permitted a limited role in monitoring repatriation of Tamils to Sri Lanka. While India joined the UNHCR Executive Committee (ExCom) in 1995 (Vijayakumar 2000: 235), it has not ratified the 1951 Convention Relating to the Status of Refugees (the Convention) nor its 1967 Protocol. The Indian Government, like many States across Asia, has consistently cited the Eurocentric focus of the Convention and its lack of focus on situations of mass influx and domestic security concerns as reason to not ratify it (Nair 2007). In addition, as Chimni (2003) has argued, the contradictory application of the Convention by industrialized states, in particular their reduced commitment to the institution of asylum, has also discouraged States across Asia, including India, from accession to the Convention.

In spite of many years of effort on the part of Indian legal advocates, and UNHCR, to develop the draft Indian Act on Refugee and Asylum Seeker Protection, India has neither adopted the Act nor implemented any other domestic legal mechanism to provide recognition or protection to refugees (Nair 2007; Saxena 2007). While some have argued (Vijayakumar 2000) that, although India is not a party to the Convention, protection could have been provided and improved for refugees if the Indian judiciary had drawn on existing constitutional law and the many international conventions to which India is signatory,[5] others have argued that this would have at best provided a limited framework for refugee protection (Saxena 2007; Sengupta 2008). For while relief has been provided in individual cases of arrest, detention and threatened deportation, this did not lead to policy shifts nor to the development of general guidelines for refugee protection. As Ranabir Samaddar has argued (in Saxena 2007: 255) 'the judicial reasoning has been mainly humanitarian and not rights based, dispensing kindness and not justice', and therefore had no significant impact on the domestic legal framework for refugees in India. However, it is important to note that the 2009 UNHCR Urban Policy makes it clear that the responsibility for providing for the rights of refugees lies both with the host state and the international community through the offices of UNHCR:

> the rights of refugees and UNHCR's mandated responsibilities toward them are not affected by their status (or lack thereof) in national legislation.
>
> (UNHCR 2009: 3)

In spite of the policy commitment, UNHCR has an extremely limited mandate in India and is authorized to work with only a small number of the total refugee population (Saxena 2007; Sengupta 2008). As mentioned previously UNHCR is not permitted access to the border areas of Mizoram and Manipur (Alexander 2009) and was only granted access to the camps in the south, hosting refugees from Sri Lanka, in early 2013 (Saxena 2007; Sengupta 2008, UNHCR 2013). The range of ethnic groups UNHCR is allowed to work with in New Delhi is also restricted and does not include refugees from Tibet, Bangladesh, Bhutan or Nepal (Saxena 2007). The major groups with whom UNHCR works in New Delhi are refugees from Burma, Afghanistan, Somalia and a smaller number from a range of other African and Middle Eastern countries (UNHCR 2011).

UNHCR works with approximately 22,000 refugees and asylum seekers in New Delhi (UNHCR 2012), with refugees from Burma being the largest group. While the number of Afghan refugees is also significant, a large percentage are Afghan Hindus and Sikhs who have been offered the opportunity of naturalization and integration in India (Bose 2006). As the naturalization process is a long and complex one, UNHCR provides assistance and follow up at each stage.

UNHCR contracts direct service delivery to a number of implementing partners. These include the Bosco Organization for Social Concern and Operation (BOSCO) and the Social Legal Information Centre (SLIC). Until 2012 when much of the programming was moved to BOSCO, the YMCA had been responsible for all refugee education and training courses since early 1996. They were also in charge of distributing and implementing the educational assistance programme to refugees recognized under the mandate of UNHCR, who were living in and around New Delhi. BOSCO assists refugees in finding employment,[6] and is responsible for the placement incentive scheme in which those refugees who have found employment through the agency are given an allowance each month to bring their income up to a basic monthly salary (Buscher 2012). BOSCO also provides psychosocial support and counselling and computer and language courses to refugees in order to broaden their skills and enhance their chances of finding employment. SLIC is responsible for providing legal assistance to all UNHCR recognized refugees in New Delhi for file renewals, naturalization, legal protection and intervention. Since 2012, in response to the high levels of sexual violence faced by refugee women in New Delhi, UNHCR has worked with SLIC to reorient their focus to include prevention, mediation and conflict resolution in addition to legal assistance. In 2012 SLIC worked in partnership with the Indian police to support a number of very well attended introductory self-defence courses for refugee women and girls. These courses were designed with the dual aims of building police awareness of the risks faced by refugee women and fostering improved relations between the local police and refugee communities.

The protection space in New Delhi

Saxena (2007) has pointed out that while incidents of arrest and deportation were relatively rare in India, many refugees continued to report discrimination at the hands of both the police and the authorities. This included being dismissed, ignored, abused or forced to pay an arbitrary fee for documents (CRR 2010, 2011; Sengupta 2008). The Urban Policy (p. 5) sets out a range of other rights to which refugees are entitled. These include protection from exploitative treatment in the labour market, access to adequate shelter and living conditions, access to education and health care, harmonious relationships with the host population, and access to the durable solutions of voluntary repatriation, local integration and resettlement. As the quotation below highlights and as is explored in the rest of the chapter, all these remain ongoing issues for refugees from Burma in New Delhi:

> Why I am telling you about a third country is because I want anywhere who cares for refugees and who shows their love to them, who sympathizes with the refugees. I want them because most of the time we are being discriminated or beaten. As a human being we want the same treatment, we want our human rights with others, here we are getting discrimination.
>
> (Refugee man from Burma, 2010)

Registration, refugee status determination and resettlement

In recent years, UNHCR has made significant efforts to streamline its refugee registration and refugee status determination (RSD) systems. In 2009, UNHCR contracted preliminary refugee registration to SLIC, and established a registration centre in West Delhi where a large number of refugees and asylum seekers live. As a result, the waiting time for initial registration was reduced dramatically, with most new arrivals undertaking registration interviews within a month of arrival (UNHCR 2011). The waiting times for RSD interviews were also significantly reduced, with most being completed within six months from registration, rather than the previous two year wait.

However, while the UNHCR registration and RSD procedures were streamlined and improved in the same period, the number of resettlement places offered by the Australian Government was reduced. While during 2007/2008, anywhere up to several hundred resettlement referrals were made each month (UNHCR 2008), in 2010 the *annual* allocation of places was less than 175. It was envisaged that in 2013 this would be reduced further; instead the number has been increased to 200.[7] This limited access to resettlement has led to increasing frustration among the refugee communities, and in particular among those from

Burma who have been in New Delhi for many years in situations of ongoing danger and uncertainty.

While instances of detention or *refoulement* from India might be few, the severe economic insecurity, high risks of sexual violence and discrimination continue to inhibit many refugees from envisaging any form of safety there. An overarching theme from the community consultations was the desire for peace and democracy in Burma so that they could return in safety to their homeland. This was reflected throughout the testimonies and in the recommendations made by the refugees. The participants of the workshops had all fled persecution and violence in towns and villages. Although their first preference was to return to their homes, the continuing conflicts made this an impossible dream. As a result, for many of the refugees who cannot return home, resettlement is viewed as the only feasible and viable solution. While this view of resettlement presents a significant challenge for UNHCR in engaging the refugee community actively in exploring local solutions to the risks they face in New Delhi, the global scarcity of resettlement places makes the development of effective protection solutions in urban areas in host countries such as India all the more imperative.

Key issues of concern for the refugees from Burma

The key issues of concern identified by the refugee community in New Delhi are:

- lack of livelihoods/poverty;
- lack of access to secure, appropriate and affordable accommodation;
- lack of access to education;
- sexual and gender based violence;
- problems with service providers;
- lack of access to health services;
- problems with registration.

In addition to being an issue of major concern in its own right, the risk of sexual and gender based violence is also a cross-cutting issue, compounding and compounded by each of the other risks faced by the refugee community. While women and girls are at risk in every aspect of their lives both in public and at work, in the streets and in parks, they are also at risk in private, including within their homes, from husbands, stepfathers and landlords. Boys and men, as well, also face risks of sexual abuse, including from other boys at school, in parks and, in a number of cases reported in 2011, by the sons of their landlords (CRR 2011). The lack of livelihoods and lack of income also compounds each of the other risks; therefore abject poverty is commonly at the root of the major risks faced by the community. Many recently arrived refugees from Chin State are malnourished, unskilled and traumatized by years of desperate poverty (Richards *et al.* 2011). They struggle to survive in New Delhi and often end up in debt, forced to use their refugee cards as security for loans from government-approved moneylenders who often charge 20 per cent interest (CRR 2011).

Livelihoods

The UNHCR (2009) Urban Refugee Policy defines an urban area as:

> a built-up area that accommodates large numbers of people living in close proximity to each other, and where the majority of people sustain themselves by means of formal or informal employment and the provision of goods and services.
>
> (UNHCR 2009: 2)

While it is acknowledged that a refugee camp situation might share some of these characteristics, it is excluded from this definition. Perhaps one of the major and key differences between most urban refugee settings and most camp settings is that in urban settings the majority of refugees are not provided food rations, access to land or shelter. Instead, they struggle to sustain themselves through either formal or informal employment. In earlier years, UNHCR had sought to address this gap by providing a subsistence allowance to most refugees in urban areas (Obi and Crisp 2000). However, as refugee numbers have increased, only those individuals and families deemed by UNHCR to be extremely vulnerable are provided with some cash assistance (UNHCR 2011).

In New Delhi, UNHCR's partner BOSCO is responsible for assisting refugees to find work in the informal sector. Along with accompanying job seekers to prospective employers, BOSCO's role is to negotiate their employment and allocate a 1,000 rupee incentive payment, bringing their wages to approximately 3,500 rupees per month (Buscher 2011). With rents ranging from Rs3,000–4,000 per month for a small room with one bathroom often shared with up to 20 people, 150 rupees for water, 70 rupees for electricity, and 1,500 rupees per head for basic food items, refugees struggle for daily survival – even before extra costs such as health care and clothing.

The majority of jobs available to refugees are poorly paid jobs in the informal sector. Most refugees work in small factories or in restaurants as dishwashers or cleaners. Women often work cutting fabric in small clothing factories. Some refugees are assisted by BOSCO to find employment; however, this can often take many months, by which time the refugees are so desperate they will take any work available. Employers are not required to process the refugees' wages through BOSCO, resulting in payments often being made late, or not at all. If a worker falls ill, their wages are cut by their employer as well as their incentives, leaving few options for the seriously ill who are supporting a family on a single wage. The refugees expressed their frustration with the flaws in this system:

> Things have been getting very difficult from 2010 because of inflation. Most of us work in factories from 8 to 9 but at the most we get Rs1,800 per month and it is not enough for the basic necessities. And from 2010 even the room rent has gone up, earlier we could get the room for five people for Rs2,000 but now it costs Rs3,000.
>
> (Refugee man, 2010)

Costs of living in New Delhi continue to rise, impacted by India's growing economy and extensions to the New Delhi metro system which moves closer to the outlying areas in which the refugees live, pushing up the costs of housing.

The refugees described appalling working conditions that included dangerous worksites, very long hours with few breaks, and factory work that involves standing for extended periods in dim light with no ventilation. Due to their desperation, women will often take positions and work hours that also put them at risk of serious sexual harassment or abuse:

> Most of our girl or ladies whenever they go to their work place they sexually harass, they treat us like animal whenever they want to play, do any sex they call us and they to play with us.
>
> (Refugee woman from Burma, 2010)

Many women recounted incidents of this harassment, and were distressed by the limited responses available when they reported them. The women also told of those who were forced into prostitution in order to support their families (CRR 2007, 2008, 2010, 2011).

A study conducted in New Delhi in early 2011 by the Women's Refugee Commission (Buscher 2011), proposed an expansion and improved targeting of income-generation training and activities. The study findings echoed the willingness of both UNHCR and BOSCO to undertake reform to address the weaknesses in the current system. In particular recommendations emphasized the importance of developing a more individually tailored approach to refugee livelihoods in New Delhi. Such an approach would build on the diverse skills within the refugee community and identify a broader range of sustainable employment opportunities, while at the same time ensuring that a programme of social support for those in need was maintained (Buscher 2011: 18–19).

Inadequate and overcrowded accommodation

Due to high rents, poorly paid work, and discrimination against the majority of the refugees, accommodation and living conditions are totally inadequate. In New Delhi, it is common for up to 15 or 20 people to live in cramped conditions in one small room, sharing one toilet, shower and cooking facilities. This leads to a range of health and safety risks and is reported by the refugees as a key factor in the high levels of family and domestic violence (CRR 2007, 2008, 2010, 2011; Pittaway and Bartolomei 2005). Landlords charge higher rents to refugees, and demand rent increases with little or no notice. Evictions were reported as a common occurrence, with regular moving having a negative effect on children's mental health and schooling.

The refugees spoke of a range of health problems arising from this overcrowding, insufficient water, and lack of fresh air or circulation. Landlords do not allow residents to see electricity or water bills, but still demand high payments despite the inadequate and intermittent supply:

Some landlords they say that they have to pay 2,500 for electricity and they have already paid that much so in the summer they do not let us buy fans or coolers and sometimes at noon the lights go off because the landlord does not want the meter to go any higher because he has already paid 2,500. In the summer the children cannot cope and some children have died because of this.

(Refugee man from Burma, 2010)

The refugees from Burma also frequently face conflicts with landlords in relation to their dietary preferences. As many Indians are vegetarian, landlords frequently object to the refugees from Burma cooking meat and using pungent fish pastes:

Sometimes when we are cooking meat they come in and pull and destroy all our food. Again, if they say don't cook this meat but as a human being we want to eat meat, so we cook it secretly but they smell it so in two or three months they stop the water or electricity. I had just had my baby, and we were very weak so we made chicken soup and the landlord stopped the water for three months and my sister had to go from here and there and she had to face sexual harassment and we had to throw away lots of clothes because of no water.

(Refugee woman from Burma, 2010)

Families and single men and women share small rooms with no opportunity for privacy. Children attending school are unable to study at home due to the number of people and lack of space. Women described how distressing it is for couples who are never alone, with women embarrassed and reluctant to have sex with their husbands. They spoke of an increase in domestic violence and marital rape caused by the stress on their relationships, and an increase in family breakdown.

Single men and women are forced to share very close quarters, often resulting in young women becoming pregnant. To avoid community shame, these women are forced into early and unwanted marriage or suffer from unsafe and expensive abortions.

The women spoke of many instances of sexual harassment of refugee women and girls, including rape by landlords.

Case study

Two widows and their families were living together, renting a small room in a two storey building with their landlord downstairs. One night the electricity went out and the families were forced to the roof of the house. One of the widows and the three daughters went upstairs to the roof, but the other widow was feeling unwell and stayed in her room. One of the daughters, a 14 year old, needed to use the toilet and so went downstairs. The landlord grabbed her, covered her mouth dragging her down the stairs into his room. The widow who had remained in her room heard the scuffle and alerted other family members. The following day, with help from their neighbours, the women reported the incident to YMCA. They returned to the house with representatives from SLIC, and the police. The police did arrest the man on this occasion, but he was not taken to the jail. In the evening the landlord held a meeting with the local people, claiming that the refugees were causing problems in the area. They demanded that the seven refugee families living in the area leave. The families, some with small children as young as two, were forced from their homes with no food or water, depending on assistance from a local Chin Christian fellowship.

While this incident is well known among community members and service providers, in 2010 many other similar cases were shared by the refugee community. During my visit to New Delhi in 2011, numerous similar cases of abuses by landlords that had occurred in the last few months of 2011 were shared during a workshop exploring community based responses to violence (CRR 2011).

Lack of access to education

There is a strong commitment in the refugee community to both primary and higher-level education. While access to government schools is open to all asylum seeker and refugee children in India,[8] the communities report numerous barriers in accessing safe and quality education. Sometimes students are barred from admission because they have missed several years of schooling when they fled their homeland. This is compounded by the children's inability to speak, read or write in Hindi, which is the medium of instruction. Teachers complain that language and cultural barriers demand more attention than they are able to give. So they resort to 'placing the child in the back of the room' to avoid disruption of the other children's learning. Other forms of discrimination are also experienced:

> From our C1 [area] side most of the children who already went to government schools now after facing many kinds of discrimination and beatings they are afraid to go again.... Their fellow students whenever they take lunch breaks they spit in their food and beat them. So after attending a short time they do not want to attend, so some go to YMCA.
>
> (Refugee man from Burma, 2010)

Those who would prefer to send their children to private schools report being unable to afford books, uniforms or transport fees, especially those who have had their subsistence allowance cut off, or those with large families or additional dependants to provide for. In some cases when families cannot afford to pay the school fees on time, the refugee children are discriminated against and mocked in front of other students. As a consequence, children are often forced to drop out and to find poorly paid work in often unsafe environments.

As a result, many parents prefer to continue sending their children to one of a number of refugee-run community schools or to the bridging programmes run by the service providers. The refugee community from Burma have run community-based schools for a number of years. Their establishment dates back to the period prior to 2009, when their children had not been permitted to attend government schools. The schools vary in size and quality, the largest offering classes in Burmese and English to over 150 students. However, the smaller schools are often staffed by one teacher working across a number of age groups in fairly rudimentary facilities. The schools lack adequate resources and generally receive what little funding they have from international church networks or community members who have been resettled. In keeping with the framework of the Urban Refugee Policy (UNHCR 2009: paras 110–113), UNHCR no longer offers any financial support to these community schools but instead directs what funding there is towards assisting children at both primary and secondary levels to access the Indian government schools (UNHCR 2011; *Basavapatna* 2011). However, it is clear from ongoing reports that significantly more work needs to be done with Indian educational authorities and local Indian schools to ensure that they provide a safe and welcoming learning environment for often highly traumatized refugee children.

Violence and discrimination against women

> Rape is common. The [rickshaw] driver doesn't recognize if they are young or old, this is the confession of the driver – we don't care if they are old or young, we just rape.
>
> (Refugee woman from Burma, 2010)

Risks of rape and sexual violence are widespread. Women and girls face danger in every aspect of their lives, on the streets, in the workplace, going to and from the night market, from landlords, rickshaw drivers, stepfathers and husbands. The risks of sexual abuse and exploitation are extremely high for children – both girls and boys – with abuses ranging from the rape of babies to sexual assault of girls and boys often tricked into going with men through offers of food or money. There have also been several reports of the rape of teenage boys and men (CRR 2010 and 2011). In all cases, these attacks occurred in a climate of almost total impunity. While many cases are reported to the police and to SLIC, UNHCR's legal implementing partner, police either fail to investigate, encourage the victims to enter into an informal agreement for monetary compensation from the perpetrator's family, or pressure the victims into withdrawing their complaints. Fear of the police that precludes some women from reporting, and difficulties in identifying the perpetrators are among the other major impediments to justice. Multiple factors increase the risks of rape and sexual abuse of refugee women in New Delhi.

One of the major risk factors identified for women and girls in urban refuge sites is the almost complete lack of access to just and effective legal processes for cases of rape and sexual and gender-based violence. Although it is claimed that the refugee communities are entitled to seek the protection of the host legal system, in practice prosecutions of the perpetrators of rape, murder and other violent acts are rare. Some of the major problems are the lack of legal documentation, lack of safe places for women to live, and lack of access to livelihood while the legal prosecution process is pursued. Local police are often unsupportive and afraid to take action against powerful figures and institutions in their own communities.[9] To date, the few cases which have been successfully prosecuted have been against refugee men from within the Burmese community. While many women from Burma reported that they were pleased with this outcome they remain distressed that when the perpetrators are Indian men, justice continues to elude them.

The problem is compounded by the low social status of women. Even accessing basic needs for their families is dangerous. Rape and sexual abuse of women when they are collecting spoiled food from the ground at night markets is common:

> While we are collecting the vegetables, the local men came and simply beat us, and sometimes they touch our breasts and also take away our purse. Sometimes we see Indian people pee on the vegetables but we still need to pick them up, despite all this we still need to look after our families.
>
> (Refugee women from Burma, 2010)

Their refugee status means they are doubly disempowered, for they lack the protection of the state, the opportunities for free movement, and the surplus income to escape situations of violence and discrimination. Upon becoming refugees, most men lost their livelihoods and income-generating capacity, and this often creates an environment of widespread frustration in which women are beaten and abused within the home. For a variety of reasons, women rarely report incidents of domestic violence. Social and religious norms within the communities which continue to discourage the public discussion of so called private family matters often keep women silent. A strong hierarchy within the community, in which significant power is held by male religious leaders, also impacts on women's abilities to speak out about the violence they face. Women are often dependent and held subservient to their husbands. The threat of losing their access to food and other aid is always a risk inherent in their decision to report incidents. The police often ignore such reports or become a part of such abuses. There are limited avenues provided for support and redress for the abused women from the perpetrator of the violence. The refugee community supports a small safe house which provides temporary shelter for women fleeing severe domestic violence. However, the refugee women who run the shelter do so at significant personal risk in their efforts to provide safety to women and to intervene to mediate an end to the domestic violence.

Other acts of violence like rape, and sexual or physical assault, are also underreported. Women face cultural and social stigmatization barriers to reporting incidences of sexual abuse. As a consequence, depression, mental health issues and threats of suicide are common. To report an incident also means risking isolation from the family or community, and therefore access to food, supplies and personal documentation. For many women the ongoing discrimination contributes to their sense of hopelessness and despair:

> Life is difficult in Burma we don't want to die, but here [in India] we want to die.
>
> (Refugee woman from Burma, 2010)

To combat these issues, women need education, empowerment and the support of society and outside parties. They need help in accessing avenues for redress, they need support in the event of leaving their husband due to violent circumstances, and they need the educational background to lobby for a better recognition of their needs. In this way, refugee women have specific issues that, in order to be addressed fairly and competently, need to be isolated from general issues facing the refugee community.

A Women's Protection Centre was established in 2006 following the consultations with the refugee women from Burma (Pittaway and Bartolomei 2005) and originally funded by the Australian government. While the aim of the Centre is admirable, its association with resettlement processing – one of its major early focuses – hampered efforts to expand its community development functions. However, the focus of the Centre has now changed as it aims to provide a safe space for counselling and referrals of asylum seeking women and men as well as a space in which to host small community meetings. This forms part of UNHCR's expanded outreach programme. In this regard, UNHCR's efforts have been critical, regarding their working more closely with the refugee women's groups including the Women's League of Burma, Women's Rights Welfare Association of Burma and Burmese Women Delhi through the framework of the Women's Protection Centre, based in Vikaspuri.

Problems with service providers

The refugees discussed the importance of having access to both responsive service provision and supportive staff. In each of the consultations held since 2005, the refugees had highlighted their frustrations with the responses of some of UNHCR's implementing partner staff in New Delhi. They reported limited access to support agencies, extended timelines for processing of applications, and inadequate provision of support. Of particular concern were the attitudes of some staff members, including security guards, who displayed a lack of cultural understanding and empathy towards the plight of the refugees and the circumstances in which they found themselves in exile. The impact of this was that service providers could exacerbate further the issues facing the refugees. With very few options available, the refugee women in particular were often deterred from reporting instances of abuse and speaking about the serious issues they face.

The refugees also reported a similar lack of support in regard to obtaining accommodation and mediating disputes with landlords. In particular, the refugees discussed the frustration of being sent from one service provider to the next, often without being able to share their problems. This often leaves the refugees feeling that the service providers are indifferent to the risks they face:

> They [local people] were threatening to kill the whole family. We go to UNHCR to tell them of this. They told us to approach SLIC, and when we approach SLIC they told us we help you once and to go to YMCA, and they told us to talk to one person who looks after housing, and he told us to wait. And we had to wait all day and the whole family was suffering and we had nowhere to go.
>
> (Refugee women from Burma, 2010)

In response to the concerns shared by the refugees, UNHCR and the implementing partners have made significant recent efforts to improve service provision in New Delhi. UNHCR hosts regular meetings with refugee representatives, has expanded their outreach programmes, and encouraged refugee groups to make use of the refugee centres that have been established in each of the areas where refugees live. They have also reviewed all current programming and approaches in the areas of income generation, vocational training, and legal and health support. In 2011 and 2012, refugee representatives shared a number of examples of improvements in service provision which were having positive impacts on their lives. In particular they referred to improvements in access to medication, decreased registration time for new arrivals and the improved quality of the sanitary napkins provided to women and girls.

Health

In the consultations, the refugees' concerns about health were threefold. First, they spoke of overcrowding and substandard living conditions. These, along with inadequate nutritional intake, led to their second concern, the increased chance of diseases and serious illness. Third, the refugees face the high costs of accessing the private system when they are denied timely access to the public hospital:

> If we admit a person to hospital the fee at least 1,500 (rupees) not including medicine. If can't pay (you get) nothing.
>
> (Refugee women from Burma, 2010)

The overcrowded accommodation and general lack of amenities exacerbate the spread of commonly reported communal diseases such as diarrhoea, tuberculosis and influenza. On visits to the refugees' accommodation during the consultations, it was common to see 15 to 20 people sharing a single room, with residents of an entire level often sharing one toilet. The overcrowding increases when children return from schools outside New Delhi during the winter and during summer school breaks when temperatures are at the most extreme, increasing the spread of disease:

> We refugees are suffering mainly from dysentery, kidney problems, kidney stones, gastritis problems. A disease such as dysentery is communicable, if one person from the family suffers, the whole family will get it. TB is a very common disease in Delhi.
>
> (Refugee man from Burma, 2010)

Malnutrition is common, with some children displaying distended bellies. The participants report that women are often most affected by malnutrition as they give priority to feeding their children. Dizziness and lack of menstruation is common. This is of particular concern for pregnant and breastfeeding women:

> a close friend was pregnant but she cannot eat proper food and also – do not take any medicine, had to work till 7 months – when she deliver baby almost die.
>
> (Refugee woman from Burma, 2010)

Children suffering malnutrition at a young age, or born of mothers suffering malnutrition, are likely to have a marked decrease in their standard of health and earning capacity throughout life. To feed their families and meet health care costs, some women resort to survival sex, increasing their chances of additional health conditions such as HIV/AIDS and depression.

The refugees reported high transport costs to and from the hospital as being a barrier to accessing treatment, particularly for those with chronic conditions. They also spoke of the overcrowding at government hospitals coupled with the lack of interpreter services, and of facing discrimination, abuse and severe rudeness from many hospital staff. While UNHCR funds BOSCO to support a small team of refugee health animators to accompany refugees to the government hospitals, their capacity to provide the necessary support is often hampered by the fact that they too face serious discrimination from hospital staff. This can lead to confusion and misunderstanding in relation to both diagnosis and treatment, which can be compounded by a lack of continuity of health staff on return visits.

This often leads to a deep mistrust of the public health system, which results in many refugees preferring to attend private hospitals where the cost of their treatment is much higher and not covered by UNHCR:

> YMCA[10] is employing a Burmese interpreter to assist patients in the hospital but sometimes many patients are there and they cannot take care of other patients sometimes we are standing in the queue while they look after patients more serious than us. When we get to the doctor we can't communicate and have to show them with hand gestures. How can we get the correct prescription when we can't communicate?
>
> (Refugee man from Burma, 2010)

In New Delhi there are two community health clinics within the community, the Yamuna clinic run by a volunteer Burmese doctor, and one run by the Women's Rights Welfare Association of Burma (WRWAB) and staffed by a volunteer nurse from Burma. Both clinics are open three half days a week and for emergencies. Generally, between 50 and 90 refugees attended the clinics each day. Most attend for assistance with gastric and respiratory problems, skin infections and kidney problems. As the clinics only receive limited financial support, mainly from refugees resettled abroad, patients are generally only provided a partial course of medicine and are left to try to find the funds to purchase the balance of the course themselves. The volunteer nurses at Yamuna also assist with delivering babies and minor surgical procedures. Housed in poorly lit and crumbling buildings, the clinics are extremely rudimentary and poorly equipped. Yet in spite of the poor conditions and limited resources, the vast majority of refugees from Burma prefer to attend these clinics rather than the Indian public hospital system. In keeping with the framework of the Urban Policy, UNHCR does not provide support to the community clinics but instead directs its resources via BOSCO to assist refugees in accessing the public hospital system by providing a daily transport service, access to some free medicines not provided by the hospitals via local pharmacies, and refugee interpreters to accompany refugees to hospital. In spite of recent improvements in access to medications, significant challenges remain in meeting both the urgent and chronic health needs of the refugees through the Indian public hospital system.

Discussion

While, tragically, many of the problems and human rights abuses that were documented during my first visit in 2005 continue, UNHCR has implemented a number of important programme and policy shifts to address some of the most urgent protection gaps. These include increased outreach to, and engagement with refugee communities, as well as faster and more streamlined registration and RSD systems (UNHCR 2011). However, the number of human rights abuses that the refugees continue to face every day in New Delhi highlights the need for UNHCR, the implementing partners and the Indian government to do much more. While, as Sengupta (2008) has stated, if refugee protection is to be achieved in India it is essential that India both 'ratifies the 1951 Convention and establishes a domestic legal framework' (2008: 4), the Urban Policy provides momentum to UNHCR's efforts to urge the Indian government and authorities to take a far greater role in refugee protection in New Delhi than they currently do (UNHCR 2009: para 30).

The significant strengths of the policy are the clear links to the human rights commitments set out in a range of key conventions other than the 1951 Refugee Convention. The policy clearly draws on both the Covenant on Civil and Political Rights (ICCPR, 1966) and the Covenant on Economic, Social and Cultural Rights (ICESCR, 1966) – both of which India had signed and ratified – to articulate the range of rights to which refugees, by virtue of being human, are entitled, regardless of where they are living. This provides a powerful base for advocacy. In India there are already a number of key examples, in particular in the areas of health and education, where the Indian government has drawn on its international legal obligations to provide services to refugees and asylum seekers. As a result of both legislative change and UNHCR's advocacy efforts, refugee and asylum seeker children have been able to access Indian public schools since 2009, as well as some places in Indian universities. In addition, the government demonstrates some commitment toward upholding the right to health for refugees by enabling them to access the public health system free of charge. Most recently the Indian Government's policy commitment to provide long-stay visas and employment rights is a significant step forward. As has been discussed, while none of these rights has yet to be fully realized, given the high levels of discrimination, this nonetheless represents progress toward the realization of some rights for refugees in New Delhi. During the consultations with the refugee community in 2010, refugees identified a number of clear recommendations which were shared directly with UNHCR New Delhi. These are discussed below within the framework of the Urban Policy. Looking back in early 2013 and as has been discussed earlier, while there is still a long way to go, it is clear that UNHCR and partners have made significant efforts to address a number of these.

Recommendations made by the refugee community to UNHCR New Delhi

Community engagement

(a) Recognize and utilize the prior and existing skills, knowledge and experi-
    ence of the refugee communities and our community-based organizations
    and include us in decision-making and service provision.
(b) Facilitate and attend regular consultations and dialogue with refugees,
    including the most vulnerable refugees, about their concerns; ensure and
    improve transparency by keeping accurate records of meetings and provid-
    ing minutes in appropriate languages to the refugees in a timely manner.
(c) Increase the number of refugee women from Burma employed as staff by
    service providers and in decision-making processes at the Women's Centre,
    providing appropriate training and remuneration.

Refugee community organizations in New Delhi play a significant role in providing protection and support to their communities. Community-based women's
groups in New Delhi have a long history of providing frontline responses to the
victims of sexual and gender-based violence. A number of these organizations
focus on advocacy and responses to women who have suffered violence.
Women's Rights Welfare Association of Burma (WRWAB), the New Delhi
based member of the regional women's network the Women's League of Burma
(WLB) (www.womenofburma.org), employs a team of trained volunteers in its

women and violence team and also runs a safe house for women fleeing domestic and family violence. Burmese Women Delhi (BWD) operate a small women and violence programme providing counselling support and small amounts of money to cover hospital transport costs to women who have experienced rape and sexual violence. This support has been provided with the most limited of resources and relied largely for its effectiveness on the commitment of small groups of often very young women volunteers who act as important interlocutors between UNHCR and the implementing partners.

The Chin Refugee Committee (CRC) also plays an important advocacy role, identifying patterns of problems as well as individuals most at risk. As was emphasized in the Urban Policy (paragraphs 39–42), the refugee communities are clearly part of the solution and need to be regarded as equal partners in finding effective responses.

Improved service provision

(d) Monitor and hold to account UNHCR New Delhi's implementing partner organizations (BOSCO, SLIC) in providing adequate assistance with health, accommodation, education, job placement, legal representation and advice.

(e) Encourage and support the UNHCR funded Women's Protection Centre to take a more active mediation role in cases of violence against women. Take an active role in promoting community driven education and awareness pro- grammes that are accessible to all members of the refugee community.

(f) Ensure the provision of accessible safe houses and adequate facilities to provide protection for the most vulnerable, with particular focus on women and their children.

Most importantly these recommendations emphasize the urgency of finding improved livelihoods, health, education, safety and housing solutions. They stress the importance of services employing workers who understand the situ- ation of the refugees and are sensitive to their needs. Others emphasize the importance of community education about refugees in schools and among local Indian communities. UNHCR is also encouraged to take a more active mediation role in cases of violence against women and to advocate for effective and timely legal solutions. The lack of action by the police in cases of rape and physical assaults is one of the most often repeated concerns of the refugee community. The lack of legal action fosters a climate of impunity and each rape or attack that goes unpunished, only compounds the community's sense of risk and vulnerability.

While this situation remains intolerable, it is essential to acknowledge that UNHCR has indeed made significant progress in recognizing and seeking to respond to the high risks of rape and sexual violence in New Delhi. This moved from a situation in 2005 (Pittaway and Bartolomei 2005) in which the risks of rape and sexual violence were not acknowledged, to one in which UNHCR recognizes it as one of the most significant risks for refugee women and girls and has taken significant steps to improve police and partner responses (UNHCR 2010b, 2011).

Registration, RSD and Resettlement

(g) Provide resources to enable quick and efficient processing of refugee applications and requests. Make regular feedback available to refugees on the progress of their application, including legitimate reasons if the outcome is unsuccessful.

(h) Actively campaign for countries to increase their refugee resettlement quota, with particular emphasis on women and girls at risk.

While UNHCR has made significant progress in reducing both registration and RSD waiting times since 2009, the limited number of resettlement places remains one of the major concerns of the refugee community. Equally, this represents a significant challenge for UNHCR to identify those most at risk and in need of resettlement when the majority of refugees in New Delhi face daily risks of abuse and violence.

In conclusion, while the long awaited revised Urban Policy does indeed give some glimmer of hope for a safer and more dignified future for refugees from Burma in New Delhi, it is a future that will only be realized through genuine political will on the part of the Indian government and the sustained commitment of UNHCR and implementing partners to implement an approach to refugee protection which respects the rights, capacities and dignity of refugees. However, as is clear in the words of the refugee below, all too often it is the quality, skill and commitment of the frontline staff upon which rests the success or failure of policy designed to enhance the protection of refugees!

> If I am chief of mission, we should have refugee law. Also, at the office and the implementing partners I would ask the staff who are having sympathy with the refugees. I would give them a job but those who do not have sympathy I would not. As a start, who love and care for the refugee, who has such a mind I would give as staff.
>
> (Refugee man from Burma, 2010)

Conclusion

In 2012 while many challenges remain, UNHCR continues its efforts to improve refugee protection in New Delhi. UNHCR has instituted additional community meetings, has offered the refugee community organizations weekend access to the refugee centres in order to coordinate their own activities, and has invited a number of proposals for community development activities directly from the Chin Refugee Committee. In response to the challenges children face attending Indian public schools, UNHCR has commissioned a study of the major barriers by the University of New Delhi. In response to the continued high risks of sexual violence, additional funding has been sought from UNHCR Geneva, and UNHCR New Delhi are building links with a local Indian NGO skilled in undertaking community safety audits. SLIC and UNHCR have expanded their programmes of trainings for the Indian police and BOSCO plans to strengthen their vocational training and job placement programmes.

However, in spite of all these positive initiatives, the refugee communities from Burma face new challenges in 2013. While some express a cautious optimism about the possibilities of positive political change, indicated both by Aung San Suu Kyi's engagement in the political process and increased international engagement in Burma, others wonder what future might await them should they be forced to return to their destroyed or desperately poor villages in Chin state. For some who faced rape and other forms of torture at the hands of the regime's soldiers the thought of return is unimaginable (Personal communication A, 2012). This new layer of uncertainty has been compounded by the retraction of funding which has long been provided to grassroots refugee groups from Burma, as the major donors from Europe and the USA shift their focus inside Burma. As a consequence, in early 2012 the two women's groups in New Delhi which both play key frontline protection roles, Burmese Women Delhi (BWD) and the Women's League of Burma (WLB), are both struggling to survive as their limited funding diminishes (Personal communications B and C, 2012). While on the one hand the political changes in Burma offer the possibility and hope of return, these new challenges mean that the refugees from Burma in New Delhi continue to face an uncertain future.

Notes

1 At the time of writing (March 2013) the number of Chin arrivals was decreasing, while Rohingya arrivals continued to increase and in early 2013 constituted the largest numbers of arrivals from Burma. The Rohingya are a stateless minority Muslim population who live mainly in Rakhine State in Burma (Refugees International, 2012).

2 Resettlement to a third country represents one of the three durable (permanent) solutions promoted by UNHCR. The others are repatriation (return to the country of origin) and local integration in the country of asylum. The durable solution of resettlement is only available to at best less than 1 per cent of the world's recognized refugee populations. Each year only 80,000 places are available, the majority in countries including the USA, Canada and Australia (UNHCR 2010a).

3 The research team over various visits included Eileen Pittaway, Kerrie James, Geraldine Doney, Rebecca Eckert and Linda Bartolomei.

4 These trainings continued in 2012 and it is envisaged that the research team will continue to travel to New Delhi each year for the foreseeable future to provide trainings should they continue to be requested by the communities.

5 The treaties to which India is a party, and which influence the treatment of refugees, are the Genocide Convention 1948; ICERD 1965; ICCPR 1966; ICESCR 1966; CEDAW 1979; CAT 1984; CRC 1989.

6 As this chapter goes to press UNHCR has expanded its NGO partnerships and made significant changes to its Livelihood programs.

7 Unexpectedly in late 2012 as a result of a shift in the policy of the Australian Government designed to make Australia a less attractive destination for asylum seekers, Australia's annual resettlement quota was increased to 20,000 places. As a result the

allocation for refugees from Burma living in New Delhi was increased to 200 places for 2013.
8 Prior to 2009, refugee children without birth certificates were denied entry to government schools. However since this time as a result of both recent legislative change and UNHCR's advocacy, refugee and asylum seeker children without birth certificates can be granted entry on provision of an affidavit certifying their age. The Right to Education Act was passed in late 2009 and entered into force on 1 April 2010 (Buscher 2011: 11).
9 The brutal gang rape and death of a young Indian woman in New Delhi in late December 2012 has had a dramatic impact on both the Indian community and police. It is hoped that the ensuing public outcry might finally challenge the impunity with which rape and sexual violence occurs in New Delhi. Some positive changes are evident across the New Delhi police force, with an increased commitment to providing effective responses in cases of rape and the employment of female officers in all police stations. It is however a great tragedy that it has taken both the death and extraordinary courage of the young woman who suffered the most brutal of gang rapes to bring about this change. Whether the current commitment to positive action is sustained and leads to increased justice and safety for women, only time will tell.
10 In 2012 UNHCR's health programming was moved from the YMCA to BOSCO and significant efforts have been made to address the problems that were highlighted by the refugee community.

References

Alexander, A. 2009. 'Without refuge: Chin refugees in India and Malaysia'. *Forced Migration* 30: 36–37.

Ananthachari, T. 2001. 'Refugees in India: Legal framework, law enforcement and security'. *ISIL Year Book of International Humanitarian and Refugee Law*, 1.

Basavapatna, S. 2011. 'The implementation of Urban Refugee Policy in Delhi'. *Refugee Watch Online*, Friday, 30 September 2011. Available at: http://refugeewatchonline. blogspot.com/2011/09/implementation-of-urban-refugee-policy.html (accessed 20 December 2011).

Bose, N. 2006. 'Afghan refugees in India become Indian, at last'. New Delhi: UNHCR. Available at: www.unhcr.org/441190254.html (accessed 18 December 2011).

Buscher, D. 2011. *Bright lights, big city: Urban refugees struggle to make a living in New Delhi.* New York: Women's Refugee Commission.

Centre for Refugee Research (CRR) 2007. *A Report from community consultations with the Women's League of Burma, New Delhi, India.* Centre for Refugee Research, University of NSW.

Centre for Refugee Research (CRR) 2008. *'We are the people who don't exist': Report on unregistered refugees in Mizoram, India* Centre for Refugee Research, University of NSW.

Centre for Refugee Research (CRR) 2010. 'Crucial support: Protecting the protector'. *Consultations with refugees from Burma: New Delhi, India.* Centre for Refugee Research, University of NSW.

Centre for Refugee Research (CRR) 2011. *'We want to be safe': Community based responses to violence against refugee women and children from Burma in New Delhi.* Trainings Report, Centre for Refugee Research, University of NSW.

Centre for Refugee Research (CRR) 2012. *Violence Against Women training for refugee men from Burma.* Training Report, Centre for Refugee Research, University of NSW.

Chimni, B.S. 2003. 'Status of refugees in India: Strategic ambiguity' in R. Samaddar (ed.)

Refugees and state: Practice of asylum and care in India 1947–2000, London: Sage Publications.

Hugman, R., Bartolomei, L. and Pittaway, E. 2011. 'Human agency and the meaning of informed consent: Reflections on research with refugees'. *Journal of Refugee Studies* 24(4): 655–671.

Human Rights Watch 2011. *World Report 2011: Burma.* Available at: www.hrw.org/world-report-2011/burma (accessed 20 December 2011).

Nair, A. 2007. *National refugee law for India: Benefits and roadblocks*, New Delhi, IPCS Research Papers, December.

Obi, N. and Crisp, J. 2000. *Evaluation of UNHCR's policy on refugees in urban areas: A case study review of New Delhi.* Geneva: UNHCR Evaluation and Policy Analysis Unit.

Personal Communication A, E. Pittaway, CRR Staff meeting minutes, 5 April 2012, on file with author.

Personal Communication B, Burmese Women Delhi, Email, 18 January 2012, on file with author.

Personal Communication C, Women's League of Burma New Delhi, Email, 16 February 2012, on file with author.

Pittaway, E. 2009. *Making mainstreaming a reality: Gender and the UNHCR policy on refugee protection in urban areas*, UNHCR. Available at: www.unhcr.org/4b0bb83f9.pdf (accessed 20 December 2011).

Pittaway, E. and Bartolomei, L. 2005. *Identification and protection of 'women at risk': Summary report.* Centre for Refugee Research, University of New South Wales.

Pittaway, E. and Bartolomei, L. 2008. *'We are the people who don't exist'.* Centre for Refugee Research, University of New South Wales.

Pittaway, E. and Bartolomei, L. 2009. 'Innovations in research with refugee communities'. *Refugee Transitions*, 21, electronic journal. Available at: www.startts.org.au/default.aspx?id=411 (accessed 20 December 2011).

Refugees International 2009. *India: Burmese Chin refugees experience sexual harassment.* Available at: www.refintl.org/blog/india-burmese-chin-refugees-experience-sexual-harassment (accessed 8 November 2011).

Refugees International 2009. *India: Close the gap for Burmese refugees.* Available at: www.refintl.org/policy/field-report/india-close-gap-burmese-refugees (accessed 8 November 2011).

Richards, A., Parveen, P. and Sollom, R. 2011. 'Life under the Junta: Evidence of crimes against humanity in Burma's Chin State'. Physicians for Human Rights Available at: http://physiciansforhumanrights.org/library/reports/burma-chin-report-2011.html (accessed 7 April 2012).

Saxena, P. 2007. 'Creating legal space for refugees in India: The milestones crossed and the roadmap for the future'. *International Journal of Refugee Law*, 19(2): 246–272.

Sengupta, I. 2008. 'UNHCR's role in refugee protection in India'. *InfoChange News and Features*, July 2008. Available at: www.infochangeindia.org (accessed 18 December 2011).

UNHCR 2008. *Newsletter New Delhi*, 31 August 2008. Available at: www.unhcr.org/48d7aff52.pdf (accessed 18 December 2011).

UNHCR 2009. *UNHCR policy on refugee protection and solutions in urban areas*, Geneva, UNHCR.

UNHCR 2010a. 'UNHCR highlights shortage of resettlement places'. *News Stories.* 5

July 2010. Available at: www.unhcr.org/cgi-bin/texis/vtx/search?page=search&docid=4c31f3826&query=Resettlement%202011 (accessed 18 December 2011).

UNHCR 2010b. *Age, gender and diversity mainstreaming, Executive Committee of the High Commissioner's Programme Standing Committee, 48th meeting.* Available at: www.unhcr.org/refworld/pdfid/4cc96e1d2.pdf (accessed18 December 2011).

UNHCR 2011. *UNHCR country operations profile – India.* Available at: www.unhcr.org/cgi-bin/texis/vtx/page?page=49e4876d6 (accessed 18 December 2011).

UNHCR 2012. 'UNHCR, refugees, others of concern and asylum seekers' (handout received from the UNHCR Office in New Delhi, 3rd January 2013).

Vijayakumar, V. 2000. 'Judicial responses to refugee protection in India'. *International Journal of Refugee Law*, 12(2): 235–243.

7 Life in limbo

Unregistered urban refugees on the Thai–Burma border

Eileen Pittaway

Figure 7.1 Drawn by a Karen urban refugee woman from Burma, Mae Sot, Thailand 2009.

Introduction

In 2011, the United Nations High Commissioner for Refugees (UNHCR) estimated that of the 10.1 million refugees recognized under its mandate, over two thirds lived in urban areas (UNHCR 2011a). However, these figures do not include the millions of people who remain hidden under multiple layers of ambiguity regarding their status as refugees, and the identities they are given or assumed as they have sought refuge from persecution. The 1951 Convention on the Status of refugees is quite clear and unambiguous in its definition of a refugee. It is someone who:

owing to a well-founded fear of being persecuted for reasons of race, religion, nationality, membership of a particular social group or political opinion, is outside the country of his nationality, and is unable to or, owing to such fear, is unwilling to avail himself of the protection[1] of that country or return there because there is a fear of persecution...

(UNHCR 1951)

However, the term is often misunderstood and is used inconsistently in everyday language. The ambiguities mentioned here have produced a number of other labels in discourse about asylum seekers, such as illegal refugees, economic refugees, illegal immigrant and irregular movers (Jacobsen 2006). While most of these labels have no basis in international law, they have become part of the 'accepted wisdom' surrounding refugees and forced migration, often propagated by politicians and media. At times these misnomers can deny asylum seekers the protection which is their right (Pittaway 2012). The definition of a refugee articulated by Ms Erika Feller, Assistant High Commissioner for Protection, is considered particularly relevant to people whose official status remains uncertain. On several occasions, Feller has clearly stated that a refugee is someone who has fled from home because of persecution or genuine fear of persecution. The act of seeking asylum and the refugee determination process merely confirms their status (Feller 2000, 2006). Until this happens, they are *unregistered refugees* and entitled to protection under international law. It is this definition that is used in this chapter.

The common picture of urban refugees is that of people living in squalid slums in major cities such as Cairo, New Delhi, Addis Ababa or Kuala Lumpur. For example, the *Journal of Refugee Studies* special edition on 'Urban Refugees' of September 2006 has a focus on refugees in major cities. However, many urban refugees live in small towns in regional and rural areas where they have sought refuge. The UNHCR Urban Refugee Policy recognizes that not all urban refugees will be located in large cities, but notes

the policy presented in this document is intended to apply to refugees in all urban areas, and not only to those in capital cities. It must be recognized, however, that UNHCR will be constrained in its ability to attain this objective in countries where refugees are scattered across a large number of urban areas.

(UNHCR 2009, para 13)

In this chapter we examine the circumstances of a large group of refugees from Burma[2] living in regional townships and villages along the Thai–Burma border (TBB). The majority of these have not been registered by UNHCR and have not received any services from them. Named as illegal immigrants by the Royal Thai Government, they themselves often accept this label. We explore the impact of this misnomer, lack of recognition and consequent lack of protection under the international protection regime on these refugees and the implications this has for them and their families. Current estimates are that over two million people labelled as illegal migrants from Burma are living and working on the TBB. In 2010, it was reported that an additional 60,500 were registered as migrant workers, the majority of whom were from Burmese minority ethnic groups, predominantly the Shan people (Interviews with UNHCR, INGOs and CBOs in Pittaway and Doney 2010; UNHCR 2013). However as Jacobsen notes 'the hidden and marginalized nature of urban refugees makes it difficult to make accurate estimates, and each "authoritative" source has its own agenda and set of reasons for the numbers it puts out' (Jacobsen 2006: 275). It can be argued that many of these migrants are, in reality, unregistered refugees who have sought refuge in Thailand from systematic and violent persecution from the ruling military government in Burma, commonly called the State Peace and Development Council (SPDC). While the specific research for this chapter was conducted in 2010, it builds on research conducted with the refugee communities on an annual basis from 2003 until 2012. As discussed below, in 2011, refugee groups reported to the author that the additional concern of forced repatriation if and when cessation occurs[3] has added to the anxiety of the population of unregistered refugees.

Research methodology

The circumstances and conditions endured by migrant workers, both legal and illegal, working in the far north of Thailand are widely known and well documented (Brees 2008; Arnold and Hewison 2005; Chalamwong and Sevilla 1996; Chantavanich *et al.* 2004). Far less is known about the unregistered refugees. Since 2002, the Centre for Refugee Research at the University of New South Wales, Australia (CRR), has conducted extensive annual community consultations with this population at the invitation of the refugee community-based organizations (CBOs). Researchers have seen the appalling conditions in which people lived, in a state often comparable to the worst refugee camps. We have visited the factories, farms, markets, bars and households where people work in often dangerous and exploitative situations. Over that time, staff from the CRR have assisted the Women's League of Burma, Shan Women's Action Network and the Karen Women's Organization by providing human rights documentation training, gender and human rights training, editing their reports and assisting in having these presented at relevant Untied Nations meetings.

The findings presented here are based on consultations undertaken with unregistered refugees and representatives of CBOs in Chiang Mai and Mae Sot on the TBB in October 2010. We also drew on data collected in 2005, 2006, 2008 and 2009–2012 in comprehensive consultations with over 400 refugees outside the camps in Mae Sot and Mae Sariang, another town hosting refugees and unregistered refugees. In addition, debate and background papers from the UNHCR annual tripartite meetings on resettlement (ATCR), Geneva, which we attended, were also utilized. We did not examine the situation of the 2,150 registered refugees living in Bangkok in 2010, which was very different to that of those living in the towns on the border.

The methodology used in the consultations is based on the 'Community Consultation' and 'Participatory Action Research' approaches, which were developed by Eileen Pittaway and Linda Bartolomei from the CRR. It grew from their work examining the occurrence and impact of systematic rape and sexual abuse of refugee women and girls in camps and refugee sites in Thailand, Kenya, Ethiopia, Sri Lanka, Bangladesh and, subsequently, in Australia.

The method uses an introduction to human rights and gender issues to provide a context to guide refugee, asylum seeker and displaced participants through an examination and articulation of issues of critical concern to their communities. The facilitators work with participants with the aim of exploring potential solutions to local problems, as well as strategies for action and advocacy. The techniques include community education strategies, the telling of stories and 'storyboarding'. The latter is a process in which participants conduct situational analyses. It is facilitated with the belief that all people have capabilities and the capacity to identify and address community problems if the resources are available to them. The visually based nature of the exercises means that they can be used with people of all levels of education, including people who are pre-literate. The focus of the method is the collection of information from often vulnerable populations in a way that is neither harmful nor exploitative but empowering, and has the potential for bringing about social change. It is ideal for use with marginalized and disadvantaged groups who have valid and historically based reasons for distrusting people in authority, including researchers, academics and representatives of governments and other institutions (Pittaway *et al.* 2010).

In Chiang Mai we worked with the Women's League of Burma, which facilitated interviews with nine unregistered refugees and provided interpreting services for the entire field trip. We interviewed representatives from the Karen Women's Organization, the Rakhine Women's Group, the La Po Women's Group and the Shan Women's Action Network, nine in all. These are all CBOs staffed by refugees.

We also conducted interviews with the Burmese Human Rights Group, two international non-government organizations (INGOs) which asked not to be identified but were staffed by refugees from Burma and international interns, and the Migrant Access Program, an NGO staffed by international staff, local Thais and workers from Burma.

In Mae Sot we interviewed four staff from the Karen Women's Organization, a community based organization (CBO) that facilitated interviews with six unregistered and four registered refugees living outside of the camps; the International Rescue Committee, an INGO that runs the Shield programme for unregistered refugees and migrants and that had just started a programme aimed at the protection of unregistered refugee children; the Jesuit Refugee Service that runs an outreach and capacity building programme with unregistered refugees, and the Migrant Access Program that runs capacity building and casework services for unregistered refugees.

We also consulted with the Head of the UNHCR Field Office and a UNHCR protection officer in Mae Sot, and spoke with UNHCR's regional senior protection officer based in Bangkok. At the request of the majority of the participants, quotes have not been attributed to specific sites to avoid identification of UNHCR offices, implementing partners (INGOs and NGOs), staff and refugees. The research was covered by comprehensive ethics procedures and all data is stored at the University of New South Wales, Australia.

Background: the situation on the Thai–Burma border

The situation on the Thai–Burma border is extremely complex. When the research for this paper was undertaken in 2010, in addition to the huge number of unregistered refugees, over 141,076 refugees from Burma were living in Thailand in camps along the border, the first of which had been established in 1984. By 2012 this had fallen to 135,873 (TBBC 2012a). Most of these refugees come from the Karen and Karenni ethnic groups, but there are also substantial populations from 10 other ethnic minority groups. Whilst the border camps on the TBB are refugee camps in every respect, the Royal Thai Government will not allow them to be called this. Instead, they are named 'Camps for people fleeing from conflict situations', in official documents and on signage outside the camps (Burma Lawyers' Council 2007). This is in deference to the wishes of the SPDC with whom the Thai Government is building a strong trade relationship (Pittaway and Bartolomei 2005). Despite internationally recognized evidence to the contrary, the SPDC claims that it does not persecute any ethnic groups, much of whose land contains a large proportion of Burma's resource wealth. The position of the SPDC is that the conflict occurs between ethnic groups themselves, or as a result of acts of aggression instigated by those groups against the SPDC (Nyein 2009; CRR 2008, 2009).

The Royal Thai Government is not a party to the UN Refugee Convention and, despite a generous record of hosting refugees including large populations from Vietnam and Cambodia in the 1970s, has refused all requests to sign and ratify the Convention. For the first 14 years of the camps' existence, the government denied UNHCR access to the population in the camps as it would not recognize them as refugees. As UNHCR was not allowed onto the TBB to offer protection to the refugees, the camps are administered by the Thai Government Minister of the Interior and run by the Thai Burma Border Consortium (TBBC), a US based NGO that was appointed to oversee the day to day running of the camps. They also assist the camp committees, comprising senior refugees, who have responsibility for managing the refugee population. For 28 years, refugees have been confined to the camps, with limited educational skills, training opportunities and no official means of earning income or gaining employment. The Thailand Burma Border Consortium (TBBC) has been the key agency responsible for providing food and shelter assistance to the refugees in these camps. Over the years it has been funded by a changing international consortium of agencies and governments from the US, Canada, Australia and the European Union. In 1998, following pressure from the international community, UNHCR was granted access to the camps to register refugees and provide some limited protection. However, they are still unable to play a role with regard to urban refugees from Burma living in Thailand.

Over the preceding decades, much has been written about the appalling conditions of the camps on the TBB, which, despite the excellent work of the TBBC, are classified as a protracted refugee situation in which thousands of people have been 'warehoused' (UNHCR 2011a; USCRI 2011). Warehousing is a term that has been used to describe the impact on refugees of living for prolonged periods in camps in which they are deprived of a myriad of human rights. Most important of these are the rights to live with dignity, and freedom of choice over most aspects of their lives as they wait for durable solutions to their situation (USCRI 2011). Protracted refugee situations are defined by UNHCR as situations in which refugees have been in camps for more than five years and have no prospect of a solution to their situation. Some refugees on the Thai–Burma border have been in camps for over 20 years (Milner 2007). Children have been born and reached adulthood in the camps without knowing any other way of life (Pittaway and Bartolomei 2005).

In 2005, an attempt was made by the international community to bring an end to the prolonged situation and to close the camps. The Royal Thai Government allowed a UNHCR resettlement programme to third countries[4] to commence from the camps, mainly to the US, Canada and Australia. However, this move coincided with renewed conflict inside Burma, including the burning of villages and crops, rape of women and girls, and kidnap of men and boys for forced labour (Nyein 2009). This was photographed and videoed by community members and has been verified by human rights activists who went inside Burma to assist (Human Rights Watch 2012; Free Burma Rangers 2009). This created a fresh influx of asylum seekers to the camps. Despite credible evidence of the continuing persecution of ethnic groups by the SPDC, the governments of Burma and Thailand blamed the arrival of more asylum seekers on the attraction of resettlement as a pull factor, rather than persecution as a push factor. The resettlement programme has therefore not achieved its goal, as the newcomers have taken the place of those resettled (CRR 2009; Pittaway and Doney 2011).

In an attempt to deter new arrivals, action was taken to suspend registration of refugees in the camps. This strategy also failed, and people continued to seek refuge in the camps. This led to situations where food rations were severely cut and malnutrition and unrest increased. Many survived only because community members shared their already inadequate rations. This forced many newly arrived refugees to move to the towns to seek basics such as food and medical services, thus increasing the numbers of urban refugees (CRR 2009; Interviews with INGOs, NGOs Mae Sot, Pittaway and Doney 2010, 2011).

Mixed migration flows

The flow of migrants from Burma across the TBB has been extremely mixed for decades, rendering the provision of services and an adequate and just response to asylum seekers difficult either to envision or maintain. In 2013, Thailand hosts nearly 85,000 registered refugees, with an estimated 62,000 still in camps. However, because of the estimated two million registered and unregistered migrants living in Thailand, of whom many are fleeing persecution, UNHCR has described Thailand as providing an inadequate protection space for people of concern. 'Refugees and asylum seekers living outside camps are regarded as illegal migrants and are subject to arrest, detention and/or deportation'. They report an improvement in detention rates in 2012, but note that arrests still continue. Many of these 'illegal immigrants form part of the estimated 500,000 stateless people currently in Thailand' (UNHCR 2013).

The mode of arrival of refugees in Thailand adds to the confusion. Some asylum seekers cross the border and go directly to camps where they apply for registration as refugees. The camps are porous and a number of people work illegally outside the camps, mainly as farm labourers. Others go directly to towns to seek employment as migrants; many do not apply for refugee status, even if they are aware that they have a valid claim. They prefer the uncertainty of life as a migrant to being warehoused in the camps. Yet another group of asylum seekers travel directly to UNHCR in towns and seek registration, but the waiting time is long and they are usually directed back to the camps to wait for assistance, even though registration had been suspended. This puts the unregistered refugees in an extremely vulnerable position. The routes to the camps have many roadblocks staffed by the Thai military and police, and if people are caught without papers they are either placed in detention or summarily put back across the border (Pittaway and Doney 2010).

The complexity of the situation also includes the mix of migrants with work permits, illegal migrants and unregistered refugees in the towns. Some Shan and Karen entered Thailand as migrant workers on short-term contracts, often to exploitative employers. The northwestern part of Thailand close to the Thai–Burma border hosts one of the biggest clusters of migrant workers, employed in agriculture, construction work, manufacturing and domestic labour (Martin 2007). The vulnerability caused by lack of not only work permits but any other form of registration leaves unregistered refugees and migrants open to all form of exploitation. Unregistered migrants are routinely paid only half of the minimum wage paid to Thai employees, and working conditions fall well below International Labour Organization standards. Their status leaves them without access to legal protection, and many employers mistreat employees with impunity, knowing that they will not report them because of their illegal status (Interview with INGO, Pittaway and Doney 2010). Many of these migrants are, in fact, refugees. Others, mainly Burmese, do cross the border purely for economic reasons (Brees 2008).

Despite a history of cross border movement, both authorized and unauthorized, and a degree of shared linguistic and cultural heritage between the Shan and some northern Thais, Shan refugees are the most marginalized group of all, as they are not even recognized as people fleeing from conflict and are officially denied access to any form of international protection. The persecution of this group, which can again be linked to the wealth of resources beneath their traditional land, is very well documented and acknowledged (Shan Women's Action Network 2003). Conditions for both illegal and indentured migrants on the border are so bad that some Shan leaders have been requesting refugee camps for the Shan people, because of the security they are seen to provide for the Karen and Karenni peoples (Pittaway and Bartolomei 2005; CRR 2009). There are a few recognized Shan communities where some minimal services are provided, but these are not comparable to the other camps and there is no access to resettlement for members of the Shan community (Interview with a Shan CBO, Pittaway and Doney 2010). Other Shan people, UNHCR and non-government service providers argue that refugee camps are such bad places that they are better off out in the community (Interviews with Shan CBO, and UNHCR, field notes 2010; Cardozo *et al.* 2004). They suggest that different ways of providing basic services such as food, water, health care and education should be identified for these urban refugees (Beyrer 2001). It is noted that some Shan people have lived in Thailand for many years, and some have been granted citizenship by the Royal Thai Government.

Additional layers of discrimination

While the exact numbers are not known, it is estimated that there are about 200,000 migrant workers living in Mae Sot alone, only half of whom are registered (Lim and Yoo 2012). Many of the local Thai people are living in poverty and working in the same factories and communities as unregistered refugees in the Mae Sot area. It is not surprising that tensions arise between the different groups around access to rights and resources (Jacobsen 2006). Migrants traditionally are employed in the very lowest wage brackets and often compete with local people for these jobs. There are many reports of illegal and harsh child labour (Robertson 2006). Refugees resent the fact that they are paid far less than the Thai workers, and have no legal redress, while local people accuse migrant workers of depressing wage rates (Interviews with INGOs, NGOs and CBOs, Mae Sot, field notes 2010).

While living conditions for unregistered refugees and migrants in Mae Sot often contravene a wide range of human rights, including the labour rights standards established by the ILO, the macro debate about the rights of refugees vis-à-vis the local population, both in Mae Sot and at a global level, is often focused on access to social, cultural and economic rights such as the right to work, health care, education and social security (Pittaway 2009; UN General Assembly, ICESCR 1966). The local poor populations in the urban areas where refugees reside usually also experience lack of access to these rights (Jacobsen 2006). It is therefore important to unpack the additional layers of discrimination and disadvantage that impact the lives of these refugees and asylum seekers in urban spaces, and differentiate them from the local poor population.

The most obvious of these is that unregistered refugees or 'illegal migrants' are routinely paid less than half the wage paid to Thai nationals and work longer hours, and suffer lack of access to fundamental human rights (Pittaway and Doney 2010; Brees 2008; Arnold and Hewison 2005). However, refugees, along with legal and illegal migrants on the TBB, also suffer from forms of discrimination and persecution in urban sites that even the poorest local people do not experience. These are those rights that come under the banner of civil and political rights (UN General Assembly, ICCPR 1966). They affect those refugees who, for a wide range of reasons, are not able to gain access to registration and therefore recognition of their refugee status; those refugees who have been rendered stateless by their country of origin; and all refugees who by definition do not enjoy active citizenship status and rights in any country in the world. These include lack of access to law and justice, freedom of movement, self-determination, and the right to participate in decision-making about their lives (Pittaway 2009). Jacobsen (2006) includes the experience of having survived conflict, torture and trauma, the danger of arbitrary arrest, and the increased chance of trafficking and sexual exploitation for women and girls as some of the many risks for urban refugees Asian Research (Centre for Migration and World Vision 2004). What this translates to on the ground is lack of access to the most basic levels of legal protection. Lack of legal protection and citizenship leaves refugees open to abuse by local people and institutions that act with impunity, knowing that the refugee/migrant population has little or no redress. Refugees from Mae Sot and Chiang Mai interviewed in 2010 stated that the Thai police and military are often hostile and unhelpful. Evidence was given of police failing to act when crimes were reported to them by refugees, in two cases of police beating male refugees who tried to complain about their treatment by an employer, and in one case, a young women sexually harassed at the police station when she went to report a rape in the workplace (Pittaway and Doney 2010). NGOs and CBOs working with refugees verified these stories and reported that refugees were very reluctant to go to the police (Pittaway and Doney 2010). UNHCR has little power and even fewer resources to enable them to provide the international protection which is meant to act as a substitute for citizenship until solutions are brokered. It is noted that those refugees who do gain registration as migrants are better protected than those refugees who live in camps.

Lack of access to these rights is what sets refugees/migrants apart from host populations. They report that they are often despised, exploited and suffer verbal and physical violence at the hands of local Thai people (Refugee from Burma, field notes 2010). They are seen to be competing for scarce resources and perceived as having no right to be there. The NSW Centre for Refugee Research (CRR) has documented several stories of Thai nationals intervening on behalf of refugees and assisting them when in trouble; however, such cases were uncommon (NGOs and CBOs, Pittaway and Doney 2010). It was reported that donor governments tended to focus on the more visible communities in camps and to ignore the needs of the ever-growing populations of urban refugees (INGOs, UNHCR, Pittaway and Doney 2010). Many have been refugees for so long that they are effectively being 'warehoused in the city'.

The UNHCR urban refugee policy

Many of the issues that were raised by refugees in consultations and interviews between 2002 and 2011 are alluded to in the new UNHCR Policy on Refugee Protection and Solutions in Urban Areas (UNHCR 2009). This document identifies many of the key problems articulated by urban refugees on the TBB, and while broad, suggests many positive steps and strategies to address some of the problems.

Key issues addressed are:

1 providing reception facilities;
2 undertaking registration and data collection;
3 ensuring that refugees are documented;
4 determining refugee status;
5 reaching out to the community;
6 fostering constructive relations with urban refugees;
7 maintaining security;
8 promoting livelihoods and self-reliance;
9 ensuring access to health care, education and other services;
10 meeting material needs;
11 promoting durable solutions;
12 addressing the issue of movement.

The policy clearly states that UNHCR has responsibility for all refugees in urban sites, both registered and unregistered, ergo the policy should apply to the unregistered refugees on the TBB. However, as noted earlier and reiterated by the head of the UNHCR office in Mae Sot in 2010, the Royal Thai Government does not recognize that there is a population of urban refugees and therefore, UNHCR does not have a mandate to work with them.

Summary of findings

The key concerns of all groups consulted were lack of registration opportunities and the long waiting times for interviews. In particular, the refugees discussed the vulnerability of women and girls, and in some cases boys and young men, to sexual abuse on the TTB, and the almost total impunity enjoyed by perpetrators (Pittaway and Bartolomei 2005).

There are many problems obtaining safe and secure affordable accommodation, as well as lack of access to justice and legal procedures. Refugees, both registered and unregistered, discussed the lack of livelihood opportunities and income security, the exploitation when work could be found, and the lack of access to health care.

Common to most refugee situations, a major issue identified throughout the consultations was the intersectionality of the areas of concern. They cannot be viewed and addressed in isolation. Lack of recognition as refugees and registration with UNHCR meant there is no formal access to basic needs, which in turn compounds risks and vulnerability. Access to safe and secure accommodation is critical to the protection of women and girls from sexual violence, but it is also crucial for the maintenance of good health, and to enable an environment in which children and adults can study. Families struggle to maintain normal familial relationships if they are not afforded some level of privacy. Refugees who cannot find safe accommodation are more likely to seek work in unsafe places that, although they provide some basic accommodation, can result in a high risk of rape, sexual abuse and other forms of violence. Women and girls who are raped face a number of health challenges, but lack of income or free health services usually means that they are not able to seek medical attention. Poor health outcomes lead to a lack of ability to protect children or to access what little employment there is – so the cycle continues (Pittaway 2009). Because of this intersectionality, it is impossible to separate out and report on individual problems. They are intimately intertwined, and any responses need to acknowledge that all aspects of the problems have to be addressed simultaneously.

Lack of access to the registration process and UNHCR services

Lack of citizenship or any other form of protection leaves refugees extremely vulnerable to police harassment and arrest. At times they are forced to resort to bribery to avoid jail or deportation. They have very low social status and are discriminated against and harassed by the local community. Some reported that they were also discriminated against by UNHCR officials and staff from implementing partners who did not always follow standard operating procedures regarding the rights of the refugees whom they are supposed to serve. Those who are unregistered or have had their claims rejected are denied access to UNHCR meetings, which might provide useful information about their rights and options. They have little avenue of appeal in the refugee status determination process (RSD) and no legal recourse when criminal acts are committed against them.

Refugees described a wide range of effects due to the lack of refugee status outlined above. These include feelings of fear, shame, humiliation, as well as a breakdown of family units, community dysfunction, domestic violence and debt. Social stigma leads to low self-esteem and loss of dignity. There are no resettlement options for asylum seekers without refugee status. The refugees also identified that the lack of legal documentation effectively denies them freedom of movement, as they are afraid that if they leave their places of employment they will be arrested, imprisoned and deported back to Burma.

The gendered nature of protection

The systematic rape and sexual harassment of women in Burma has been well documented by women from many of the minority ethnic groups (Shan Women's Action Network 2002; Karen Women's Organization 2004, 2007,

2010) and has been cited as one of the main reasons that many women and girls seek refugee in Thailand. However, for many, sexual violence does not cease once they flee Burma (Consultations 2005, 7, 8 and 9, Interviews with Shan Women's Action Network, Women's League of Burma (WLB), Chiang Mai and Karen Women's Organization, Mae Sot, Pittaway and Doney 2010). Rape and sexual abuse of women and girls are endemic in all aspects of their lives as unregistered refugees. Out of desperation and the need to feed themselves and their children, some women are forced to engage in 'survival sex'. They work in great danger, often employed by pimps, and are forced to engage in unsafe sexual practices. Sexual harassment and threats from employers are the everyday reality of urban refugees' lives. The inequality in family relationships and the lack of women's participation in community and family decision-making exacerbate social stigma. Women who are known to have been sexually abused or raped are often ostracized by the local community. This leads to feelings of shame, humiliation and helplessness, which in turn can result in the neglect of children. The emotional effects of family abuse and breakdown suffered by children can lead them to mirror these negative patterns of behaviour in their own lives.

The physical impacts of sexual abuse, including the effects of sexually transmitted diseases, impact livelihoods. Early and unwanted pregnancy results in dangerous, self-induced abortions, babies born as a result of rape, or deaths of young women too small to bear children. Their experience is encapsulated in a report produced by the Burmese Women's Union, entitled 'Caught between two hells' (BWU 2007). In 2010, refugees, CBOs and NGOs all reported major child protection issues and lack of legal process to address these. They included allegations of child abuse and neglect by parents forced to work long hours, leaving small children to fend for themselves. Local people also abuse children, both sexually and through exploitation in the workforce (Consultations Mae Sot, Chiang Mai and Mae Sariam 2005, 7, 8 and 9, Interviews 2010).

One of the major risk factors identified for women and girls in urban refugee sites around the world is the almost complete lack of access to just and effective legal processes for cases of rape and sexual and gender-based violence (UNHCR 2011c). Refugee women in Mae Sot gave detailed accounts of rape by employers, landlords and even the police. Many are severely traumatized, and yet are ashamed to disclose what has happened to them for fear of stigma and community exclusion. One Chin woman commented, 'if my husband finds out what has happened to me he will put me out and no-one will speak to me again.' A young Karen woman stated, 'If people find out what they did to me, no-one will ever marry me – I will be alone all of my life.' One Karen man talked of the powerlessness felt by himself and his friends who know these things happen but can do nothing to prevent it (Consultations 2009, Pittaway and Doney 2010). Prosecutions of the perpetrators of rape, murder and other violent acts are rare. Local police are often unsupportive and afraid to take action against powerful figures and institutions in their own communities:

If women are sexually harassed they have nowhere to report it to, because our existence itself is illegal. It is best not to approach the police because otherwise they will arrest us and send us home.

(Karen refugee woman, Mae Sot 2010)

Specific risks for men and boys

The impact of the abuse of the women on husbands and fathers, sons and brothers is enormous. They share the shame, powerlessness, and knowledge that there is no justice for their female relatives or for their communities. The knowledge that whole communities are unable to respond and can neither provide for, nor protect, the women and children causes enormous anguish. While women and girls are the primary targets of sexual and gender-based violence, boys and men are also raped and targeted for sex tourism and trafficked to Bangkok.

Many of the men are also carrying the physical and psychological scars of torture and the horror of war, and what is happening in urban situations is a repeat of what has happened during conflict. Instead of fleeing to safety, they have fled to continuing violence. In addition, they suffer from racism and the erosion of their identity as breadwinners and protectors of the family. Lack of access to political participation, autonomy and decision-making over their own lives perpetuate feelings of isolation and powerlessness. It is the resilience of the refugee population in the face of ongoing discrimination that maintains their communities, but after 20 years, the resilience is wearing thin and hope for a peaceful return to their homelands is fading (Interviews with CBOs and refugees, Mae Sot 2010).

Inadequate housing and accommodation

The lack of affordable and suitable accommodation leaves many refugees homeless. Single young men and women, often under the age of 18, are forced to share cramped accommodation and women are frequently raped. There are major problems with landlords who exploit the refugees, especially if they are not registered (Interviews with CBOs, NGOs, Mae Sot, Pittaway and Doney 2010). The very low wages paid to undocumented workers means they are unable to afford the rent. The refugees made strong connections between protection from further violence and secure shelter. They reported that when communities attempt to establish safe houses for women who have suffered rape and violence, they are often targeted by local police and forced to keep moving from one space to another (CBO interviews, Pittaway and Doney 2010).

Health

While refugees and migrants are allowed to use government hospitals, unregistered refugees and migrants are often afraid to approach these institutions because of their lack of legal status (interviews with refugees, Pittaway and

Doney 2010). Lack of access to medical services has multiple long-term impacts on refugees. Unregistered refugees are denied access to medical and hospital services, either through fear of accessing them without documents or because they cannot afford to pay for services. They are forced into debt in order to obtain essential medication. Refugees reported low health status and decreased life expectancy because of insanitary living conditions and insufficient food, causing malnutrition and health deficiencies and increased vulnerability to preventable diseases and illness. Children suffered stunted growth, resulting in long-term complications (Interview with refugee paramedic, Pittaway and Doney 2010).

> Many people are suffering from malnutrition due to lack of good food, their bodies are thin and bloated, their skin wrinkled, and many including children are having problems with their eyesight.
> (Karen refugee woman working as a paramedic, Mae Sot, 2010)

Many refugees suffer from conflict and flight-related trauma, exacerbated by conditions in urban areas. They suffer from mental health issues including depression, suicide and attempted suicide. Many children are the sole carers for sick adults, causing isolation and loneliness. Increased drug and alcohol abuse were reported to be much higher than in their previous lives, and there are very few services for refugees with a disability on the TBB. Health conditions affect refugees' ability to access a livelihood through employment:

> People are dying because we cannot access health services. If there is an emergency we have to go to the hospital but if we do not have the money to pay we are sent away.
> (Karen refugee woman, Chiang Mai 2010)

A major source of support to refugees and migrants both registered and unregistered, in Mae Sot is the Mae Tao Clinic, founded by a Burmese doctor, Dr Cynthia Maung. It is a community-based hospital financed by donation and overseas aid that has provided good quality health care to the Burmese refugee population in western Thailand since 1989 (Mae Toa Clinic 2012). However, since 2009 funding sources are becoming limited and donors focus more on service provision within Burma, causing problems for the thousands of people who rely on the services provided by the clinic (Interview, KWO Mae Sot, 2012).

Access to livelihoods

The discussion of livelihoods within this context is problematic. As mentioned earlier, as long as the unregistered refugees are not recognized as such, UNHCR and NGOs are not permitted to offer them any livelihood opportunities as an alternative to employment in the highly exploitative informal labour markets. There are also grey areas between the formal and informal labour markets, which are used to advantage by some employers. Weak legislation addressing labour rights in Thailand, with often poor enforcement, leaves migrant workers extremely vulnerable (HRW 2010; Voice of America 2013). It effectively precludes the introduction of independent income generation and opportunities for refugees to become self-sustaining (Interviews with UNHCR and INGOs, 2010 field notes). It is therefore not surprising that livelihoods and access to income security are key issues for all refugees who attended consultations. Yet access to livelihoods is the best protection women can have against sexual and gender based violence and can assist them to provide adequate protection for their children. It is also equally important that men can access safe livelihoods to enable them to survive, continue supporting their families, and maintain their dignity.

While the recently instituted system of worker registration provides some protection from arrest, it is limited to a small number of work categories. It does not cover those unregistered refugees working for community organizations and NGOs in the health, social or education sectors. Registration is also expensive and many are simply unable to afford it. Unregistered refugees are unable to advocate for fairer wages, as their employers can refuse to pay them at all or threaten them with arrest due to their illegal status (Interviews with refugees and CBOs, Pittaway and Doney 2010).

Informants also identified a range of ways in which refugees are exploited by Thai employers, which include having to pay fees to employment brokers to find them work. In such cases, women are often trafficked into employment in factories or brothels and burdened with a debt to the broker or trafficker (Chantavanich *et al.* 2004). Men are trafficked into hard manual labour, which has very high health risks. It was reported that they are sometimes killed in accidents and their deaths concealed by employers who bury the bodies illegally. In other cases, the employers retain the refugees' labour cards and charge them a high penalty if they wish to leave and seek alternative employment. Women who become pregnant are victimized by their employers who either force them to have abortions or report them to the police who then deport them as illegal immigrants. Refugees are often forced to pay bribes in order to gain employment or to stop employers or work colleagues from reporting their illegal status (Pittaway and Doney 2010):

> If women are sexually harassed they have nowhere to report it to, because our existence itself is illegal.
>
> (Kachin refugee woman, Chiang Mai, 2010)

Children are forced to work in dangerous conditions for long hours, causing them to miss schooling, and families are forced to separate to find work. Refugee women bear the double burden of having to work long hours for poor wages as well as working at home and caring for and feeding a family:

> I sometimes think I was better in the jungle – I could toilet without being seen, I could hide under a bush when men came looking for girls. Here there is nowhere to hide – but there was no food and no work in the jungle.
>
> (Karen refugee woman, Mae Sot, 2010)

Migrant workers v. refugees: registration and labour rights

Over the years a number of CBOs have developed and worked with the unregistered refugees and migrants. Some international NGOs also provided limited assistance, but not openly, as the Thai Government has not approved this. The work of the CBOs has been the main and most important support for refugees and migrants over two decades. They have worked tirelessly and fearlessly to deliver services to their target community and to advocate for them at national and international levels. It was due to their work that the unregistered refugees and migrants gained some significant rights.

Since 2002, progressively small changes have been achieved and there is now a limited system of migrant registration, some access to basic health care, and the children of migrants are allowed to attend primary schools (Pinkaew 2008). In consequence of the work of the CBOs, life chances have increased significantly for many of the unregistered refugees or migrants on the TBB. However, an unintended consequence is that in their naming of this group of people as migrants – either legal or illegal – and in using a broad spectrum of human rights instruments to advocate for their rights, they inadvertently changed the perception of them away from that of unregistered refugees and firmly to that of migrants. This creates a dilemma, because if and when the Royal Thai government decides that it neither wants nor needs these people in the workforce or when the Burmese military government demands that they be returned to Burma, the unregistered refugees will have no recourse to refugee protection and assistance. While they are accepted and treated as migrants, no one is fighting for their rights as refugees.

Many of the problems obtaining secure, safe and sustainable livelihoods are linked to the lack of the right to work in host countries. However, some preliminary studies show that if refugees were legally absorbed into the labour market, paying for business licenses and tax, their labour could have significant advantages for the host government. They often contribute both legally as consumers, and sometimes illegally, as employees and business owners to the local economy in the areas in which they live, and the benefits of this participation could be enhanced to spur further economic development in impoverished urban settlements (Porter *et al.* 2008).

It has to be noted, however, that these exploited and cheap migrant workers have become an essential part of the local economy. It was reported in a interview with the MAP Foundation, Chiang Mai, 2010 that when one group of migrant workers held a protest rally aimed at the government in 2010 because they were being threatened with forced repatriation to Burma, some employers joined with them. However, the fact that the migrants felt confident enough to hold a rally indicates that the rights gained for them by the activities of the CBOs and their growing realization of the importance of their cheap and exploited labour to the Thai economy encourage them to refuse to stay in such a marginalized position. This in itself could backfire, because winning the fight for improved conditions and wages means the employers might not see an advantage to employing a migrant worker over a local Thai. Demand for their services could drop, again making forced repatriation a reality. The fluctuations of the world economy also directly impact this group of people. For example, in economic downturn, when the demand for labour drops, there have been frequent roundups and deportations of redundant migrant workers (CBO Mae Sot, Pittaway and Doney 2010).

Challenges to service provision

While there are groups of CBOs striving to assist the unregistered refugees, these organizations also suffer from constraints because of their illegal status. Working on the issue of labour laws, and undertaking tasks such as documenting rapes of migrant workers or forced abortion by employers is very dangerous, as the workers face incarceration, possible abuse and deportation with very little access to due legal process. The different ethnic groups speak different languages and lack of suitable interpreters renders communication extremely difficult. Despite these challenges, the CBOs offer counselling, legal advice and some material support. The majority of INGOs and NGOs who provide assistance to migrants, often covertly, are situated in the centre of Mae Sot, while many of the migrant workers live on the outskirts or in smaller cities and towns in Tak State. Many unregistered refugees are unaware of these services or unable to travel to them.

While many CBOs are very effective and productive, others lack the infrastructure and resources to function effectively. Despite the excellent and untiring work of the Karen Women's Organization and other CBOs in Mae Sot it is evident that they do not currently receive the resources and support required to enable them to respond fully to the needs of the large numbers of women and girls at extreme risk. The few safe houses are desperately underresourced and overcrowded, and constantly have to change premises to avoid detection by the police. Community-based women's groups are also at risk from employers and perpetrators of domestic violence if they become known as having provided shelter. Poor communication between the CBOs and INGOs, and the lack of recognition by some humanitarian aid workers of the capacity and expertise of the refugees, exacerbate the problems. CBOs reported their extreme frustration about being excluded from meetings at which critical decisions about their communities were being made. This is taking on a greater importance as the reality of camp closure and return becomes high on the agenda (Pittaway and Doney 2010, 2011, 2102). Refugees report that there is a perceived lack of confidentiality amongst all key personnel involved in responding to labour exploitation, child protection issues and violence against women. This prevents some members of the community from seeking assistance even when it is available (Consultations Mae Sot, Chiang Mai and Mae Sariang 2005, 7, 8 and 9, Interviews 2010 Pittaway and Doney 2010, 2011).

The politics of recommendations

The refugee protection regime is based on the belief that eventually a durable solution will be found for all refugees. It is argued that whichever of the three durable solutions are found to be most appropriate for individuals, families and communities, the international community has a vested interest in improving the level of services to a standard that promotes a positive outcome or at the least causes no further harm to refugees. While this is an admirable goal, it only applies to those refugees who are recognized as such by the international community. Unregistered refugees are not considered for durable solutions. They are rendered invisible by the ambiguities discussed earlier. They have been named by UNHCR Geneva some of the most vulnerable refugees in the world (ARRA 2010, 2012).

Suggesting recommendations for actions to assist the unregistered refugees living along the Thai–Burma border is complex because of their ambiguous situation. Until and unless they are recognized and named as unregistered refugees by all key stakeholders, their claims to protection come under the same umbrella as other migrants, both legal and illegal, with whom they live and work.

The first recommendation would be for the Royal Thai Government to allow UNHCR to reopen registration of refugees and extend this to cover not only refugees fleeing recent conflict inside Burma and newly arrived in the camps but also those unregistered refugees who have been living illegally in border towns, sometimes for decades. This would then enable them to claim the rights to which they are entitled under international law (David and Holliday 2012). The answer might be a joint approach which at the same time recognizes this group as refugees so that they are protected from *refoulement* when their labour is not required, but grants them the right to work in decent conditions for a decent wage when needed. Many additional recommendations about the implementation of the UNHCR Refugee policy on Urban Refugees have already been made in other countries (Pittaway 2009). Most of these could easily be transferred to this cohort.

Karen, Karenni, Shan and Chin refugees interviewed in 2010, and who took part in a UNHCR sponsored consultation in 2011, have highlighted the need for stronger labour laws in Thailand to avoid exploitation and for the protection of migrant workers (Pittaway and Doney 2010; UNHCR 2011c). This includes reducing the cost of the labour card so that it is affordable for all migrant workers. A previous revision of the current system of labour permits had allowed migrants the right to change employers, to work while pregnant and to refuse forced abortions, at least on paper. This needs to be fully implemented and to be available to unregistered refugees. In an ongoing effort to register all illegal migrant workers, commenced by the Royal Thai Government in 2009, employers were asked to pay between 3,000 and 4,500 baht (US$100–150) to register existing illegal employees, with a large fine if caught employing illegal migrant workers. The unregistered workers themselves also faced large fines if caught. While a step in the right direction, the scheme is complex and cumbersome, open to corruption and manipulation (Democracy for Burma 2011). The scheme had limited success, and in 2013 the government extended the registration date for the estimated two million migrants still unregistered for four months. There are doubts that this target can be fulfilled, and also doubts that the scheme will ever be totally successful. Many employers are unwilling to pay the newly introduced minimum wage of 300 baht (approximately $10) per day and continue to flout the law (Voice of America 2013).

Refugees highlighted the need for human rights groups to document cases of forced abortions, as well as trafficking for employment and deportations, in particular when deportations have split families. The refugees also emphasized the importance of legal documentation for CBOs engaged in social, health and education services and their employees. Work needs to continue with the Royal Thai Government to improve the living conditions of migrant workers and to ensure their access to Thai legal processes and social services. Realistically, this is unlikely. As detailed earlier, one reason that the refugees have been tolerated for so long is because of their important role in the Thai economy as cheap illegal migrants (Martin 2007). UNHCR staff are in a very difficult situation. They have a very limited mandate on the Thai–Burma border, and are so under-resourced that they struggle merely to deal with the existing caseload. They could not handle the processing of potentially millions more applications for refugee status. It is also unlikely that the Executive Committee of UNHCR would sanction this action because of lack of resources and political pressure (Interview UNHCR 2010). Since 2010, the changing political landscape within Burma has also shifted the focus of UNHCR, governments and NGOs to preparing for an anticipated end to the conflict situation and closure of the camps on the border (ARRA 2012). Much of the funding previously received by the INGOs and NGOs working with the TBBC has been diverted to programmes within Burma, leaving serious gaps in service provision for the refugees both in camps and in the urban sites (TBBC 2012b; Pittaway and Doney 2012).

This has led to a further complication: that of the ongoing threat of forced repatriation, often used as a big stick by the Royal Thai Government to control the refugee populations. CBOs and refugees struggle to understand these threats, how to separate rumour from reality and how to plan for their communities. In November 2011, members of refugee CBOs on the TBB requested the author to run a workshop on repatriation. The organizations understand that there is no immediate plan to return refugees to Burma, but they assess from political developments that repatriation will occur at some time in the foreseeable future. They wish to work collaboratively with authorities to develop a framework to protect refugees' rights and ensure their safety in the planning and implementation of repatriation. Rumours abound in the camps and in the urban refugee population that the SPDC has requested the Royal Thai Government to return all illegal migrants to Burma, as their labour is needed on the many new economic development projects which the SPDC is planning in collaboration with multinational organizations and other governments (Burma Partnership 2012; Interviews with NGOs, CBOs, Doney and Pittaway 2010, 2011, 2012). Critical to a successful end to this refugee situation is improved communication between all key stake-holders, including the refugee communities.

One of the major fears of the unregistered urban refugees is that they will be forcibly returned to Burma and, because of their ambiguous status, will once again have no international protection. Unlike the registered refugees, their return and security will not be monitored by UNHCR, nor will they receive international aid to assist in rehabilitation. Unless they are recognized as refugees and the international community considers them when determining future relations with the government of Burma, once again they will be invisible and their fate unknown and unacknowledged. Unregistered refugees expressed their concern, that if cessation is declared by UNHCR and Burma requests the Royal Thai Government to return all refugees, nobody knows what will happen to them (Pittaway and Doney 2010, 2011, 2012). To use an expression from Burma, the unregistered refugees are caught between a snake and a tiger (Pittaway 2008).

Notes

1 Protection is defined as all activities aimed at obtaining full respect for the rights of the individual in accordance with the letter and spirit of the relevant bodies of law, namely human rights law, international humanitarian law and refugee law (UNHCR 2011b).
2 While the ruling Junta in Burma, the SPDC, has changed the name of the country to Myanmar, this name change is rejected by the refugee groups. They see it as a further attempt to eliminate their status as specific ethnic groups within Burma. They therefore request that supporters of their cause refer to their homeland as Burma. They also reject the nomenclature 'Burmese refugees' as this also takes away their ethnic identity. Instead they refer to themselves as 'refugees from Burma', usually adding their ethnic group to the title.
3 Cessation is the term used when the circumstances in the country of origin have changed such that UNHCR declares it safe for refugees to return home.
4 Resettlement to a third country is one of the three UNHCR designated 'durable solutions' for refugees. It is intended as a last resort for those refugees who are unable to gain protection or to integrate into the country of first asylum, or to return to their homeland (UNHCR 2005).

References

Arnold, D. and Hewison, K. 2005. 'Exploitation in global supply chains: Burmese workers in Mae Sot'. *Journal of Contemporary Asia*, 35 (3).

ARRA (Australian Refugee Rights Alliance) 2010. *Report on the UNHCR Consultation with NGOs, Geneva*, CRR UNSW, Sydney, Australia.

ARRA 2011. *Report on the UNHCR Consultation with NGOs, Geneva*. CRR UNSW, Sydney, Australia.

ARRA 2012. *Report on the UNHCR Consultation with NGOs, Geneva*. CRR UNSW, Sydney, Australia.

Asian Research Center for Migration and World Vision 2004. *Response to trafficking of persons, especially women, youth, and children along the Thai–Burmese border*. Asian Research Center for Migration, Chulalongkorn University, Thailand.

Beyrer, C. 2001. 'Shan women and girls and the sex industry in Southeast Asia: Political causes and human rights implications'. *Social Science and Medicine*, 53 (4): 543–550.

Brees, I. 2008. 'Refugee business: Strategies of work on the Thai–Burma border'. *Journal of Refugee Studies*, 21 (3): 380–397.

Burma Lawyers' Council 2007. 'Analysis on the situation of the refugee camps from the rule of law aspect'. *Lawka Pala Legal Journal of Burma* (Spring 2007), reprinted in *Thailand Law Journal* Spring 2008, 11 (1).

Burma Partnership 2012. 'Refugees deserve right to choose when to return home'. Available at: www.burmapartnership.org/2012/06/refugees-deserve-right-to-choose-when-to-return-home/ (accessed 28 January 2013).

Burmese Women's Union 2007. *Caught between two hells*. Thailand: BWU Chiangmai.

Cardozo, B.L., Talley, L., Burton, A. and Crawford, C. 2004. 'Karenni refugees living in Thai–Burmese border camps: Traumatic experiences, mental health outcomes, and social functioning'. *Social Science and Medicine*, 58: 2637–2644.

Centre for Refugee Research 2006. *Consultation Report, Mae Sot* (ed.) L. Bartolomei, CRR UNSW, Sydney Australia.

Centre for Refugee Research 2007. *Consultation Report, Mae Sariang* (ed.) E. Pittaway, CRR UNSW, Sydney Australia.

Centre for Refugee Research 2008. *Consultation Report, Mae Sot and Chiangmai* (ed.) G. Doney, CRR UNSW, Sydney Australia.

Centre for Refugee Research 2009. *Consultation Report, Mae Sot* (ed.) G. Doney, CRR UNSW, Sydney Australia.

Chalamwong, Y. and Sevilla, R.C. 1996. 'Dilemmas of rapid growth: A preliminary evaluation of the policy implications of illegal migration in Thailand'. *TDRI Quarterly Review*, 11 (2).

Chantavanich, S., Premjai, V. and Samarn, L. 2004. *Migration and deception of migrant workers in Thailand*. World Vision Thailand in collaboration with Asian Research Centre for Migration, Chulalongkorn University, Thailand.

David, R. and Holliday, I. 2012. 'International sanctions or international justice? Shaping political development in Myanmar'. *Australian Journal of International Affairs*, 66 (2): 121–138.

Democracy for Burma 2011. Available at: http://democracyforburma.wordpress.com/2011/05/13/thailandregistration-for-illegal-migrant-workers-starts-june-15/ (accessed 28 January 2013).

Feller, E. 2000. 'Statement by the Director, UNHCR Department of International Protection, to the 18th Meeting of the UNHCR Standing Committee'. 5 July 2000, *International Journal of Refugee Law*, 12 (3): 401–406.

Feller, E. 2006. 'Asylum, migration and refugee protection: Realities, myths and the promise of things to come'. *International Journal of Refugee Law*, 18 (3/4): 509–536.

Free Burma Rangers 2009. *Annual Report.* Available at: www.freeburmaranegrs.org (accessed 12 February 2013).

Human Rights Watch 2010. 'Thailand: Migrant workers face killings, extortion, labor rights abuses'. Available at: www.hrw.org/news/2010/02/22/thailand-migrant-workers-face-killings-extortion-labor-rights-abuses (accessed 14 November 2011).

Human Rights Watch 2012. 'Burma: New violence in Arakan State'. Available at: www.genocidewatch.org/myanmar.html (accessed 31 January 2013).

Jacobsen, K. 2006. 'Refugees and asylum seekers in urban areas: A livelihoods perspective'. *Journal of Refugee Studies*, 19(3): 273–286.

Karen Women's Organization 2004. *Shattering silences*. Thailand: KWO Mae Sariang.

Karen Women's Organization 2007. *State of terror*. Thailand: KWO Mae Sariang.
Karen Women's Organization 2010. *Walking amongst sharp knives*. Thailand: KWO Mae Sariang.
Lim, R. and Yoo, H.B. (eds) 2012. *Migrant and refugee: Field study in Mae Sot, Thailand*. John Hopkins Bloomberg School of Public Health and Medipeace USA.
Mae Tao Clinic 2012. Available at: maetaoclinic.org (accessed 3 February 2013).
Martin, P. 2007. *The economic contribution of migrant workers to Thailand: Towards policy development*. ILO Subregional Office for East Asia, Malaysia.
Milner, J. 2007. *Towards solutions for protracted refugee situations: The role of resettlement*, Annual Tripartite Consultations on Resettlement, UNHCR Geneva, 29 June.
Nyein, S.P. 2009. 'Ethnic conflict and state building in Burma'. *Journal of Contemporary Asia*, 39 (1): 127–135.
Pinkaew, E. 2008. *Good practices to protect and promote migrant workers' rights in Thailand: Lessons learnt from NHRCT and its counterparts through people's capacity building and networking for enhancing human rights mechanisms*. National Human Rights Commission of Thailand.
Pittaway, E. 2008. 'Between a tiger and a snake: The Rohingya refugees in Bangladesh: A failure of the international protection regime', in H. Adelman (ed.), *Protracted displacement in Asia: No place to call home*. UK: Ashgate Press.
Pittaway, E. 2009. 'Making mainstreaming a reality: Gender and the UNHCR Policy on Refugee Protection and Solutions in Urban Areas: A refugee perspective paper'. Commissioned by UNHCR for the December High Commissioner's Dialogue on Urban Refugees.
Pittaway, E. 2012. 'Citizens of no-where: Refugees, integration criteria and social inclusion', in N. Steiner, R. Mason and A. Hayes (eds), *Migration and insecurity: Citizenship and social inclusion in a transnational era*. London: Routledge. pp. 169–189.
Pittaway, E. and Bartolomei, L. 2005. 'Risks for refugee women on the Thai–Burma Border'. Centre for Refugee Research Occasional Paper, UNSW Sydney.
Pittaway, E., Bartolomei, L. and Hugman, R. 2010. '"Stop stealing our stories": The ethics of research with vulnerable groups'. *Journal of Human Rights Practice*, vol. 2, no. 2, pp. 229–251.
Pittaway, E. and Doney, G. 2010. *Field report: Urban refugees, Mae Sot and Chiang Mai, Thailand*. CRR, UNSW, Sydney Australia.
Pittaway, E. and Doney, G. 2011. *Field report: Repatriation, Mae Sot, Mae Hon Son, and Chiang Mai, Thailand*. CRR, UNSW, Sydney Australia.
Pittaway, E. and Doney, G. 2012. *Field report: Human rights documentation and UNSCR 1325 Trainings, Mae Sot Chiang Mai, Thailand*. CRR, UNSW, Sydney Australia.
Robertson, S. Jr (ed.) 2006. *The Mekong challenge: Working day and night. The plight of migrant child workers in Mae Sot, Thailand*. The Federation of Trade Unions, Burma (FTUB) Migrants Section.
Shan Women's Action Network 2002. *Licence to rape*. Chiangmai. Available at: www.shanwomen.org/publications.html (accessed 29 January 2013).
Shan Women's Action Network 2003. *Dispelling the myths: Chiangmai*. Available at: www.shanwomen.org/publications.html (accessed 14 November 2011).
Thai Burma Border Consortium (TBBC) 2012a. 'TBBC history'. Available at: www.tbbc.org/camps/history.htm (accessed 28 January 2013).
TBBC 2012b. *TBBC programme report: January–June 2012*. Available at: www.unhcr.org/refworld/country,,TBBC,,MMR,,506bfe262,0.html (accessed 20 January 2013).
UN General Assembly 1966. *International Covenant on Civil and Political Rights*

(ICCPR), 16 December 1966, United Nations, Treaty Series, 999, p. 171. Available at: www.unhcr.org/refworld/docid/3ae6b3aa0.html (accessed 14 November 2011).

UN General Assembly 1966. *International Covenant on Economic, Social and Cultural Rights (ICESCR)*, 16 December 1966, United Nations, Treaty Series, 993, p. 3. Available at: www.unhcr.org/refworld/docid/3ae6b36c0.html (accessed 14 November 2011).

UNHCR 2005. *Resettlement Handbook.* Available at: http.//unhcr.org/pages/4a2ccba76. html (accessed 14 November 2011).

UNHCR 2009. *Policy on refugee protection and solutions in urban areas.* Available at: www.unhcr.org/refworld/docid/4ab8e7f72.html (accessed 8 December 2011).

UNHCR 2011a. 'Background notes for the roundtables, the intergovernmental event at the ministerial level of member states of the United Nations on the occasion of the 60th anniversary of the 1951 Convention relating to Refugees, and the 50th anniversary of the 1961 Convention on the Reduction of Statelessness. Geneva'. December, Geneva.

UNHCR 2011b. *Notes on international protection.* Available at: www.unhcr.org/refworld/type/UNHCRNOTES.html (accessed 20 January 2013).

UNHCR 2011c. *Survivors, protectors, providers: Refugee women speak out.* UNHCR. Geneva.

UNHCR 2013. *Country program, Thailand.* Available at: www.unhcr.org/refworld/type/UNHCRNOTES.html (accessed 12 February 2013).

USCRI 2011. *Help end human warehousing.* Available at: www.refugees.org/our-work/refugee-rights/warehousing-campaign/ (accessed 18 January 2013).

Voice of America 2013. Available at: www.voanews.com/content/thailand-extends-migrant-worker-registration-deadline/1586395.html (accessed 28 January 2013).

8 Urban refugees and UNHCR in Kuala Lumpur

Dependency, assistance and survival

Gerhard Hoffstaedter

Introduction

Malaysia has long been a place of intra- and intermobility, people moving across the Malay world and beyond via this *entrepôt*, strategically located along major trade routes. National borders continue to be acts of fiction rather than material delineation of the state's authority and sovereignty, especially on Borneo. Even on the peninsula people move, are smuggled and are trafficked daily, often through locally licit rather than legal avenues (Kalir and Sur 2012). Thus movements continue into, through and out of the country. Malaysia has largely forgotten its history of a regional mobility and cosmopolitanism (Hoffstaedter 2011a). Today the Malaysian government, like most governments, is more focused on constricting mobility and flows of people through a range of disciplinary techniques, such as detention, and border protection mechanisms, including biometric passports and more border patrols. However, the terrain of the border remains porous and the sea especially offers many alternative routes to and from Malaysia. As a result, Malaysia has become home to numerous groups of peoples from the Malay Archipelago, with most Malays being able to trace some of their kinship ties to modern day Indonesia or some other regional country. This allows other foreigners to blend into Malaysian society quite easily – a fact of great significance for many refugees.

The situation for urban refugees in Kuala Lumpur (KL) in Malaysia is complex. There is no single experience for the urban refugees of KL; rather, individuals from varied ethnic and religious backgrounds have diverse experiences of refugee life in Malaysia's largest city. This chapter will provide a glimpse into the lives of urban refugees in Malaysia by examining briefly the history of refugee populations in Malaysia, the ethnic and national make-up of the refugee population, and the Malaysian government's position vis-à-vis refugees, while also describing the services available to refugees, either through local service providers or their own communities. The chapter discusses the UNHCR policy on urban refugees and its effects on the refugee populations in Kuala Lumpur. Fieldwork for this chapter was undertaken in June-July 2012, during which time I interviewed refugees and asylum seekers from a broad range of backgrounds, as well as representatives from the United

Nations High Commissioner for Refugees (UNHCR) and civil society actors. I also lived with the Chin Refugee Community for one month.

At the end of December 2012, there were some 101,080 refugees and asylum seekers registered with UNHCR in Malaysia (UNHCR 2012). The vast majority of them (92,560) are from Burma, comprising 32,900 Chins, 25,580 Rohingyas, 10,530 Myanmar Muslims, 6,820 Rakhines, 3,630 Mons and minorities from other ethnicities in Burma. Seventy per cent of refugees and asylum seekers are men, while 30 per cent are women. There are also some 21,810 minors below the age of 18. In addition there are thousands of persons not registered with UNHCR for a variety of reasons, some because they do not know about the refuge determination process, others because they do not have the funds available to travel or are too scared to travel to Kuala Lumpur for registration. Numbers are difficult to gauge beyond the UNHCR figures, but refugee communities speak of tens of thousands, while conservative UNHCR estimates put unregistered asylum seekers at about 15,000. The majority of refugees and asylum seekers reside in the Klang valley stretching from the port of Klang to Kuala Lumpur and its satellite cities.

The largest group within the refugee world of Malaysia is made up of people who have fled from Burma. More than 90 per cent of registered refugees are from Burma, with Chins the largest group, followed by the Rohingyas. Refugees from Burma number more than 90,000, with many additional asylum seekers who are either waiting to be processed by UNHCR or have not yet been able to register. These numbers are high, as is the diversity of the Burma refugee population in terms of ethnic identity, religion, clan and kinship networks. The majority have fled Burma due to the ongoing civil wars and unrest, especially in the north and west of the country. Following independence from British colonial rule in 1948, Burma experienced several decades of internal conflict – due to the desires for political autonomy by various ethnic minority groups – and more recently military rule (Streit 2007: 10–11). Many people in Burma, particularly the ethnic minorities, have suffered human rights abuses including forced labour, forced relocation and mass killings (US Congress 2009: 6). These military incursions, torture, dispossession, forced labour and other human rights abuses have led many to flee across the border to Bangladesh, India and Thailand. From there human trafficking and smuggling rings provide passage to Malaysia via jungle treks across the Thai-Malaysian border. Most arrivals have family, friends or contacts living in Malaysia who will link them into existing refugee networks.

The Rohingya present a special case, because they have much in common with the majority Malay community in Malaysia and have resided in Malaysia for a prolonged time, yet integration remains an elusive goal that both UNHCR and the Rohingya themselves hope for. The Rohingya are a Muslim minority ethnic group from Burma. They have effectively been stateless since 1974 when they were stripped of their nationality. In 1982, this was enshrined in the Burma Citizenship Law, which declares Rohingyas as 'non-national' or 'foreign residents' (Refugees International 2009: 36). This view is based on a particular view of migration during British rule of Burma. During the colonial period migration into the areas now inhabited by Rohingya increased, leading to the charge that Rohingya are immigrant Bengalis (ibid.: 36). Having been made stateless, many fled to Bangladesh and later to Malaysia, where they were initially well received and even obtained six-month work permits in 1992 (Human Rights Watch 2000). While other refugees from Burma were not treated so well, the Malaysian government led UNHCR to believe that local integration for Rohingyas was a reasonable goal. Negotiations had proceeded to a point that in 2004 a plan to issue 10,000 temporary work visas was announced in parliament. However, by early 2006, just days into its implementation, the plan was stopped amidst corruption claims (Cheung 2012: 5). Apparently the Malaysian authorities had not cooperated with UNHCR and UNHCR registration records, working instead with self-proclaimed Rohingya leaders (The Equal Rights Trust 2010). Leadership in the Rohingya community has been a major issue for UNHCR too, with no single community organization able to act as a representative.

The legal status of urban refugees in Kuala Lumpur

Malaysia is not a signatory to the 1951 Convention on refugees, or its 1967 Protocol, and as such does not officially recognize refugees nor provide any rights traditionally afforded to refugees (USCRI 2009). However, Malaysia, like most of its regional neighbours, is party to the Convention on the Rights of the Child, which means it has a responsibility to provide assistance and protection to refugee children. Legal aid organizations and civil society actors in Thailand are probing whether this convention could be fruitfully activated to provide some legal standing vis-à-vis the state for refugee minors. In Malaysia, NGO shadow reports make use of the Convention on the Rights of the Child and the Convention on the Elimination of All Forms of Discrimination against Women (CEDAW); however, no formal legal challenge has been explored to date.

Currently the most powerful Malaysian law that pertains to refugees is the Immigration Act that makes their presence unlawful, although the Minister can exempt people or groups and provide IMM 13 visas to allow temporary residence and provide work rights.[1] So far this has been applied to, for example, Moro refugees residing in Sabah, and Bosnians and Acehnese in the peninsula. On the whole, however, the government has generally regarded refugees and asylum seekers as illegal immigrants and they are thus subject to treatment as such. If caught, they face caning, fines, imprisonment as well as detention and deportation (US Congress 2009: 12). UNHCR has had assurances from the Malaysian government that registered refugees in possession of their UNHCR card would be permitted to remain in Malaysia until their resettlement (see further on). However, the Malaysian government has had an inconsistent approach to allowing refugees to stay in Malaysia and providing work rights for them.

UNHCR and the government's approach to urban refugees

Since the exodus of South Vietnamese refugees in the 1970s and 1980s there have been no refugee camps in Malaysia and most refugee populations live amongst Malaysians in rural as well as urban settlements and cities. Many squatter settlements in jungle camps and amongst plantations have been razed recently, pushing refugees into the anonymity of Malaysia's major cities, especially Kuala Lumpur, Johor Bharu and Penang.

UNHCR Malaysia spends most of its resources on refugee registration and refugee status determination (RSD). The agency also has a close working relationship with most refugee community organizations, which collaborate and provide data on new arrivals for registration to UNHCR. This has been a deliberate policy to strengthen (especially Burma refugee) community groups as a first port of call. Owing to the large numbers of refugees from Burma, they are no longer allowed to go directly to the UNHCR compound to get an appointment. Instead their community organization registers them and sends the names and contact details to UNHCR. This has in effect outsourced some of the burden on UNHCR and cut down the number of people queuing outside the office every day. It has also created a dangerous precedent of relying on other organizations to document these details accurately and forward them. There have been substantiated allegations of fraud; for instance, members of some organizations have charged refugees substantial amounts of money to register. UNHCR has acted on some of these allegations and put data gathering for some of those organizations on hold.

Refugee organizations agree that the RSD process is slow and occasional reports from refugees claim the RSD process suffers from unprofessional conduct by the case officers or translators. With limited funding RSD can slow down, especially neglecting those refugees outside of Kuala Lumpur as mobile registrations are cut back. The Australian government has filled some shortfalls and recently increased funding to regional Southeast Asian UNHCR offices, especially Malaysia, in part to stabilize the regional refugee population. Some of the funding has been tied to expediting registrations, which, UNHCR officials told me, allowed UNHCR to register an additional 15,000 refugees in 2010.

The Malaysian government allows UNHCR to operate but does not engage further with refugees, as it shuns any legal liability or responsibility for them. There is no consistent policy framework, but mostly verbal and ad hoc pronouncements that often do not translate through the implementing and executive government bodies. Refugees registered with UNHCR are afforded some minimal protection, because police have been instructed not to detain them. The Attorney General issued written directions to this effect in 2005, but this is not known to all police officers, nor implemented across the police and auxiliary forces, such as the 500,000 strong People's Volunteer Corps (RELA) (Zimmermann et al. 2011: 177; see also Nah 2011; USCRI 2009). There remains a major issue of rent-seeking on the part of low-paid immigration and police officials as well as RELA officers. Most refugees spoken to for this project have had to pay a bribe or been accosted for one, to avoid being detained or arrested. The legal limbo refugees live in, or their illegality vis-à-vis the Malaysian immigration law, allows for this rent-seeking to go largely unchecked as refugees are too scared to report such incidents to the police or to anyone else. Only the granting of temporary resident visas would ameliorate this situation, because it would allow refugees to stay in Malaysia legally. On occasion, the Malaysian government has granted temporary residency and work permits to some refugee communities, such as the Acehnese and Moro (from the Philippines) in East Malaysia.

In 2011 the government launched the 6P amnesty programme, or the *Program Pemutihan* (Legalization Programme), which allegedly allowed illegal immigrants to apply for work permits or face deportation. According to Deputy Prime Minister Muhyiddin, this would give 'everyone the opportunity to come clean' (Mazwin Nik Anis 2011a). However, individuals registered with UNHCR are explicitly ineligible for the permit (Kementerian Dalam Negeri 2012). When asked why this was so, the Human Resource Minister replied that it is because 'According to the UNHCR law', their residence in Malaysia was temporary, pending resettlement to another country (*The Star Online* 2012). He added that, 'These refugees are not allowed to work as their welfare is being taken care of by UNHCR' (*The Star Online* 2012). This is not the case, as UNHCR ceased providing rental and welfare assistance in the mid 1990s and has since focused on community welfare programmes such as providing assistance to set up and run schools, health services, community organizations and income generating endeavours.

The 6P registration process began in July 2011. By September 2011, 1.3 million illegal immigrants had registered for the permit. In October of 2011, the government announced that they would begin cracking down on those employers who continued to hire illegal immigrants, which includes refugees, who had not registered for the permit under the 6P programme (Mazwin Nik Anis 2011b). The 6P programme was extended and registration made mandatory for refugees in August of 2011. They were sent home with slips that said '*Slip Pendaftaran PATI*' (PATI registration slip) and '*Tujuan: Pulang Ke Negara Asal*' (Purpose: Return to Country of Origin) (Fernandez 2011). Many refugees registered, whilst others did not. Some refugee communities and advocacy NGOs tried to find out what the process was and how refugees would be affected. As soon as they realized that asylum seekers and refugees could be refouled under this plan they warned those who had not already registered not to do so.

Subsequently, Malaysia's national news agency Bernama (2011) reported that talks between UNHCR and the Malaysian government yielded an agreement to begin registering UNHCR refugees and asylum seekers in January 2012 (see also UNHCR 2011a). This move has produced a massive biodata collection drive and seems to suggest an integration of this information into government databanks. However, as this was not deemed to be part of the process of registration as UNHCR refugees (i.e. refugee status determination), many refugees and asylum seekers did not attend or provide their data. Data collection was ongoing when I was in the field in mid 2012 and refugees often told me that the exercise was not deemed advantageous, especially by those refugees with jobs, and not worth losing a day's work or even one's job for. In addition, some refugees were fearful of the new data collection, because its intentions had not been adequately communicated to them. UNHCR, meanwhile, are hopeful that the registration will increase the protection space for registered refugees, reducing the risks they face of arrest or deportation, and also combat the use of fraudulent UNHCR identity cards (Muthiah *et al.* 2011).

Current urban refugee programme in Kuala Lumpur

Refugees in Kuala Lumpur continue to face arrest, detention, harassment and rent seeking. Refugees continue to feel anxious about major crackdowns by police, immigration and RELA that endanger and disrupt their everyday lives (Nah 2011; CRC 2013). The government does not officially recognize their presence and they face extortion, violence and other forms of oppression from Malaysians. Reports of gangs holding up refugees and stealing wages, wallets and mobile phones are heard daily in most refugee communities. Refugees do not feel empowered enough to report these incidents for fear of the authorities. Furthermore, police are generally not interested in assisting non-citizens, much less illegal immigrants. Thus these crimes go unreported and undetected, becoming an added tax on the lives of refugees, physically, mentally and financially. Médecins Sans Frontières reported that 26 per cent of incidents of violence committed against refugees in their (admittedly small) sample of 100 were conducted by 'ordinary Malaysians' (Médecins Sans Frontières 2007). Women are especially vulnerable, as many of the jobs they do require them to walk to public transport in the very early or late hours of the day. These are significant issues for urban refugees living in a foreign country's cheapest and most dangerous parts of town.

UNHCR 2009 urban refugee policy and 2012 review

As a response to the growing number of urban refugees compared to camp based ones, in 2009 UNHCR issued a policy document that realigned priorities and practices with regard to refugees in urban areas (UNHCR 2009: 1–29). The policy had two main objectives: to ensure refugee rights in urban settings, including their right to reside there, and 'to maximize the protection space available to urban refugees and the humanitarian organizations that support them' (UNHCR 2009: 5). This policy relies heavily on the host country's support (financial and otherwise), or at least tolerance of UNHCR and its programmes, and as a result has had varying degrees of success in its implementation. The policy emphasizes practicality in dealing with refugees in urban settings, which can be disparate and changeable depending on a variety of factors. This is extended by the policy's focus on sensitivity to diversity of the refugee populations. In line with other UN programmes such as the Millennium Development Goal 8, partnerships feature prominently, both in respect to advocating for host states to take an active interest in and responsibility for refugee welfare, as well as partnerships with non-state actors. A part of this engagement push with organizational structures is creating and supporting community-based refugee organizations that are supposed to be conduits for UNHCR support and interactions. Thus this represents a shift from interacting with refugees individually and addressing their needs on a case-by-case basis to one that encourages a 'mobilization' of refugee groups. Another policy feature is the emphasis on self-reliance and empowerment through employment or self-employment, increased access to services such as education and health care, and ongoing skills or vocational training. As ever, the desire to find 'durable solutions' for refugees, such as voluntary repatriation, integration or resettlement, remains a key referent. The policy document is largely aspirational in a country that not only has not signed the refugee treaties but refuses to acknowledge refugee rights.

UNHCR conducted a review of its urban refugee programme in Kuala Lumpur, which was published in 2012. Its report outlined a raft of challenges, opportunities as well as criticisms and commendations of the Malaysian UNHCR urban refugee policy. In the following section I will summarize and contextualize some of the 2012 review (Crisp *et al.* 2012: 1–38).

In terms of challenges, the review team identified a series of problems with the implementation of the policy as well as structural problems relating to the host country and refugee populations themselves. Whereas the new urban refugee policy aims to get the government more involved in refugee issues, refugees in Malaysia continue to be thought of as the responsibility of UNHCR (Crisp *et al.* 2012: 2, 11, 17). Meanwhile, the number of refugees arriving in Malaysia is increasing steadily. However, UNHCR funding has been reduced drastically and though the UNHCR staff size is already at capacity, they are increasingly unable to cope (Crisp *et al.* 2012: 20, 23). While the registration of asylum seekers and the granting of refugee status have been beneficial to refugees, the process is also inefficient and refugees face high risks while waiting to be registered (Crisp *et al.* 2012: 22). There is also a concern that refugee community organizations may be corrupt, undemocratic and male-dominated (Crisp *et al.* 2012: 22, 36). Divisions within the refugee community along ethnic lines are also a point of concern (Crisp *et al.* 2012: 37). These problems may negatively impact the new community-orientated approach to urban refugees in Malaysia.

The report outlined a series of opportunities and positive outlooks on the situation for refugees. First, the treatment of refugees in Malaysia has improved slightly, particularly as the new Malaysian government appears more concerned with its international image (Crisp *et al.* 2012: 13). In the region the report made reference to the failed Australian 'refugee swap deal' and whilst making no value judgment on the proposed deal the authors note that it has been beneficial in focusing public attention on refugees and UNHCR in Malaysia. Thus they encourage UNHCR to take advantage of this with more awareness raising campaigns (Crisp *et al.* 2012: 17). Another positive development is the general high level of organization amongst refugee communities, often facilitated by the new urban refugee policy and its funds, which emphasize self-sustainability. Finally, even though the

Malaysian government does not officially recognize refugees, it does give UNHCR some degree of approval or tolerance and often enters into 'tacit agreements' (Crisp *et al.* 2012: 21, 26) with UNHCR. For example, refugees registered with UNHCR may be arrested but are usually released (Crisp *et al.* 2012: 14); UNHCR was allowed to conduct a 'large scale mobile registration exercise of refugees and asylum seekers' in 2009 (Crisp *et al.* 2012: 21); since 2006 refugees have been allowed a 50 per cent discount on health care fees and have been granted access to maternal and child health clinics (Crisp *et al.* 2012: 28).

In the assessment, the review team is generally impressed with UNHCR Malaysia activities and organization. However, the team remains pessimistic that its new urban refugee policy, which prioritizes finding long-term 'durable' solutions, is attainable. According to the report, 'there are no solutions in sight for the majority of refugees, even in the medium or longer term' (Crisp *et al.* 2012: 3). The report finds that UNHCR has been generally successful in adapting to the unique conditions of operating in a non-signatory country in which refugees are not recognized or confined to camps, but face different problems such as risk of arrest and deportation, and a lack of access to basic services. The review further finds that UNHCR's policy of concentrating on seeking resettlement opportunities for all its refugees exacerbates rather than mitigates the refugee issue in Malaysia, particularly the Malaysian government's treatment of them as bodies in transit to be handled by UNHCR. They recommend focusing on securing work and resident permits instead (Crisp *et al.* 2012: 16). This remains very aspirational, as no traction has been achieved on this goal, except for limited periods of time for specific refugee cohorts (see above, e.g. Acehnese refugees receiving IMM13 visas). Access to an income or sufficient funds to support themselves is the biggest challenge and source of anxiety and hardship faced by refugees in Malaysia. This is exacerbated by the fact that the urban refugee policy does not address refugees' access to accommodation (Crisp *et al.* 2012: 25). In terms of other major changes to the urban refugee policy, UNHCR Malaysia was lauded for service provision, such as refugee schools, but improvements need to be made (Crisp *et al.* 2012: 26–27) and self-reliance programmes such as the Social Protection Fund seem successful, but without more funding cannot meet the demand (Crisp *et al.* 2012: 38–39). The report also found that the recent switch in focus from an individual to community-based approach seems effective, given the unique conditions in Malaysia. These are the official views from UNHCR so it is encouraging to see the review take issue with some key problems, but it remains to be seen how well UNHCR Malaysia can respond to structural and internal issues either not of their making or impossible to remedy.

Community outreach and partnerships

Malaysian civil society and refugee community groups

Working with refugees in Malaysia poses a series of problems and issues. First and foremost is the small and strained civil society service provision sector in

Malaysia. This is the result of a government campaign to limit (especially progressive) civil society. Historically the state has only recently intervened with a heavy hand, with Giersdorf and Croissant (2011: 7) describing an early period of relatively relaxed attitudes towards civil society in general between 1957 and 1969. In the 1970s universal concerns came to the fore with the establishment of several women's and human rights groups (Brown *et al.* 2004: 12; Case 2003: 46; Giersdorf and Croissant 2011: 8). The 1980s were marked by a resurgence in Chinese ethnic concerns, particularly around vernacular Chinese schools. This led to tensions within the ruling Barisan Nasional coalition. The tensions and instability led to Operation Lalang in 1987, a police crackdown on civil society groups in Malaysia, involving the arrest of over 100 activists and politicians under the Internal Security Act and the temporary ban of several newspapers (Brown *et al.* 2004: 12; Giersdorf and Croissant 2011: 9). This event was aimed at restricting and intimidating civil society in Malaysia with the government keeping a tight grip on who can and who cannot register as a society[2] and using the powers of the Internal Security Act to detain people indefinitely without charge. As a result, many civil society organizations are registered as private companies. However, civil society organizations have expanded since then with most based in urban centres of peninsular Malaysia. The most vocal and powerful NGOs, such as the Consumers Association of Penang, women's rights NGOs, such as AWAM, WAO and Sisters in Islam (SIS) and Aliran, a social justice movement, are concerned with local and universal rights issues.

The role and efficacy of civil society in Malaysia is largely circumscribed by a number of factors. Legally, civil society organizations (CSOs) have to contend with a number of laws that restrict what they can say, the actions they can take as well as who they can work with or for. The amended federal constitution and Sedition Act circumscribe a series of 'sensitive' topics for discussion. These include 'the status of Islam as the official religion', 'the status and power of the Malay Rulers' and – perhaps significantly in the context of refugees – 'citizenship rights for non-Malays' (Brown *et al.* 2004: 6). Meanwhile non-governmental organizations have to operate in an environment of increased surveillance by the state and a strictly policed media, although the internet has generated a range of alternative media outlets that cover refugee issues with less bias. The government is particularly prone to curtail human rights organizations rather than charitable ones. Thus, the scant resources available tend to focus on Malaysian political and human rights. Tenaganita, an NGO that speaks out for migrant and refugee rights, is one of the few organizations that specifically work in this area, often advocating for refugees with UNHCR and acting as advisor and conduit for refugee concerns. However, tight budgets and staff shortages mean this area remains under-resourced and inadequately serviced. SUARAM, a human rights organization, also works with refugees and acts on their behalf, but their resources are also strained. Their work on domestic human rights often brings them in direct opposition to the government, which in turn harasses them and limits their operations.

As a result of this state intervention, progressive civil society remains focused on domestic politics and human rights battles in Malaysia. This focus results in a curtailed pool of service providers for refugee and also implementing partners for UNHCR. The latter is also due to the strained relationship some NGOs have with UNHCR, which often acts to minimize the Malaysian government's exposure to criticism in order to maintain good relations. One case in particular is the lack of legal aid for refugees. Although there are some very committed lawyers from the Bar Council who act as advocates for refugees, there is no centralized organization that provides legal aid.

Refugees are often left to find their own way and rely on their personal and group networks for advice and guidance. For most refugees from Burma, that place is their community refugee organization. Refugee community organizations have been established largely along ethnic or national lines. The two biggest Chin groups have a membership in excess of 50,000, made up of refugees currently residing in Malaysia and some who have been resettled, as records depend on members self-reporting. The ethnic minority groups of refugees from Burma created an umbrella organization called the Coalition of Burma Ethnics, Malaysia (COBEM) in 2007 to share information and pool resources, as well as coordinate interactions with UNHCR. COBEM includes the Kachin Refugee Committee (KRC), Organization of Karenni Development (OKD), Malaysia Karen Organization (MKO), Chin Refugee Committee (CRC) together with Alliance for Chin Refugees (ACR), Mon Refugee Organization (MRO), Arakan Refugee Relief Committee (ARRC) and Shan Refugee Organization (SRO). COBEM is arguably the best organized entity within the refugee community at large and represents the majority of refugees in Malaysia. UNHCR has a relatively good working relationship and efforts have been made to replicate this sort of community organization in other refugee groups, as it makes interactions easier for UNHCR if they have a legitimate representative as a counterpart.

Each community organization runs its own programmes to improve the living and working conditions of its members. Some have extensive links to factory, restaurant and plantation owners and other employers in the city and can place newly arrived refugees into a workplace very easily and quickly. Often they also check the workplace agreement and follow up with the employer if there are complaints from the refugee workforce. Thus they can provide a range of sophisticated services for their members ranging from employment to shelter to schools for refugee children and training facilities for adults. Most of these facilities and programmes are funded by UNHCR and donations and support from the diaspora community and membership fees. The average membership fee for employed refugees is around 50RM (US$16) for the membership card, which is often the first piece of identification an asylum seekers receives and offers a modicum of protection. In addition, many refugee organizations charge 10RM (US$3) per month. Most other refugees are not as highly organized and do not have membership organizations, although UNHCR is encouraging them to replicate this model.

The success of many of these organizations has taken considerable pressure off limited UNHCR resources, but it has also problematized the relationship between them. Refugee community organizations depend on UNHCR for registration, but little else, as many run their own education, work and health programmes, often in partnership with Malaysian NGOs and some UNHCR implementing partners. As the organizations mature and find effective leadership they increasingly hold UNHCR and other organizations to account for their services and demand better and more effective ones. For instance, in a recent survey of Chin refugees, the Chin Refugee Committee reports slow, unprofessional and inadequate support from UNHCR (Chin Refugee Committee 2013: 49). This shows that refugees are far from the passive victims often conjured up by the media and also aid organizations, particularly in the West (Harrell-Bond 1999).

Income, shelter and self-reliance

The biggest challenge refugees and asylum seekers face is finding adequate income streams to provide for themselves and their families. Most refugees and asylum seekers are self-reliant, although income streams can be fractured and inconsistent. Generally they have no work permits, thus any work they find and conduct is illegal work and if their workplace is raided by authorities they will face punishment in the form of detention and possibly a fine and even caning for immigration offences. There has been a continued call from NGOs and the Bar Council for work rights for refugees, but the government has hitherto not responded favourably, except for some cases, such as those for the Acehnese and Bosnian refugees for a limited time. Rohingya have been hoping for their IMM 13 visas for decades and there is no solution to this situation in sight.

UNHCR can provide limited support for very vulnerable refugees, but this is a last resort due to limited funding and ever-greater need. Some NGOs provide micro-loan services, which refugees can use to start small businesses. Tech Outreach (*Pertubuhan Kebajikan Tech Menceria*) focuses on empowering women through a system of micro-credit financing modelled on that of the Grameen Banking System, first introduced in Bangladesh. They regularly grant loans to women interested in starting their own small businesses. Partners in Enterprise Berhad, a Christian NGO, also provide micro-finance loans and are involved in some community development projects. Hartford Academy is a private organization that specializes in providing IT related education and training skills. They have partnered with the UN on several projects including the SPF Computer projects and a project that enabled almost 500 refugees to open savings bank accounts.

Access to shelter

When refugees first arrive in Malaysia many have contacts, friends and family they can stay with or their refugee community organizations take them in for a short period of time. The next step is to find a job, which may come with attached accommodation (often employers pay lower wages and offer a shared room to their workers) or will allow them to pay the rent for a shared room with other refugees. There are some shelters available to refugees, although the infra-structure on the whole is weak and places are extremely limited. In theory, refugees are allowed to access the Malaysian welfare system, mostly run by NGOs; however, they are often not admitted or pushed out. Some local NGOs or church organizations run shelters that are also open to refugees and, in some cases, asylum seekers. The latter are especially problematic and there have been cases where asylum seekers have been turned away by shelters, even when they were in need of post-operative care. In such cases, shelters feared exposure to Malaysian authorities for harbouring an asylum seeker (read illegal immigrant) as this could result in their losing their license to operate.

Access to health care

There are three main service providers in Kuala Lumpur who have provided, and continue to provide, most of the free or heavily subsidized health care to UNHCR registered refugees[3]: A Call To Serve (ACTS), Taiwan Buddhist Tzu-Chi Foundation Malaysia and Health Equity Initiatives. Each of these organiza-tions has a particular focus and clientele depending on their location and affiliation. There are other support services, ad hoc medical clinics and outreach programmes run by individuals, church groups and others, but these tend to be tied to specific funding, timing and individuals, whereas the three aforemen-tioned organizations have established themselves as specific health care provid-ers first and foremost.

   Health Equity Initiatives (HEI) is a Malaysian NGO primarily concerned with refugees in Malaysia and their access to health care services. They provide a range of health care services to refugees, especially psychiatric and psychologi-cal treatment. They have a permanent centre in Brickfields, operate mobile clinics and offer home visits. HEI also train refugee health care workers for early health issue detection and intervention. Their progressive and also activist work in this area is in high demand. Their path-breaking report on Afghan refugee mental health in Malaysia is the first in-depth look at an understudied and under-served area that has wide implications for the well-being of families and entire communities (Health Equity Initiatives 2010).

   ACTS (A Call To Serve) is a Christian NGO that provides health care ser-vices to refugees in Malaysia and champions their right to basic health care. They are an implementing partner of UNHCR and are situated in Brickfields, close to UNHCR and transport links. Some of their projects include awareness campaigns, the Arrupe Clinic (a clinic catering for refugees) in Brickfields, various mobile clinics and convalescent homes around Kuala Lumpur. The con-valescent homes are an important addition to providing frontline health services, as there are no facilities available to refugees who have no family or friends willing or able to look after them following an operation or prolonged illness. ACTS runs two homes, one for women and one for men. Clients include some with severe injuries from road accidents, muggings and workplace injuries. The latter is a major issue, as refugees have no work permit. Therefore, they have no access to any compensation from their employer and, if injured, face the double burden of health care costs and loss of income. ACTS also run mobile clinics to far-flung areas across the peninsula and to immigration detention centres.

The Taiwan Buddhist Tzu Chi Foundation Malaysia has become a central node for refugee health care services. The Tzu Chi Clinic in Pudu is a modern clinic that has just been upgraded to also provide rehabilitation (physiotherapy) facilities to refugees. Tzu Chi is a Buddhist organization from Taiwan with a growing membership in Malaysia that organizes various programmes and campaigns relating to humanitarian aid, access to medical care, access to education and environmental awareness. Initially, they were working on international humanitarian aid projects when UNHCR approached them in 2004 to involve them in providing medical aid in Malaysia. In time, Tzu Chi took over many of the functions Médecins Sans Frontières had served after coming to Malaysia following the 2004 tsunami in Aceh. MSF stayed to work with urban refugees and build up local capacity until 2007. Tzu Chi's impressive two-floor operation in Pudu is close to large refugee communities. They offer refugees a range of health care services, from Western medicine to psychosocial services as well as dental work on a limited basis. Tzu Chi also runs mobile clinics in other parts of Kuala Lumpur and to some detention centres in collaboration with UNHCR. On occasion, they also help refugees with financial aid for operations at government hospitals.

There is a range of other NGOs and civil society actors engaged in providing health care services on a pro bono or subsidized basis. For instance, several of the refugee community organizations under COBEM have doctors and/or nurses who staff a makeshift clinic once or twice a week. These are often very rudimentary clinics that act as a first port of call, with patients referred to other refugee clinics or a state clinic or hospital. They can deal with minor injuries and ailments and provide few medications. MSRI, which was set up to help Palestinian refugees, has extended its work to help Iraqis and others and maintain a small clinic in Ampang. Meanwhile there are several organizations that offer mobile clinics in Kuala Lumpur and further afield. Mercy Malaysia (Malaysian Medical Relief Society) used to provide mobile medical clinics, especially to Rohingya and Myanmar Muslims around Kuala Lumpur, but owing to budget constraints no longer does so. In theory, refugees have access to public hospitals and clinics, but many do not make use of this because they are scared of detection by authorities, do not have the funds to pay for treatment or even consultation, or do not have access for other reasons. UNHCR registered refugees receive 50 per cent discount on the foreigner fee charged at public hospitals (JRS Asia Pacific 2012: 62). This is an improvement on paying the full fee, but still a heavy burden on low or no-income earners. The situation is even worse for asylum seekers, who do not receive the discount and face the threat of arrest in government facilities if they were to be reported to police. While the hospital staff should not do this, I came across several reports where hospital staff had themselves reported asylum seekers to the police and immigration authorities.

Access to education

Refugee and asylum seeker children have no access to Malaysian state schools, as they are not registered residents. There have been several attempts to remedy this situation, with occasional support from parts of the education ministry, but at present the legal situation remains unchanged. Many long-time refugees, especially Rohingya who reside in smaller settlements and speak Malay, find ways around the legal problem by, for instance, having their children adopted by a Malay family for a fee. This ensures that their children are allowed into the state education system, but its cost can be prohibitive. Other initiatives have included international schools taking in a small number of refugee children, but neither is a workable and sustainable solution for the over 20,000 affected children under the age of 18 currently residing in Malaysia. As a result proactive refugee community organizations, with the help of UNHCR and local partners, have started their own schools, mostly delivering basic English, mathematics and computer skills. Most UNHCR accredited community organizations have received funding under the Social Protection Fund (see below) to start up schools. These are open to asylum seekers and refugees independent of their UNHCR status. In addition some charitable organizations run schools for refugees across Kuala Lumpur, usually staffed by volunteers and relying on community support and donations. Some of the community refugee schools have boarding facilities – usually mattresses are laid down in the classrooms at night – for refugee children whose parents work far from Kuala Lumpur and can pay the boarding costs.

Harvest Centre is one of the bigger refugee schools that offer pre-school through to secondary schooling. They also train teachers for smaller refugee schools and run UNHCR certified programmes that teach technical skills like computing, cooking and organizational skills. Harvest is run by the Dignity for Children Foundation, which in turn was started by a minister from the New Covenant Community Church, and is an official UNHCR implementing partner. Tzu Chi (see earlier in health care) is another major UNHCR implementing partner running four schools for Rohingya children, which are fully funded by UNHCR, in part because they believe the Rohingya community has the least capacity and therefore requires additional support. A range of other organizations and community groups run schools for specific refugee groups. Most are monocultural, catering for specific ethnic groups, even though the main curriculum is supplied by UNHCR and is therefore relatively standardized. Better staffed and funded schools are able to provide more than the basic mathematics and English from UNHCR textbooks.

Assistance and durable solutions

UNHCR provides a range of programmes aimed to improve the lives of refugee communities in Malaysia. These programmes provide funding for projects that develop the capacity of refugee populations to serve their own communities. The most successful programme is the Social Protection Fund (SPF)

Scheme. The SPF scheme is a fund organized by UNHCR for the purpose of aiding non-governmental organizations or community-based groups that seek to provide 'self-help' or empowerment opportunities to refugee communities. The fund recognizes the importance of self-reliance and community development amongst refugees in Malaysia. Amongst others, it provides grants for projects that allow refugees to gain the skills needed to support themselves financially, projects that provide opportunities for refugees to make an income, projects that provide education opportunities to refugee children and projects that give refugees a safe environment to socialize with each other. Applications are open to most groups with very few conditions and are assessed on a case-by-case basis before grants are awarded. These funds are usually accessed by community organizations, which means that poorly organized groups find it much harder to gain access to funding. UNHCR has worked on this issue by providing help to set up refugee community organizations for those refugee groups that are not yet adequately organized. This has been ineffectual with the Rohingya, but the Tamils have been supported to form a community organization that UNHCR now works with and funds, the Sri Lankan *Tamil* Refugee Organization of *Malaysia* (*STROM*). Thus access to assistance remains an issue across refugee communities.

The SPF has mainly funded the setting up of refugee community schools, equipment for computer training, and smallholdings. The latter is a project taken up predominantly by COBEM refugee groups. The leased land is worked by refugees who then sell the produce within their community and also to markets in the city. Most of the farm projects will not be continued, however, as refugees have not found them to be economically beneficial. Many refugees working on the farms found they could earn more money in the city, also preferring to be closer to friends and UNHCR.

UNHCR considers three options as durable solutions: repatriation, local integration and resettlement. The ongoing conflicts in the majority of source countries, such as Burma, Somalia and Afghanistan, mean that repatriation is not a realistic solution for most refugees in Malaysia. Local integration has been discussed for Rohingya, but the Malaysian government has backed away from this option, at least in any systematic fashion. There are occasionally accounts of individuals gaining a MyKad (Malaysian personal ID card) and thus citizenship status. For the majority of Rohingya, or any other refugee group, a Malaysian government led integration is far from realization. This leaves resettlement as the only durable solution for UNHCR to pursue. With limited annual international resettlement quotas, it would take decades to resettle the current caseload alone. In addition, resettlement countries are scaling down resettlement places in a global economic climate that does not see the global refugee crisis as a priority. In 2011, UNHCR resettled 8,370 refugees to third countries, a number that is going to increase and make Kuala Lumpur the biggest resettlement quota for UNHCR offices worldwide (UNHCR 2011b). Whilst this offers some hope to refugees, the limited numbers also spell prolonged hardship for the vast majority.

Inter-agency cooperation

The UNHCR compound in Kuala Lumpur stands on a large block of land on the hill behind the old *istana* (palace). The large number of caseloads and the extensive range of partners for the growing resettlement programme means that the compound has outgrown its usability. The Malaysian government, however, does not give official status to any organization involved in refugee work, which means international organizations also encroach on the limited space available in the UNHCR compound, and there are no plans for expansion. The International Organization for Migration (IOM), International Rescue Committee (IRC) and International Catholic Migration Commission (ICMC) all have small offices in the UNHCR compound. IOM and IRC assist refugees with resettlement processes. IOM handles the logistics for resettlement to third countries, the health checks and actual accompanying refugees from Kuala Lumpur to their new countries of residence. ICMC is working on a gender-based violence programme for refugees.

Conclusion

UNHCR in Kuala Lumpur has been fairly successful in maintaining its presence in Malaysia, in the face of a non-cooperative government, and in protecting key refugee interests. By outsourcing some responsibilities to organized refugee communities they have been able to increase registration capacity and service provision. Outsourcing, however, may also undermine the UNHCR operation by allowing refugee communities to carry out operations according to their own practices, which may run counter to UNHCR standards and protocols. For instance, for refugees from Burma the only way to get an appointment with UNHCR is through refugee community organizations, most of which charge a membership fee. This system raises a serious issue given that UNHCR services should be free. The legal non-recognition of refugees in Malaysia continues to be the greatest hurdle for improving the well-being and protection space for refugees, especially around work rights, along with the continued operations to arrest and detain refugees.

The urban refugee experience in Malaysia is dominated by one's ethnic identity, as it is this key identifier that determines one's community support networks. This can be beneficial for some and detrimental for others, especially those with smaller or less organized community groups. Refugees from Burma are best organized and represent the highest numbers of refugees. This allows individuals to fall back on ethnically-based kin networks in times of need and provides a limited safety net independent of the UNHCR. The resilience, entrepreneurialism and strong community bonds of many refugees from Burma has mitigated some vulnerabilities and offered limited support to these refugees. The Malaysian government remains disconnected from most refugee issues, abandoning responsibility for this to an already overburdened and under-resourced UNHCR country office. UNHCR in turn has outsourced service provision to the refugee communities themselves and a small Malaysian civil society. This arrangement is not sustainable, even with additional funding provided to Malaysia from regional partners such as Australia, which is in turn intent on outsourcing its refugees to the region, or at least stabilizing the refugee population in Malaysia, in order to diminish push factors which motivate many irregular migrants to attempt travel to Australia by boat.

Notes

1 Both Nah (2007) and myself (Hoffstaedter 2011b) have theorized and attempted to make sense of the shifting nature of the legal and illegal in Malaysian discourses of the refugee as either inhabiting a 'judicial indeterminacy' (Nah) or being a 'radical other' (Hoffstaedter).
2 Non-governmental organizations have to register with the Registrar of Societies in order to operate legally. This registration is often under threat of being revoked (Jesudason 1995: 338; Barraclough 1985: 809).
3 Only in special circumstances can they accept asylum seekers who are not yet recognized by UNHCR. They are considered as second line treatment costs and may not be covered by UNHCR funds.

References

Barraclough, S. 1985. 'The Dynamics of Coercion in the Malaysian Political Process'. *Modern Asian Studies* 19(4): 797–822.
Bernama 2011. 'Refugees to be Registered From Next Year'. *Malaysiakini*. Available at: www.malaysiakini.com/news/180681 (accessed 14 June 2012).
Brown, G.K., Siti Hawa Ali and Wan Manan Wan Muda 2004. 'Policy Levers in Malaysia'. *CRISE Policy Context Paper* 4. University of Oxford: Queen Elizabeth House.
Case, W. 2003. 'Thorns in the Flesh: Civil Society as Democratizing Agent in Asia'. In Schak, D. and Hudson, W. (eds) *Civil Society in Asia*. Hampshire: Ashgate, pp. 40–58.
Cheung, S. 2012. 'Migration Control and the Solutions Impasse in South and Southeast Asia: Implications from the Rohingya Experience'. *Journal of Refugee Studies* 25(1): 50–70.
Chin Refugee Committee (2013) *Chin Refugee Committee Annual Report 2012* Kuala Lumpur: Chin Refugee Committee.
Crisp, J., Obi, N. and Umlas, L. 2012. ' "But when will our turn come?" A review of the implementation of UNHCR's urban refugee policy in Malaysia'. *UNHCR Evaluation Reports* May 2012 (02). UNHCR Policy Development and Evaluation Service. Available at: www.unhcr.org/4faa1e6e9.html (accessed 29 June 2012).
Fernandez, I. 2011. 'Halt '6P' registration of refugees immediately'. *Malaysiakini*. Available at: www.malaysiakini.com/news/174040 (accessed 14 June 2012).
Giersdorf, S. and Croissant, A. 2011. 'Civil Society in Competitive Authoritarianism in Malaysia'. *Journal of Civil Society* 7(1): 1–21.
Harrell-Bond, B. 1999. 'Refugees' Experiences as Aid Recipients'. In Ager, A. (ed.) *Refugees: Perspectives on the Experience of Forced Migration*. London: Pinter, pp. 136–139.
Health Equity Initiatives 2010. *Between a Rock and a Hard Place: Afghan Refugees and Asylum Seekers in Malaysia*. Available at: http://refugeerightsasiapacific.org/pdf/Afghan_RNA_final per cent20_report_June per cent20_2010.pdf (accessed 26 June 2012).

Hoffstaedter, G. 2011a. *Modern Muslim Identities: Negotiating Religion and Ethnicity in Malaysia.* Copenhagen: NIAS Press.

Hoffstaedter, G. 2011b. 'Radical othering: Refugees, displacement and landscapes of difference in Malaysia'. *TASA, Newcastle.* Refereed conference proceedings.

Hotli Simanjuntak 2008. 'Malaysia Extends Permits for Aceh Refugees'. *The Jakarta Post.* Available at: www.thejakartapost.com/news/2008/07/29/malaysia-extends-permits-aceh-refugees.html (accessed 13 June 2012).

Human Rights Watch 2000. 'Living in Limbo: Burmese Rohingyas in Malaysia'. *Human Rights Watch* 12(4c). Available at: www.unhcr.org/refworld/docid/3ae6a8743.html (accessed 13 June 2012).

International Federation for Human Rights (FIDH) 2008. *Undocumented migrants and refugees in Malaysia: Raids, detention and Discrimination.* Available at: www.unhcr.org/refworld/docid/47f0b03b2.html (accessed 21 June 2012).

Jesudason, J.V. 1995. 'Statist Democracy and the Limits to Civil Society in Malaysia'. *The Journal of Commonwealth and Comparative Politics* 33 (3): 335–356.

JRS Asia Pacific 2012. 'The search: protection space in Malaysia, Thailand, Indonesia, Cambodia and The Phillipines'. Available at: https://jrsap.org/Assets/Publications/File/The_Search.pdf (accessed 29 June 2012).

Kalir, B. and Sur, M. (eds) 2012. *Transnational Flows and Permissive Polities: Ethnographies of Human Mobilities in Asia.* Amsterdam: Amsterdam University Press.

Kementerian Dalam Negeri 2012. 'Brochure – Program Pumitihan PATIH'. *Portal Rasmi Kementerian Dalam Negeri.* Available at: www.moha.gov.my/images/stories/pdf/Pekerja_Asing/brochure6p.pdf (accessed 14 June 2012).

Lee, R. 2011. 'Mahathir blames police over Ops Lalang'. *Malaysiakini.* Available at: www.malaysiakini.com/news/155542 (accessed 8 July 2012).

Mazwin Nik Anis 2011a. '6P programme starts on July 11'. *The Star Online.* Available at: http://thestar.com.my/news/story.asp?file=/2011/10/5/nation/20111005170727&sec=nation (accessed 14 June 2012).

Mazwin Nik Anis 2011b. 'DPM: Enforcement against those hiring illegals to start immediately'. *The Star Online.* Available at: http://thestar.com.my/news/story.asp?file=/2011/10/5/nation/20111005170727&sec=nation (accessed 14 June 2012).

Médecins Sans Frontières 2007. '"We are worth nothing": Refugee and asylum seeker communities in Malaysia'. Brussels: Médecins Sans Frontières. Available at: www.msf.ch/fileadmin/msf/pdf/2010/04/MSFbriefingpaper.pdf (accessed 8 July 2012).

Muthiah, W., Lee Y.M., Wong, P.M. and Farah, F.Z. 2011. 'Working for refugee rights'. *The Star Online.* Available at: http://thestar.com.my/news/story.asp?file=/2011/11/7/nation/9853469&sec=nation (accessed 21 June 2012).

Nah, A.M. 2007. 'Struggling with (il)legality: The indeterminate functioning of Malaysia's borders for asylum seekers, refugees and stateless persons'. In Rajaram, K. and Grundy-Warr, C. (ed.) *Borderscapes: Hidden Geographies and Politics and Territory's Edge.* P Minnesota: University of Minnesota Press, pp. 35–64.

Nah, A.M. 2011. 'Legitimizing violence: The impact of public 'crackdowns' on migrant workers and refugees in Malaysia'. *Australian Journal of Human Rights* 17(2): 131–157.

Pragalath, K. 2012. 'Keep your "ISA" word, Najib told'. *Free Malaysia Today.* Available at: www.freemalaysiatoday.com/category/nation/2012/04/09/keep-your-isa-word-najib-told/ (accessed 8 July 2012).

Refugees International 2009. 'Nationality rights for all: A progress report and global survey on statelessness'. Available at: www.unhcr.org/refworld/pdfid/49be193f2.pdf (accessed 26 June 2012).

Rieffel, L. 2010. 'The moment'. In Rieffel, L. (ed.) *Myanmar/Burma: Inside Challenges, Outside Interests.* Washington DC: Brookings Institution Press.

Robinson, W.C. 1998. *Terms of Refuge: The Indochinese Exodus and the International Response.* London: Zed Books.

Setudeh-Nejad, S. 2002. 'The Cham Muslims of Southeast Asia: A historical note'. *Journal of Muslim Minority Affairs* 22 (2): 451–455.

Streit, C. 2007. *The Problem of Ethnic Insurgencies and Its Impact on State Building in Myanmar.* Germany: GRIN Verlag.

Teoh, S. 2011. 'Najib announces repeal of ISA, three emergency declarations'. *The Malaysian Insider.* Available at: www.themalaysianinsider.com/malaysia/article/najib-announces-repeal-of-isa-three-emergency-declarations/ (accessed 8 July 2012).

The Equal Rights Trust 2010. *Trapped in a Cycle of Flight: Stateless Rohingya in Malaysia.* Available at: www.equalrightstrust.org/ertdocumentbank/ERTMalaysiaReportFinal.pdf (accessed 21 June 2012).

The Star Online 2012. 'Myanmar refugees not eligible for 6P programme'. *Online.* Available at: http://thestar.com.my/news/story.asp?file=/2012/2/24/nation/20120224121629&sec=nation (accessed 14 June 2012).

To, Q. 2011. 'IMM13 documents not issued to illegal immigrants, says Nazri'. *FreeMalaysiaToday.* Available at: www.freemalaysiatoday.com/category/nation/2011/10/10/imm13-documents-not-issued-to-illegal-immigrants-says-nazri/ (accessed 14 June 2012).

UNHCR 2009. 'UNHCR policy on refugee protection and solutions in urban areas'. Available at: www.unhcr.org/4ab356ab6.pdf (accessed 29 June 2012).

UNHCR 2011a. 'Registration of UNHCR document holders to start in 2012'. *Komunitikini.* Available at: http://komunitikini.com/kl-selangor/kuala-lumpur/registration-of-unhcr-document-holders-to-start-in-2012 (accessed 14 June 2012).

UNHCR 2011b. 'A new beginning in a third country'. Available at: www.unhcr.org/pages/4a16b1676.html (accessed 1 January 2013).

UNHCR 2012. 'Figures at a glance'. Available at: www.unhcr.org.my/About_Us-@-Figures_At_A_Glance.aspx (accessed 1 January 2013).

United States Committee for Refugees and Immigrants 2009. *World Refugee Survey 2009 – Malaysia.* Available at: www.unhcr.org/refworld/docid/4a40d2adc.html (accessed 13 June 2012).

US Congress 2009. *Trafficking and Extortion of Burmese Migrants in Malaysia and Southern Thailand.* Washington, DC: US Government Printing Office.

United States Committee for Refugees and Immigrants 2009. *World Refugee Survey 2009 – Malaysia.* Available at: www.unhcr.org/refworld/docid/4a40d2adc.html (accessed 14 June 2012).

Zimmermann, A., Dörschner, J. and Machts, F. 2011. *The 1951 Convention Relating to the Status of Refugees and Its 1967 Protocol: A Commentary.* Oxford: Oxford University Press.

9 The Japanese pilot resettlement programme

Identifying constraints to domestic integration of refugees from Burma

Saburo Takizawa

Introduction

In December 2008, Japan announced that it would start a three-year pilot programme to resettle 90 Burmese refugees from Mae La camp in Thailand. Each year, 30 refugees will be selected and resettled in Japan. The announcement came as a pleasant surprise to the humanitarian community in and out of Japan because it was viewed as a sign that Japan is changing its hitherto restrictive refugee policy. As Japan was the first Asian country to start a resettlement programme, it raised hope that other Asian countries may follow suit to address the problem of the 'protracted refugee situation' (Loescher *et al.* 2008) in the region. However, the pilot programme has been facing difficulties from the start and the initial quota of 90 will not be met. In May 2012, the Government decided to extend the pilot programme to five years. In the meantime, previously resettled refugees are concentrated in the metropolitan cities and adjacent areas, forming part of what one may call 'metropolitan refugees.'

Based on the author's interviews with Japanese Government officials, interviews with refugees in Mae La and other refugee camps in Thailand in November 2007 and August 2011, and records of the Resettlement Experts Council, as well as experience in setting up a civil society organization to assist resettled refugees in Matsumoto City in central Japan, this chapter first describes recent asylum trends in Japan and then addresses three questions: (1) why and how the pilot resettlement was started; (2) how the pilot programme has been implemented, and (3) why the programme has been facing difficulties.

Conceptual framework: global public goods

In attempting to provide answers to these questions, this chapter uses a 'global public goods' approach as presented by Kaul *et al.* (1999). Global public goods must meet two criteria. The first is that their benefits have strong qualities of publicness – namely non-rivalry in consumption and non-excludability. Non-rivalry means that use by one individual does not reduce availability to others (my use does not affect yours). Non-excludability means that individuals cannot be effectively excluded from use (it is available for everyone to use or enjoy).

The second criterion is that their benefits are quasi universal in terms of countries (covering more than one group of countries), people (accruing to several population groups) and generations (extending to both current and future generations). This property makes humanity as a whole the beneficiary of global public goods. Because they are non-rivalrous and non-excludable, public goods tend to be under-supplied, the main reason being *free-riding*. As it is difficult to preclude anyone from using a public good, those who benefit from the good have an incentive to avoid paying for it and free ride on the supply of the good provided by others, collectively causing its under-supply (Kaul *et al.* 1999: 2–6).

From a public policy point of view, three weaknesses in the current arrangements of providing global public goods can be identified:

1   The jurisdictional gap, which is the discrepancy between a globalized world and national, separate units of policymaking. Policymaking at the national level often does not take into account *externalities* or *spillovers* and this creates undersupply of public goods (or oversupply of public bads) at the global level. To fill the gap, national governments should *internalize externalities* by increasing production of public goods (or refraining from production of public bads). The roots of public goods (or bads) are at the national level; international cooperation starts 'at home'.

2   The participation gap, which means that international cooperation is mainly an intergovernmental process and excludes the people, civil society, the private sector and voiceless groups (such as refugees). Without participation of those national actors who could share responsibilities and make contributions as well as enjoy benefits, supply of global goods will be limited.

3   The incentive gap. International cooperation today has moved from between-country and at-the-border issues to behind-the-border issues, making implementation of international agreements more important. Yet the incentive mechanism is limited to financial aid, and other policy options such as compensation payments or use of market mechanisms are not considered (Kaul *et al.* 1999: xxiv–xxxiv).

The Global Refugee Regime as global public good

International protection of refugees can be considered as a global (or regional) public good insofar as the provision of protection benefits a number of states, irrespective of whether they themselves contribute to providing protection. Once protection is provided by a state, it will benefit a number of states in terms of providing security and fulfilling humanitarian goals, irrespective of whether those states actually contribute to its provision. As a result, states are likely to free-ride on other states' contributions to protection, and protection will be under-provided in comparison to what would have been collectively desirable (Betts 2009: 81).

International protection of refugees in the form of global public goods is provided by the Global Refugee Regime, which consists of Refugee

Conventions, states that are primarily responsible for refugee protection including financial provisions, UNHCR as a supervisory body, as well as non-governmental organizations that provide direct services to refugees. The aim of the Global Refugee Regime is to protect and empower refugees in such a way as to bridge the protection gaps when national protection fails, and to find durable solutions to the refugee problem (voluntary repatriation to the home country, local integration in the country of asylum and resettlement to a third country).

From the viewpoint of a state, there are two ways in which it can contribute to refugee protection: either by providing protection to refugees who reach its territory (asylum), or by contributing – through resettlement or financial contributions – to the protection of refugees who are on the territory of another state (burden sharing) (Figure 9.1).

In the case of asylum, the refugee regime sets out a strong normative and legal framework, underpinned by the principle of *non-refoulement*. In contrast, in the case of burden-sharing, the regime provides a very weak normative and legal framework, setting out few clear norms, rules, principles or decision-making procedures. Burden sharing represents the degree of international cooperation in the Global Refugee Regime (Betts 2009: 87).

Refugee resettlement is an international policy practised under the Global Refugee Regime. The aim of resettlement is twofold: to eliminate dangers to the most vulnerable refugees, and to share the burden or responsibilities of protection among states. Its implementation by states will contribute to the protection of refugees and is desirable collectively. As a global public good, it benefits all states in terms of promotion of human security of refugees and international burden/responsibility sharing. Being a global public good it has two attributes: non-rivalry in consumption and non-excludability from use; hence resettlement tends to be undersupplied, particularly because it is not mandatory but optional, not a legal but a political option available to states willing to make contributions to refugee protection. In response to the 'Agenda for Protection' in 2002 that called for states to expand resettlement and requested UNHCR to intensify its efforts to expand the resettlement base (UNHCR 2003), some 30 states are now providing resettlement with varying numbers of places (or quotas). Japan is the latest country to join the 'resettlement club' (the Geneva-based Working Group on Resettlement).

Resettlement as a national public good

Resettlement is a global public good but is also a national public good, which is produced within a national border. The resettlement process entails identification

Refugee protection as Global Public Good	Asylum	
	Burden sharing	Resettlement
		Financial contributions

Figure 9.1 Two ways of refugee protection.

of needs, selection of candidates, pre-departure orientation, transfer to a destination country, post arrival orientation followed by support for local integration. While the host government could provide services up until the refugees' transfer to the destination country, integration support requires many national actors and a long-term engagement.

Adoption of a resettlement policy by a state provides refugees with territorial protection and 'freedom from fear'. However, that does not by itself provide 'freedom from want', another component of human security. To support refugees' self-reliance and their integration into a new state/society, a multitude of integration support services have to be provided, such as assistance in accommodation, employment, language training, children's education, becoming a member of the community and eventually in obtaining citizenship. These services have to be provided not only by the central Government but also by local governments, civil society comprising NGOs, corporations and media as well as existing refugee communities. The roles of non-state actors are crucial for successful implementation.

These support services benefit a number of communities, irrespective of whether they themselves contribute to providing integration support. They are 'national public goods', which show non-rivalry in consumption and non-excludability from use. As such, their provision is subject to possible 'free riding' and under-production. If one local community accepts a large number of (resettled) refugees and provides integration assistance, other local communities can 'free-ride' without assuming costs, and integration support may be under-provided in comparison to what would have been collectively desirable.

As a public good, national production of integration support for resettled refugees faces thee gaps:

1   The jurisdictional gap, which is the discrepancy between national concerns and local, separate units of policymaking. Policymaking at the local level does not take into account externalities or spillovers and this creates undersupply of public goods at the national level. To fill the gap, local governments should 'internalize externalities' and increase production of public goods. The roots of national public goods are at the local level; international cooperation ultimately starts at the local level.

2   The participation gap, which means that national policy on implementing resettlement is decided by central ministries and excludes the people, civil society, the private sector and voiceless groups (such as refugees). Without the participation of those local actors who could share responsibilities and make contributions as well as enjoy benefits, supply of integration support will be limited.

3   The incentive gap. The incentive mechanism is limited to financial aid and quantity is limited (or non-existent). Other policy options such as compensation payments, use of market mechanisms or non-monetary rewards are not available.

With the above theoretical framework, this chapter separates the issues into two stages: international adoption by the state of the resettlement policy, and its implementation by domestic non-state actors. Although the Japanese government adopted a resettlement norm, its effective implementation requires adoption by other national actors that provide services, and internalization by domestic institutions and actors that must be 'socialized' to perceive such norms as legitimate.

This chapter points out that domestic implementation of the resettlement programme has been found much more difficult than its international acceptance, due to the conflict between international standards and the domestic or local norm and capacity, as well as the gap between camp life and urban life, and suggests the need to create a new incentive mechanism at the local level in the host communities.

Japan's refugee policy

To understand the significance of the decision to start a resettlement programme and the difficulty in its implementation, it is useful to situate the issue in the context of Japan's asylum policy in recent years. On the one hand, Japan has been accepting much smaller numbers of refugees compared to other industrialized countries. Figure 9.2 shows the trend in the number of asylum seekers, recognized refugees and humanitarian status holders in Japan. There is an upward tendency in all the numbers. An important reason is the revision of the Immigration Control and Refugee Recognition Act in 2004, where the 60 days rule was removed and the Refugee Adjudication Counsellors (RACs) System was introduced.[1] These changes made Japan's refugee status determination process less restrictive and more transparent.

In 2011, 1,867 asylum applications were made (491 from Burma, 251 from Nepal and 234 from Turkey as well as lower numbers from other countries). Out of the total applications, 21 cases were recognized as refugees and 248 were granted humanitarian status. The total asylum applications filed between 1981, when Japan joined the 1951 Refugee Convention, and November 2011 came to 11,754, of which only 598 were granted refugee status. The overall 'recognition rate' is 5 per cent. Most of the asylum seekers are from Asian countries: of the total asylum seekers, 4,215 were from Burma and 1,489 were from Turkey. Most of those recognized are also from Asian countries: 307 Burmese nationals, 69 Iranians, 59 Vietnamese, 50 Cambodians and 48 Laotians. In addition, 1,994 special permits to stay on humanitarian grounds (humanitarian status) were issued, most of which were from 2008 onward primarily to Burmese nationals (see Ministry of Justice 2012). The small number of asylum seekers granted refugee status has caused some 'Japan bashing' where Japan is depicted as closing doors to refugees fleeing from persecution, and 'free riding' on the asylum provided by other states.

On the other hand, Japan has been very generous in contributing to UNHCR and other UN agencies for refugee-related programmes worldwide. Japan has

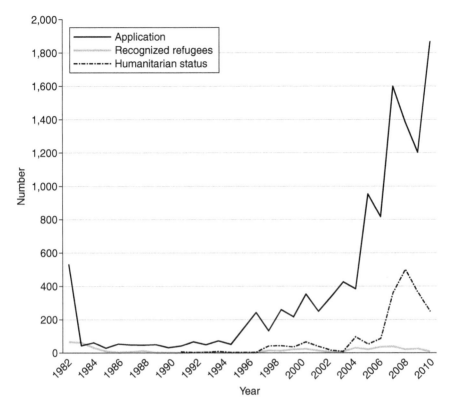

Figure 9.2 Asylum trends in Japan (source: compiled from data provided by Ministry of Justice).

been the second highest financial contributor to UNHCR after the United States and in the last several years, Japanese contributions have consistently exceeded US$110 million. In 2011 Japan contributed US$226 million, covering 10.8 per cent of UNHCR's total expenditure, and in 2012, it contributed US$185 million by September (see UNHCR 2012). Japan also spends some US$8 million for assisting asylum seekers and recognized refugees in Japan. Admittedly, the high value of contributions has been leveraged by a strong Japanese Yen in the currency market in recent years. In fact, much of the increase in Japanese contributions to UN organizations has been due to the valuation of the Japanese Yen. During June 2007 and June 2012, the Yen appreciated by 33 per cent from 120 Yen to 80 Yen per US Dollar, which is the budgetary currency of UNHCR (UN Treasury n.d.). At the same time, it is noted that Japanese official development assistance (ODA) has been halved from 1997 to 2011 when it was reduced to 537 billion Yen or some US$6.3 billion (MOFA 2011).

One can say that Japan's refugee policy from the viewpoint of contributing to the Global Refugee Regime is characterized by a strong focus on financial burden sharing and less emphasis on asylum. However, this profile combination has been criticized as 'cheque book diplomacy' – an approach with the implicit message that 'we will give you money, please take care of refugees outside of Japan'. Some argued that this policy mix represents 'shifting, not sharing responsibility to protect refugees' (Dean and Nagashima 2007). Sadako Ogata, a former High Commissioner for Refugees from 1991 to 2000, claimed that 'Japan lacks humanity' in dealing with refugees (Kingston 2007).

These criticisms are based on a mistaken assumption that Japan has a consolidated refugee policy on balancing asylum and financial burden sharing. In fact, however, refugee acceptance is the responsibility of the Immigration Bureau, Ministry of Justice (MOJ), while financial contributions are made by the Ministry of Foreign Affairs (MOFA). The main concern of MOJ is a proper refugee status determination process, while MOFA is concerned about the international image of Japan abroad. The two ministries do not coordinate respective refugee-related policies with a view to designing an integrated Japanese refugee policy. Bureaucratic sectionalism prevails and policy coordination by the Cabinet Secretariat is not strong. The mandate of the Inter-Ministerial Working Group on Refugee Issues (IMWG) established under the Cabinet Office is to coordinate policies on refugees already in Japan among 11 ministries,[2] and it does not interfere in the contribution policies of MOFA.

In light of contributions to the Global Refugee Regime, 'Japan bashing' seems to be unbalanced in that it focuses exclusively on asylum while ignoring significant financial burden sharing. Also, it appears that it is not so much that the Japanese Government does not let refugees in as asylum seekers as that refugees are not coming to Japan. One criticism of Japan's refugee policy is that the 'recognition rate' is too low. However, while the overall 'recognition rate' of 5 per cent is lower than those of traditional immigration countries, it is not necessarily lower than those of European States, which have been narrowing protection space in recent years. In Japan, rejected asylum seekers can repeat applications without limit and in recent years two-thirds reapply (in 2011, 1,719 out of 2,001 rejected asylum seekers, or 86 per cent, reapplied). If one takes into account the repeated applications, the net recognition rate is higher. If the increasing granting of 'humanitarian status' is factored in, the effective 'protection rate' in the last five years has been between 14 per cent and 38 per cent.[3]

What is conspicuous is the small number of asylum seekers in Japan in comparative terms. In 2011, an estimated 441,300 asylum applications were registered in the 44 industrialized countries, 20 per cent more than in 2010. The 2011 level was the highest since 2003 when 505,000 asylum applications were lodged (UNHCR 2011a). However, only 1,867 or 0.01 per cent of asylum seekers in the world applied in Japan. Notably, in 2011, out of some 27,000 Chinese who applied for asylum worldwide, only 27 applied in Japan. Only 278 asylum seekers were from neighbouring China during the last 30 years. The same goes for Russia from where some 20,000 asylum seekers leave every year. No Russian sought asylum in Japan in 2011. Even North Korean refugees do not choose Japan as a country of asylum. This suggests that an important reason for 'Japan passing' is the historical and political relations between Japan and Asian countries of origin. The memories of Japan's invasions into Asian countries during World War II have not been forgotten and mistrust of Japan's intention to accept refugees lingers. During the Cold War era, Japan did not accept political refugees from communist countries. Refugees from those countries tended not to seek asylum in Japan, either. In the late 1990s, the then Director of the Immigration Bureau, MOJ, expressed his wish that no refugee would come to Japan, arguing that Japanese people were opposed to accepting refugees (Hatano 2010).

Another root cause of 'Japan passing' is the multiple barriers to social integration of refugees, comprising a jurisdictional gap, participation gap and incentive gap at the national and local level. The following sections examine these issues.

The pilot resettlement programme

In December 2008, the Japanese Government announced that it would start a three-year pilot refugee resettlement programme from 2010. Each year, 30 Karen refugees from Burma living in Mae La camp in western Thailand will be resettled in Japan. This decision came as a pleasant surprise to the humanitarian community, which is used to 'Japan bashing'. It was surprising because, unlike the acceptance of Indochinese refugees during the 1980s that was 'imposed' on Japan by the United States (Mukae 2001: 241–243),[4] this time Japan took the decision without foreign pressure. As it was the first time in Asia, the decision was hailed not only as a turning point in Japan's exclusionary refugee policy, but also as a humanitarian decision that significantly changed Asia's image. The move conveys a message to the international community that there is a country in Asia welcoming asylum seekers and refugees (Kipgen 2008). It was a welcome development at a time when European States have been narrowing protection space.

For years, UNHCR tried to persuade the Japanese Government to start a resettlement programme but was unsuccessful (see Takizawa 2011). Following the 'Agenda for Protection' in 2002 that called for the expansion of resettlement, UNHCR intensified efforts to expand the resettlement base, including Japan (UNHCR 2003). Since 2006, High Commissioner António Guterres has taken up the issue of resettlement with the Ministers of Foreign Affairs and Justice during his annual visit to Japan. In February 2007, with the arrival of the first Japanese Representative who used to work at MOJ (the author), the UNHCR Office in Japan started a two-pronged approach to promote resettlement: public advocacy for a resettlement programme, and quiet diplomacy via informal meetings with key officials of the Immigration Bureau in the MOJ. Concurrent efforts were made to improve relations between UNHCR and the MOJ, which had been strained following deportation of two Turkish asylum seekers in 2005 who were recognized by UNHCR as 'mandate refugees'; this incident caused international condemnation of Japan. Initially, the MOJ officials were mute on the idea of starting a resettlement programme. In an interview with the author (the then UNHCR Representative), Toshio Inami, the then Director-General (DG) of the Immigration Bureau, responded to the suggestion of a resettlement programme that, while he understood the significance of Japan taking such an initiative, he considered that politicians and the general public would not support the idea. Time was not mature, he remarked. He understood the difficulty of initiating a discourse on resettlement in the face of the existence of the jurisdictional gap, participation gap and incentive gap surrounding the issue.

Nonetheless, Inami took an initiative and established a 'study group' to understand the refugee resettlement programme. The members were three middle-level managers of MOJ, MOFA and the Cabinet Secretariat. Their activities were kept in strict secrecy to avoid stirring up political opposition and also reflecting Japanese bureaucratic culture not to disclose any policy proposal when it is still being formulated. The group researched the policies and practices of a few resettlement countries with assistance from UNHCR, while sharing information with the Inter-Ministerial Working Group on Refugee Issues (IMWG). In June 2007, a Japanese diplomat attended the Resettlement Working Group meeting in Geneva for the first time, raising hope that Japan could become a new resettlement country. The warm welcome extended to the delegate was an incentive for Japan to engage with the issue of resettlement. The study group completed its work towards the end of 2007 and proposed to establish a 'Consultation Group on Resettlement.' In Japanese bureaucratic practice, which emphasizes the importance of consensus decision-making, a 'consultation group' is a way to overcome jurisdictional barriers and to ensure wide participation of stakeholders of concerned ministries, and the establishment of a 'consultation group' also implies that a positive consensus has been achieved in favour of the issue concerned. The establishment of the consultation group was officially conveyed to High Commissioner Guterres by the Justice and Foreign Ministers in November 2007 when he visited Japan. The announcement was reported in the media.

In parallel, DG Inami advocated for mobilizing support of concerned officials and parliamentarians in his attempts to dispel concerns and sell the benefits of a resettlement plan. The then Minister of Justice Kunio Hatoyama supported the initiative. At the Parliamentary debates in the spring of 2008, Hatoyama stated that he considered that a resettlement programme would be Japan's contribution to the international community and even made a personal commitment that he and his staff at the MOJ would lead the discussion on the introduction of the resettlement programme.[5] This was a strong drive for Japan to start a new burden-sharing initiative in the form of resettlement in addition to its financial contribution. In an interview with the author conducted in 2011, Hatoyama said that he trusted the judgment of DG Inami and his staff in starting a resettlement programme and believed that accepting refugees is a form of Japan's international contribution.

In the first half of 2008, research and discussion took place in the IMWG meetings and by summer 2008, when the budget for 2009 was prepared, an implementation plan and budget had been prepared by MOFA, the Ministry of Science and Education and the Ministry of Welfare and Labour. The decision was fast by Japanese bureaucratic standards; by fall 2008, the pilot programme was endorsed at the top level of the Government and the Cabinet endorsed the programme by issuing a Cabinet Understanding on 18 December 2008. The Understanding and the detail of the plan were immediately made public on the Cabinet Office website.[6] The Government decision was formally conveyed by Prime Minister Taro Aso to Guterres, who was visiting Japan. In fact, the Cabinet decision was set so that Guterres could receive the good news in person. Assisted by the improved working relations with the MOJ, UNHCR was able to sell the 'benefits' of the resettlement programme to reform-minded Government officials, who in turn persuaded parliamentarians that it was in Japan's 'national interest' to dissipate the negative image of a country closed to refugees or a country that lacks humanity, through a resettlement programme. Such a programme would be in line with Japan's identity as a humanitarian power and enhance its reputation in international society, while its scale would be very small and associated costs and risks would be low.

MOFA officials were aware of the value of a resettlement programme in improving Japan's image and identity. The resettlement programme is a concrete form of international responsibility/burden sharing and is in line with the notion of human security, a major principle of Japanese ODA. MOFA officials have repeatedly stated that the resettlement programme is a form of Japan's international contribution. MOJ officials were also persuaded of the 'benefits' of a resettlement programme from MOJ's own perspective. MOJ could dispel the long-held image of an excessively restrictive refugee policy caused by stringent refugee status determination (RSD) procedures. Under a resettlement programme, MOJ can choose those refugees who it deems better fit for resettlement to Japan and increase the number of refugees in Japan, without going through the complex and lengthy RSD process, including litigations. As for MOJ's concern about security risks, UNHCR would recommend for resettlement only those who have no security risks. Thus MOJ would have nothing to lose by starting a resettlement programme. However, not all ministries were supportive of the programme. The Ministry of Internal Affairs and Communications (MIC), which is responsible for economic and social activities including local administration and related tax systems, was opposed to the idea. The National Police Agency (NPA) was also not supportive as it was concerned about criminal activities committed by foreigners. There is, however, no hard proof that foreigners, including refugees, commit more crimes than Japanese nationals in relative terms. It is likely that the NPA stance reflects the Government's policy of reducing the number of foreigners without regular status.[7] The Cabinet Secretariat, which chairs the IMWG, eventually managed to reconcile the divergent views and interests, and succeeded in bringing about a compromise agreement to start a small pilot resettlement programme lasting three years.

On the political side, senior parliamentarians including former Prime Minister Yoshiro Mori of the then ruling Liberal Democratic Party (LDP) joined the group of supporters. Komeito, a political party affiliated with the religious organization Soka Gakkai, established a project team on refugee issues and actively supported the resettlement initiative. The introduction of the resettlement programme was also helped by the debate on immigration. Given the prospect of a rapidly shrinking and ageing population in Japan, some opinion leaders started campaigns to open Japan for immigration. In 2008, the LDP issued a policy paper urging Japan to establish an Agency for Migration and accept up to 10 million immigrants in the next 50 years (Liberal Democratic Party 2008). UNHCR Japan, along with the IOM (International Organization for Migration) office in Japan, adopted an issue-linkage tactics whereby refugees are presented as 'humanitarian migrants' who need preferential treatment. The LDP paper eventually included a proposal to increase the number of humanitarian migrants (refugees) to 1,000 per year, at a time when only 20 to 30 asylum seekers per year were granted refugee status in Japan. While this proposal did not result in any concrete measures (partly due to the defeat of the LDP at the 2009 general election), it had a significant symbolic value in that the then ruling party officially published such a policy paper and initiated a discourse on a very sensitive issue of migration. A similar view is reflected in the report on immigration submitted to the Prime Minister by an influential body of conservative opinion makers (Japan Forum on International Relations 2010). One can say that a shared view and consensus has emerged among key decision-makers and opinion leaders that it is high time for Japan to accept more refugees than before.

Finally, the national media played a major role by reporting positively on the resettlement issue. Reports and editorials in the national dailies as well as TV programmes have been supportive and encouraged the Government's resettlement initiative, with a proviso that the Government has to provide resettled refugees with sufficient language and skills training as well as social integration support so that they can become members of Japanese society as soon as possible, avoiding the repetition of the difficulties the Indochinese refugees experienced. Such urging in turn encouraged officials and parliamentarians who consider media reports and editorials would reflect popular thinking. DG Inami admitted the very important role played by the Japanese media in forging positive public opinions on resettlement.[8]

Inami and Hatoyama *internalized* the resettlement issue by linking it to Japan's positioning in the world and helped break down the jurisdictional barrier. Governmental stakeholders' participation was ensured by the IMWG. UNHCR was not allowed to join the IMWG but played a catalytic role through advocacy and provision of technical information. Different incentives and interests were offered to the key stakeholders. The three gaps have thus been narrowed and this facilitated the decision to start a resettlement programme. As only a limited number of Government officials were involved in the decision, the decision to start the pilot programme itself was relatively easy and quickly made by Japanese standards. However, this offered no assurance that implementation would be as easy as the decision itself. Indeed, domestic implementation of the resettlement was found to be much more difficult than expected as it requires the participation of a number of national actors, their buy-in, and incentives to ensure their commitment. These were largely missing, and the three gaps at the local level would soon be exposed.

Implementation of the pilot resettlement programme

Resettlement as a national public good requires filling the three gaps: the jurisdictional gap, participation gap and incentive gap. First, local governments and people have to *internalize* externalities, that is, they have to understand refugee issues (virtually none does so in Japanese cities and towns) and find value in accepting and assisting refugees in their localities. However, the value of international resettlement in local cities and towns was neither understood nor shared among concerned actors. No answer was provided to questions like: Why should we accept refugees in our town? Isn't it the responsibility of the central Government? No internalization took place. Second, actors other than the Government ministries did not participate, or were barred from participating in local integration support activities. Initially even information on the refugees was withheld by the Government and communication was limited. In the decision leading to the start of resettlement, local municipalities, NGOs, the private sector and importantly refugees were not consulted. Infrastructure for their local integration was slow to build and done in an ad-hoc manner. Refugees' voices were largely ignored. Finally, the central Government did not provide incentives for those who support refugees in their localities. Only partial financial compensation was made to schools that had to assume inevitable additional costs such as additional teachers.

In summer 2009, the Government requested UNHCR to prepare a list of 60 refugees recommended for resettlement in Japan. The Government asked UNHCR to take into account the Japanese selection criteria: Karen refugee families (parents and dependent children, or a nuclear family) in Mae La camp who have a capacity to adapt to Japanese society and who are expected to get employed.[9] The reason for targeting Mae La camp was that it was the largest camp where active resettlement activities were taking place. The Karen ethnicity was chosen as the majority of the camp residents were Karen whose livelihood was agriculture, hence it was thought that they would adapt to Japanese culture and customs without much difficulty. The family criterion was applied because, compared to single people, families would have less difficulty in adapting to Japanese society.[10] In retrospect, the criteria were based on impressions and untested assumptions.

After information campaigns, UNHCR submitted a list of candidates who met its selection criteria: legal and/or physical protection needs; survivors of torture and/or violence; medical needs; women and girls at risk; family reunification; children and adolescents at risk; and lack of foreseeable alternative durable solutions. UNHCR did not officially take into account whether the recommended refugees have 'ability to settle' because resettlement should not be determined on the basis of 'integration potential' or other non-protection criteria.[11] This created a discrepancy with the Japanese selection criteria that emphasized the ability to integrate. This discrepancy is a reflection of the jurisdictional gap, where Japan's domestic interest differs from a global interest as reflected in UNHCR's selection criteria.

When the resettlement plan was announced in Mae La camp, to the surprise of the Japanese Government it did not attract much interest from the refugees and the number of applicants did not reach the quota of 30. Selection interviews were conducted in Mae Sot town in February 2010 by an MOJ official in charge of RSD based on the Japanese selection criteria mentioned above. As the quota of 30 was not filled, a second interview was conducted and eventually five families consisting of 27 adults and children were selected as the first group. After attending a one-month IOM pre-departure orientation programme, the first group arrived at Narita International Airport on 28 September 2010 and was met by a large number of media reporters, Government officials, NGO members and Karen community members. TV reports and magazine articles welcomed the arrival of the refugees while cautioning that the Government should strengthen the measures to support their integration.

Upon arrival, the refugees attended a six-month language and cultural adaptation programme organized by Refugee Headquarters (RHQ), a semi-governmental organization specialized in refugee assistance in Japan, in a training centre in Shinjuku, downtown Tokyo, where many people from Burma live and work. The programme consisted of 570 hours of Japanese language training, 120 hours of social adaptation guidance as well as job placement assistance. Information on the refugee families, including the location of the training centre, was withheld by the Government on the grounds that the refugees' privacy needs to be protected. The first group was effectively cut off from the local community and NGOs who intended to provide support. A precious opportunity to turn the high level of interest into effective participation of various national actors who were willing to support the refugees was lost. The Shinjuku municipality office, where the training centre was located, was asked to accommodate the training centre and refugees but no incentive or compensation was given by the Government.

Before the five families left the centre, RHQ assisted them in finding employment, work place adaptation and finding schools for the children. RHQ found a corporate farm in Chiba prefecture, about 20 kilometres from Shinjuku, to undertake on-the-job training for two of the five refugee families. From the beginning, the refugees complained about difficult living and working conditions, long commuting distances to work and schools (they live in one city and work in another), absence of day-care services for infants, as well as communication problems with the employer and RHQ social workers. They had not been informed of the conditions of life and work practices in Japan before they left the camp for Japan. RHQ had given priority to finding jobs for the refugee parents. First employers were identified then the municipal government was informed of the arrival of the refugees. There was no local NGO that provided support to the refugee families. The wide gap between refugees' expectations and the reality in Japan, including the expectations of the Government officials, RHQ staff and local community members, was exposed at last. A source close to the families said that one refugee doubted if coming to Japan was the right decision, and another refugee frequently shared by phone her dissatisfactions and complaints with friends and relatives remaining in Mae La camp, initially spending up to US$800 a month in phone bills. It can be easily imagined that such communication spread a negative image of Japan as a resettlement country among refugees in Mae La camp, who were anxious to know what happened to the first group. On the other hand, the farm employer criticized the Government for leaving at the farm refugees who spoke poor Japanese, adding that the six-month training is too short to acquire mechanized agricultural know-how. Negotiations among the employer, refugees and RHQ staff did not succeed, communications broke down and the refugees asked lawyers to intervene to represent their interests. The lawyers publicly asked MOFA to improve the assistance scheme, and the lawyers' intervention was reported in the media, generating perceptions that the resettlement programme was facing serious difficulties (Japan Lawyers Network for Refugees 2011). Eventually the two families abandoned the on-the-job training and moved back to Tokyo. One of the two families is receiving welfare benefits as the husband's income is insufficient to support a large family.[12] Reliance on welfare benefits is the last thing the Government wants to see as it gives an impression that refugee acceptance is only a burden on society and there has been much controversy on possible abuses of welfare benefits. Low salaries, expensive housing and schooling cost are creating pressure for the two families.[13] Their children are attending elementary and junior high schools and reported to be doing fairly well, while parents are struggling with Japanese language and suffer from social isolation from both the local community and Burmese ethnic communities (both Karen and others) (Yamashita 2013).

The remaining three families of the first group have settled in Suzuka City in Mie prefecture in central Japan, where the adults have been working at a large mushroom farming firm. Masaharu Nakagawa, a Member of Parliament and then Minister of State, helped the refugees find employment through his personal network. The owner of the firm and his wife have been committing considerable time and resources to assist and support the refugee families in learning Japanese, finding accommodation, learning farming skills, negotiating with schools to make special arrangements for the refugee children, all without compensation from the Government. As there are thousands of foreign factory workers in automobile companies in Suzuka City, city officials and school teachers have been cooperative in taking initiatives such as opening an evening Japanese class for refugee families. The children are quick at learning the Japanese language and are doing well in school. However, the adults are having difficulties in learning the language. After 18 months of life in Japan, some of the adults still lack basic Japanese language skills and work-related instructions are given by examples and gestures (Japan Broadcasting Corporation 2012). On 30 June 2012, Suzuka City organized a symposium on refugee resettlement entitled 'Refugee Settlement and Multicultural Community' where the Mayor attended as a speaker. Participation by the city government is notable. As a whole, the resettlement of the three families appeared to be going well, thanks to the understanding and participation of the employer, schools, and city officials. However, at the end of February 2013, one of the three families moved to the Tokyo area despite persuasion by support groups. According to sources close to the family, they have always wished to live in the Tokyo area where there are Karen communities. Proximity to the same ethnic group is an important factor in the choice of place of residence and explains the reason for the concentration of Karen in the metropolitan area.

The resettlement of the second group started in the spring of 2011. As in 2010, there were not many applicants. The refugees had received information on the difficulties faced by the first group in Japan. While six families were selected for resettlement following two rounds of interviews, two of them withdrew just before departure for Japan in September 2011 because they were advised by the selection mission that, since the cost of living in Japan is very high, both husband and wife will have to work to earn a living. As one of the women had a baby and another was expecting delivery in February 2012, they became uneasy and gave up the idea of resettlement to Japan.

The four families, consisting of 18 individuals, arrived at Narita airport at the end of September 2011 and were met by a smaller number of reporters and Japanese supporters than in 2010. Although the date of arrival was known, no members of the Burmese community were at the airport to welcome the arrival of the second group, suggesting the existence of difference of opinions among the ethnic community.[14] The second group received somewhat better services from RHQ than the first group. In the face of criticism that the assistance pro-gramme is overly protective and secretive, MOFA and RHQ released more information on resettlement than in 2010. The programme contents have been improved and RHQ started contacting possible employers as early as November 2010. Eventually, the four male refugees found employment in a shoe factory in Tokyo while their spouses worked in a linen shop in Misato City, located 15 kilometres from downtown Tokyo. In March 2012, all families found apartments in a housing complex to start a new life. A group of volunteers are helping the families integrate into the local community. Proximity to the Burmese community in Tokyo, working and living together and incentive based work, among other things, seem to facilitate their integration process in Japan.

The Government was alarmed by the unexpected unpopularity of the Japa-nese resettlement programme, as the credibility of the programme is at stake. In the summer of 2011, when criticism arose against the handling of the two fam-ilies in Chiba, the Government started taking action. Information campaigns were made, assisted by UNHCR in Mae La camp screening a DVD on the life of families resettled in Japan. At home, the Government used the internet and TV to carry an information campaign for the Japanese public on the refugee resettle-ment programme. RHQ also improved information disclosure on the web: for the first group, there were only two news briefs, but for the second group, there are a dozen news announcements on the activities of the RHQ and the refugees. The Government also started an outreach programme. In November 2011, 12 officials of the IMWG took an initiative to visit Suzuka City and Matsumoto City to assess the situation and exchange information on resettlement in the provincial cities. Furthermore, to find ways to improve integration support mechanisms, the Government held two consultation meetings (December 2011 and January 2012) with interested NGOs to seek advice on the resettlement programme. However, these information campaigns were not enough to encourage participation by local governments as they had been kept uninformed, and there are no incentives provided to engage in international cooperation.

The selection interview for the third and the last group was conducted in February 2012. Amidst widely reported difficulties encountered by the first group (particularly the two families initially resettled in Chiba), only two families consisting of 10 individuals applied for resettlement. Through telephone and email, resettled refugees informed relatives and friends that conditions surrounding resettled refugees in Japan are more challenging than expected. In a press interview, Mr Tun Tun, the chairman of the refugee committee of Mai La camp, said that some of the first refugees who moved to Japan (presumably a refugee family who moved to Chiba) told their relatives in the camp that it is tough adjusting to Japanese culture (*Asahi Shimbun Digital* 2012). Meanwhile, the wives of the two families settled in Misato city, who had never experienced putting children in school, babies in nursery or working in a factory, have been complaining that life is tough in Japan, their efforts are not recognized, they would not have come to Japan had they known the difficulties facing them, and even that they wish to go back to their refugee camp.[15] Another factor is the rapid political changes in Burma that gave rise to the hope that it might be possible to return there.

At the end of March 2012, the Government (IMWG) issued a report assessing the progress of the pilot programme. The report highlighted the Japanese language as the biggest challenge to local integration, and introduced several improvements in the pilot programme such as an enhanced information campaign in the camps, enrichment of the language courses and the placement of Integration Supporters (*Chiiki Teijuu Shien-in*) in the communities where refugees live. The Government also decided to extend the pilot programme for two more years until 2015 as it was too early to make any meaningful assessment of its success or failure. The extended pilot programme will now target refugees at Umpiem Mai and Nupo camps in addition to Mae La camp. Furthermore, the Government decided to establish a Resettlement Experts Council (*Yuushikisha Kaigi*, hereafter REC) under the IMWG to find better management of the pilot programme and its prospects after 2015. However, the selection criteria and implementation modality remain within the framework of the Cabinet Understanding and related procedures announced in December 2008.

To increase the number of selected families, the Government conducted another selection interview in June 2012 targeting refugees in Nupo and Umpiem Mai camps, and one family of six was selected. At the symposium commemorating

World Refugee Day on 20 June 2012, a representative of the Immigration Bureau of the MOJ admitted that he was surprised by the unexpected unpopularity of the Japanese resettlement programme. This suggests the existence among government officials (and the general public) of a strong but untested assumption and a mindset that Japan is sought after by refugees as a country of asylum.

By August 2012, the Government decided to accept three families consisting of 16 members, and in September they started a four-week pre-departure orientation programme organized by IOM in Mae Sot. However, on the second day of the programme the two families from Nupo camp received a phone call from their parents in the camp pleading with them not to go to Japan and, despite persuasion by UNHCR, they decided not to participate in the resettlement programme. The remaining one family from Umpiem Mai camp did not want to go alone and also withdrew. All three families went back to their respective refugee camps. Eventually one family was selected for resettlement in the United States.[16]

As a result, there was no refugee resettlement in 2012 against the target of 30. At the end of the three-year pilot period, out of 163 refugees (35 families) recommended by UNHCR, only 45 refugees, comprising 18 parents and 27 children, have been resettled against the target of 90.[17] The refusal of the selected refugees to move to Japan was a big shock not only to the Government but also to civil society. It was an embarrassment to the Government, that intended to use the resettlement programme to enhance the image of Japan, and a humiliation to those who boast that 'The first phase (of resettlement) may be difficult but Japan has a lot to offer, this is an exciting and enjoyable place. Once they are in Japan, we will extend assistance for this entire generation'.[18] Importantly, it has shattered the assumption and mindset that many asylum seekers wish to come to Japan, and forced people to realize that Japan may be de-selected by refugees as a country of destination. The episode reflects how wide and deep is the gap between Japanese self-image and the images held by refugees.

This gap in perception is particularly ironic because, at the political level in November 2011, both the House of Representatives and the House of Councillors unanimously passed a resolution commemorating the thirtieth and sixtieth anniversaries of Japan's accession to the Refugee Convention and the creation of UNHCR respectively. The Resolution concerning Japan's continued commitment to refugee protection and search for solutions reads as follows:

> The year 2011 will mark the 60th anniversary of the 1951 Convention relating to the Status of Refugees and the 30th anniversary of Japan's accession thereto. As a member of the international community, Japan has, for 30 years since its accession to the 1951 Convention, made an active engagement with humanitarian assistance for refugees and IDPs worldwide, emphasizing the concept of human security and peace building. Launching the pilot resettlement programme that accepts Myanmar refugees from Thailand, Japan became the first resettlement country in Asia in 2010. Japan has made every effort to ensure the transparency and efficiency of its

refugee status determination process with a view to further enhancing and developing its domestic asylum system.

Building on the above and respecting the basic international refugee protection principles and values, Japan wishes to commit, through strengthened coordination with international organizations, civil society and non-profit organizations that support refugees, to develop a comprehensive asylum process in Japan and to make every effort to further develop the current resettlement pilot project. Japan also reaffirms its commitment, based on its foreign policy, to continue assisting and supporting refugees and IDPs worldwide, thereby undertaking a leadership role in Asia and in the world to strengthen and improve refugee protection and to identify durable solutions for refugee/IDP problems worldwide.[19]

It was the first time that a national parliament passed such a resolution committing the country to the international protection of refugees and IDPs, and it was much welcomed by UNHCR Headquarters in Geneva. In fact, the UNHCR Office in Japan played an important role behind the scenes in realizing the adoption of the Resolution in coordination with the Parliamentarians' League for UNHCR. The resolution set a direction and parameters for the Japanese Government in the formulation and implementation of future refugee policies.

Also, at parliamentary debates held in the spring of 2012, the Minister of Foreign Affairs replied that efforts would be made to widen the target group to other refugee camps, relax selection criteria, involve NGOs as well as local governments, and to review the performance of the RHQ and its management structure. On 20 June 2012, at the symposium organized by UNHCR Japan on the resettlement programme, Minister of State Nakagawa (in charge of the programme) stated that the resettlement programme is a national policy that is aimed at making Japan a multicultural nation. It is part of efforts to determine a 'new shape of Japan.' With these political developments and an issue-linkage to the future shape of Japan, it will be impossible to discontinue the resettlement programme without damaging Japan's credibility. The stakes have been raised considerably. Yet, these lofty announcements have not been backed up by concrete measures on the ground to facilitate refugees' integration in Japan.

The Government was at a loss facing the withdrawal of three refugee families. At the sixth meeting of the Resettlement Experts Council of 17 November 2012, a senior official admitted that the Government did not foresee such an eventuality and could not really understand why the refugees withdrew. She hinted that the strict selection criteria may be the main reason.[20] Earlier, when the news of the refugees' refusal was known, the former Minister of State Nakagawa instructed the Government to review the selection criteria and integration support measures.[21] UNHCR on its part made a four-point recommendation to the Government. They are: (1) to expand the definition of family from a nuclear family in such a way as to allow single persons or a family reunion, (2) to expand the target refugee camps from only Mae La camp to all nine camps in Thailand and to urban refugees in Malaysia, (3) to expand the target group from only Karen to all ethnic groups, and (4) to improve integration support measures such as involving NGOs, volunteers and refugees. The reactions of the ministries to these recommendations were quite negative. They question that there is no assurance that more liberal criteria would lead to more applicants; more data and analyses are needed on the reasons for the unpopularity; additional costs are expected if more refugees are accepted, and so forth. In general, one can detect a sense of confusion, distrust of UNHCR's recommendation and anxiety about the possible increase in cost. There were rather acrimonious exchanges of views at the REC itself.[22] At its seventh meeting, in connection with UNHCR's recommendations to relax the selection criteria, members agreed on tentative recommendations. They are: (1) to allow family reunification, (2) to expand the target group to ethnic groups other than Karen, (3) to expand the programme to urban refugees in Malaysia.

Why is the Japanese resettlement programme facing difficulties?

To understand the implementation difficulties of the pilot programme, it is useful to put the programme in a wider historical, political, social context and reflect on the huge differences between the camp society and Japanese society. The armed conflicts between the Burmese military and ethnic minority groups including the Karen people in the eastern provinces of Burma have been going on for decades. Most of the minority militant groups have entered into ceasefire agreements with the government forces, including the Karen National Union (KNU), which agreed to a ceasefire with the government early in 2012. While in Burma, members of these ethnic groups used to live a self-sufficient and very poor life in eastern rural and mountainous areas. The military strategy was to target the civilian support base, which undermined human and food security and has impoverished large parts of the civilian population (Loescher and Milner 2008: 305). The army forcibly confiscated land and relocated civilians from some 3,000 villages to new government-controlled villages to obtain free labour and other resources. A large number of ethnic minority groups have been used as forced labour for military and development purposes. Food insecurity, lack of education and basic health services and the outbreak of major health crises (malaria, cholera, HIV/AIDS) have resulted in a major humanitarian crisis in Burma. Up to half million people have been internally displaced and many have fled to Thailand. There are estimated to be as many as three million migrants/ migrant workers in Thailand, of whom at least 80 per cent are believed to be from Burma. Many are de facto refugees, having left their homes due to the same circumstances as those living in the camps (TBBC 2011: 9).

Camp life is not much better. The Thai military have placed security forces around the camps and enforced severe restrictions on the more than 100,000 refugees living there. Refugees cannot move freely between camps or beyond the camp perimeters and are not allowed to work locally on Thai farms or as day labourers, although some manage to do so. Refugees have become entirely dependent on international aid and, in effect, have been 'warehoused' (USCRI 2011). Refugees in the camps are in a 'protracted refugee situation' – a long-lasting and intractable state of limbo, in which their lives may not be at risk, but their basic rights and essential economic, social and psychological needs remain unfulfilled after years in exile and they are often unable to break free from enforced reliance on external assistance (UNHCR 2004). Having been 'encamped' for 10 to 15 years and obliged to depend on welfare assistance provided by international NGOs for food, schools and medical care, they have lost whatever skills they used to have, such as agricultural skills, which has weakened their ability to live self-reliant lives. In the late 1990s, the Royal Thai Government consolidated smaller refugee camps into nine large-scale camps and imposed strict restrictions on refugee movement in and around camps. This resulted in the loss of opportunities to earn a living outside the camps and refugees became almost totally dependent on international assistance (Bowles 1998). Without opportunities to work to earn a living, they have been forced to remain in a situation of 'aid dependency' (Brooks 2004). Refugees have little opportunity to maintain or develop skills, as job opportunities in the refugee camps are very limited and they have been disempowered. Most adults have not received formal schooling or formal education and some remain illiterate and innumerate. It is not surprising that some have developed a 'dependency mentality', which does not help resettlement of refugees in industrialized and urbanized countries like Japan.

Japan is at the other end of a camp life–urban life spectrum. It is not a multi-ethnic country. Out of a population of 127 million, only two million are non-Japanese, of whom some 800,000 are Koreans who have been living in Japan for decades and who are virtually indistinguishable from Japanese. Generally, people are not hospitable to foreigners from Asian or African countries. Japanese society is highly urbanized and organized and there is strong pressure for conformity. Group consensus is appreciated while individualism or 'being different from others' is frowned upon. Almost 100 per cent of the population receives secondary education and over 50 per cent of high school graduates enter universities. Illiteracy and innumeracy are virtually unknown. Due to a 20-year old recession, the relative poverty rate rose to 16 per cent in 2010 and over two million people are on social welfare (Ministry of Health, Labour and Welfare 2010). Competition for jobs is tough and even Japanese college students have difficulty in finding jobs. Unskilled foreign workers who have poor Japanese language skills are not competitive in the job market. From 1998 to 2012, over 30,000 people committed suicide every year, roughly a quarter for economic reasons (National Police Agency 2011). Due to rapid population decline and ageing, the sustainability of the national social security system is in question. The safety network for the most vulnerable is inadequate; self-reliance is the dominant social norm and social pressure and expectation for 'self-help efforts' is strong. People are expected to be hardworking, disciplined and purposive to overcome barriers. While improving, public understanding of refugee issues is low.

Three gaps

Because of the huge differences between a protracted refugee situation and urbanized Japan, and because of the profiles of the resettled refugees, integration of resettled refugees in Japan has been facing several serious challenges. The challenges can be analysed in terms of the three gaps under the conceptual framework; the jurisdictional gap, participation gap and incentive gap.

First, there is a jurisdictional gap. Although the stated policy goal of the resettlement programme was global (humanitarian assistance and international burden sharing), the primary goal was national: to improve the image and identity of Japan in international society. Resettlement offers a convenient way for both MOFA, that wishes to improve Japan's image, and MOJ, which wishes to increase accepted refugees without RSD procedures. In developing the policy, refugees' motivation for resettlement, related needs and capacities were not adequately assessed. The Government policy was based on the assumption that there will be many applicants for resettlement in Japan. The restrictive selection criteria are based on that assumption. However the assumption was found to be unrealistic. As early as in November 2007 when the Japanese Government had been looking into the possibility of starting a resettlement programme, the author visited Tham Hin camp, Nan Mai Nai Soi camp and Mae Surin camp in the north of Thailand to learn about conditions in the camps and the wishes of refugees. In a meeting with representatives of refugees in Tham Hin camp, the author asked if anybody was interested in resettlement in Japan. The answers were generally negative. Other than the fact that refugees had a vague notion about today's Japan, some pointed out that during World War II, Japan invaded Burma, which was then a British colony. The British mobilized the Karen while the Japanese Imperial Army solicited the support of the Burmese to fight against the British. The Karen people still remember the atrocities committed by the Japanese soldiers. Others mentioned that Japan has been providing Official Development Assistance to the present Burmese Government that has been oppressing the Karen and other ethnic minorities. The historical and political relations between Japan and Burma negatively affect refugees' choice of Japan as a destination country. Lack of interest in resettlement to Japan was also clear in August 2011 when the author visited Mae La camp. Three years after the announcement by Japan that it would resettle Karen refugees, refugees were not enthusiastic about this option. The Government's targeting of Karen refugees was based on the reasoning that the majority of refugees in Mae La camp are Karen and did not take into account this historical background and the collective memories of Karen people.

It is also important to keep in mind that refugees can apply for any of the 13 resettlement countries operating in the camps, and Japan is only one of the latecomers, a less competitive and least experienced country. Notably, Mr Tun Tun, Chairman of the Refugee Committee of Mae La refugee camp, who knows Japan fairly well, told the author that he would not choose Japan as a country of resettlement: he prefers the United States. In his view, Japanese selection criteria are too restrictive and inflexible. His preference (and those of other refugees) is partly based on the feedback of refugees who have been resettled in Japan. When the camp leadership does not favour the Japan option, one cannot expect fellow refugees will choose Japan as a country of resettlement. A resettlement policy that strongly focuses on domestic concerns and lacks global perspective (including target refugees' needs and expectations) is flawed. In retrospect, UNHCR Japan should have advised the Government of issues pertaining to Japan's positioning in the resettlement circle as well as refugees' identity, their collective memories and expectations. Internalization of the global perspective or norm was not sufficient.

Furthermore, Japan's national selection criteria diverge from global standards as reflected in UNHCR's *Resettlement Handbook*. According to the handbook, resettlement is provided for humanitarian reasons for those who need special protection when no other solutions are available. UNHCR's resettlement submission categories include: (1) legal and/or physical protection needs; (2) survivors of torture and/or violence; (3) medical needs; (4) women and girls at risk; (5) family reunification; (6) children and adolescents at risk; and (7) lack of foreseeable alternative durable solutions. The Japanese selection criteria are explicitly defined in the Cabinet Decision as follows: (1) those who are recommended by UNHCR as needing international protection, and (2) those who have the capacity to adapt to Japanese society and who are employable to earn a living. In addition, selection is made on a family basis, namely parents and dependent children (a nuclear family), which is typical in today's Japan.[23] Cultural adaptability reflects the Government's wishes to minimize social costs by rapid assimilation rather than integration. The employability criterion reflects the Government's wishes to minimize economic costs. Selection is made on a 'nuclear family' basis and excludes single persons and family reunification because the Government believes that a nuclear family would cost less (socially and economically) than single persons or extended families that would include elderly parents. In 2007, UNHCR Japan also recommended the Government to start a small scale resettlement programme on a family basis, not realizing that in camps housing refugees from Burma most refugee families are 'extended families' consisting of children, parents, grandparents and relatives. This strict criterion makes it impossible for a family reunification to take place later for those who are left behind in the camp, making the Japanese programme less attractive.

The Japanese criteria are not in line with the UNHCR policy that resettlement should not be determined on the basis of 'integration potential' or other non-protection criteria. Fortunately, UNHCR's criterion (7) can be used for refugees who do not require resettlement for immediate protection needs but when it opens strategic possibilities for a comprehensive solution, and this category was used for the Japanese programme. For UNHCR, the start of a resettlement programme in Japan, even if 90 refugees in three years is a drop in the ocean, has a strategic and symbolic value, as it could be a catalyst in increasing the number of resettlement states in Asia. For Japan, its benefits are big enough to add to the country's reputation yet associated costs were believed to be small enough to prevent opposition and to keep costs low. Japan can contain a free-rider criticism by a cheap initiative. For UNHCR the relatively small size of the programme was not an issue as it could be expanded later. UNHCR's strategy was to adjust to the interests of the donor. UNHCR and Japan have a common and strategic interest in starting a resettlement in Japan.

The Japanese criteria emphasizing ability to settle are very restrictive and reflect domestic concerns rather than global needs to reduce the number of those who need resettlement. Also, it is getting more difficult to find those who have the 'ability to settle' in the camp as the target population has become smaller. Since 2006, 73,775 refugees have left the camps for resettlement in third countries, including 56,968 to the United States. Probably at least 75 per cent of the most skilled registered refugees have left the camps (TBBC 2011: 8). Those who have education, have command of English and work experience with international NGOs operating in the camps, or the 'best and the brightest', are the first to be resettled to countries where they could utilize their potentials to become self-reliant and make contributions (Banki and Lang 2008). Japan is not alone in applying 'ability to settle' criteria but few other countries, if any, make such criteria explicit and official. In the eyes of refugees, Japan is less attractive than the US, Canada, Australia and European resettlement countries. The selection criterion (only nuclear families) is a disincentive as it would break up the 'extended family' where parents, children, grandparents and other relatives live together or support each other. Finally, resettlement of a nuclear family is not necessarily less expensive than individuals as it excludes young, single, ambitious and talented refugees who could adapt to Japan faster. In fact, all the refugee families who have been selected by Japan have four to five dependent children who will not be able to work and earn income for 5 to 10 years. Until that time, support NGOs, municipal governments or the host community will have to bear the additional costs.

The second gap is the participation gap. Integration is a continuum with the starting point well before the refugees arrive in the country of resettlement and the end point years after arrival. Local integration involves not only language and employment but the entire spectrum of integration process covering economic, legal and cultural/social dimensions (UNHCR 2011b). In terms of time frame, there are three phases of integration: short term (immediately after arrival); medium term assistance phase; and long term support phase towards self-reliance.[24] Effective local integration is a national public good and its production requires the participation of a number of actors other than the central Government, such as municipal governments, employers, NGOs and community residents. The Japanese programme does not have a mid-term and long-term perspective. Japan adopts a six-month intensive orientation and settlement assistance system in Tokyo provided by RHQ. Other possible service providers could not participate even if they wish to do so. Compared to the short-term support programme, refugees' concerns are of a long-term nature such as stable employment, medical insurance, education of children, old age pension and permanent residency or naturalization, which allow them to make a life plan.[25]

RHQ reapplies the short-term assistance system used for the Indochinese refugees accepted in the Japanese resettlement programme between 1978 and 2005, although the resettled refugees coming from a protracted refugee situation have different backgrounds, capacities and expectations from the Indochinese. While the resettled refugees intend to stay in Japan indefinitely, many of the Indochinese refugees initially intended to stay in Japan only temporarily before moving on to the US and other countries.

The experience of the Indochinese refugees demonstrates the inadequacy of a short-term centralized assistance programme. In 2008, the UNHCR Office in Japan and the UN University in Tokyo conducted a study on the integration of 11,000 Indochinese refugees accepted in Japan with a view to providing the Government with lessons learned from Indochinese experiences so that a better integration support system can be designed by the Government (UNHCR 2009). The study found that there have been several barriers to social integration experienced by the refugees, the Japanese language being the main obstacle. Mastering Japanese requires years of continuous learning, yet the Indochinese refugees were offered only four-month training at reception centres, which is utterly inadequate for them to become self-reliant in Japan. A small number of volunteer groups offered ad hoc language courses, but no long-term training was offered by the local governments.

Lacking language proficiency, they had few job opportunities except dirty, difficult and dangerous (3D) jobs. Such jobs were available in the 1980s when the Japanese economy was booming, but since the early 1990s, due to the long economic recession, even such jobs have become fewer. Lacking stable jobs, refugees found it difficult to find decent housing. As an apartment for a family of five would cost at least US$1000 per month, many were obliged to share small sub-standard rooms. Education of children has been a particularly serious concern. Few schools were ready to accept foreign born (refugee) children and fewer provided for special teachers for foreign students. With very high costs of schooling, opportunities for higher education are limited. The cost of completing education up to high school is up to US$120,000 and if one proceeds to university the cost could be up to US$180,000, which few refugees can afford to pay.[26] Refugees can join the national social security system, but to qualify for the national pension 25 years of annual contribution is needed by age 60 and some adult refugees will not meet this condition. Many refugees could not afford to join the medical insurance system. As to the acquisition of nationality, the requirements for Japanese citizenship are so high that many refugees have given up the idea. As of 2004, a quarter century after the influx of Indochinese refugees started, only 5 per cent of the refugees who responded to a survey had obtained Japanese nationality (Yamada 2007). Finally, refugees have suffered from social isolation in the face of the language barrier and in the absence of ethnic communities. There was no national integration support system that encourages participation of potential supporters at the local level. In the eyes of refugees, Japan is not a hospitable country as a multi-ethnic and open culture does not exist. Although many refugees appreciated the fact that Japan has provided them with protection, freedom and a level of safety, many refugees remain in the lower socio-economic bracket of society as opportunities for upward social mobility are very limited. They are aware that refugees who have been resettled in other countries like Australia and Canada have much easier access to higher education, skill acquisition and citizenship. In a survey conducted in 2004 by a group of researchers with regard to 163 Vietnamese refugees, 12 responded that they regret having come to Japan (Yamada 2007: 143).

Despite the inadequacy of the integration support system for the Indochinese refugees, it was reused by the RHQ for the new resettlement pilot programme without much improvement. The initial training period was extended from four months to six months and on-the-job training of six months was added. However, 570 hours of Japanese language training is too short for refugees who are illiterate and/or innumerate to be self-reliant. The cultural adaptation programme has been improved to include practical training, such as lessons on how to ride a bicycle. However, after the six-month training, RHQ's support is reduced to a minimum and refugees are expected to be on their own. The recent introduction of Integration Supporters is a step forward. One problem of a centralized six months training is that afterwards refugees have to move out of the metropolitan Shinjuku area where the training centre is located; thus forfeiting whatever connections they may have developed with the community during the training period. Refugees have to start a new life in a completely new environment and this has a destabilizing effect. One alternative is, as many resettlement countries do, to resettle refugees directly to the municipalities that accept refugees and provide customized integration support from the beginning with the participation of all stakeholders and potential service providers. This has not been done.

To minimize the participation gap, it is important to listen to the voices of the refugees, both in the camps and in Japan, to understand their interests, expectations, concerns and social norms and to encourage them to participate in designing the integration support system. A pre-departure survey by IOM shows that 45 per cent of the refugees are concerned about new languages, 19 per cent about different people and cultures and 8 per cent about neglect and loneliness. Around 17 per cent of the respondents stated that they have no particular concerns.[27] However, the last figure needs to be assessed in light of their unassertiveness or social norm: they would say 'I have no problem' even when they do have problems, in order to avoid troubling others.[28] As a whole, the Japanese programme is biased towards material assistance (jobs, housing, cash assistance, etc.) while psychological support to address refugees' identity, aspirations, fears, expectations, capabilities, customs, culture and social norms is weak. Participation of those who understand the importance of the psychological dimension was not sought until the problems started to emerge.

Except for the RHQ, there are few service providers that are capable of providing, or are willing to provide, integration support in areas where refugees settle. For a resettlement programme to be effective:

> Integration programmes require coordination, cooperation and collaboration. Investments need to be made at an early stage to ensure that sound

coordinating infrastructures and processes are established; that cooperative relationships are fostered between players, and relevant personnel have opportunities to develop their expertise in integration programme development and implementation. At the local level, communities must be prepared to welcome and support resettled refugees and opportunities to bring newcomers and their new community members together to build relationships and identify and address issues are critical to the programme's success. These tasks require an investment in time, resources and expertise.

(UNHCR 2011b)

Such conditions have been missing from the Japanese pilot programme. Official services have been provided exclusively by the semi-governmental RHQ and its services outside Tokyo are limited. The Government throws all the support programmes into the lap of RHQ, which is easy from an administrative point of view. NGOs have effectively been excluded from participating in the provision of services. MOFA's Invitation to Tender requires that a successful contractor provides assistance to both resettled refugees and Convention refugees (some 30 per annum in total). Only a few NGOs could meet such a requirement. Due to its long service history and closeness to the Government, only RHQ has an infrastructure to assist both groups.[29] As a result, few NGOs have submitted a proposal. On their part, NGOs have not coordinated their activities to compete with RHQ, such as creating a consortium of NGOs.

The Forum for Refugees Japan (FRJ) is a network of NGOs and other organizations that supports refugees and asylum seekers who have fled to Japan, and those allowed to remain in Japan with humanitarian status. FRJ was established in 2004 with 12 NGOs and UNHCR as members. It is a loose network of NGOs: each has its own mission and policy and FRJ does not have its own operational capability. As regards the pilot programme, some NGOs consider that improving the asylum system is more pressing than starting a resettlement programme. FRJ's role has been limited to advocacy such as issuing a joint press release urging the Government to change the selection criteria and improve the integration support projects.[30] One of the member NGOs, JAR (Japan Association for Refugees) has been most active in promoting the pilot programme. It has organized a number of national and international symposia and seminars on resettlement to learn the best practices of other resettlement countries and provide support to the resettled families. However, JAR has been overloaded by the recent sharp increase in the number of asylum seekers in Japan and cannot compete with RHQ on the resettlement front.

The Government took the position that refugees are free to choose the place of resettlement, and RHQ gave priority to finding employers who would employ refugees. It was only after employers were identified that the search for accommodation was started and the municipalities were contacted for providing support. Municipalities are not involved until they are informed that resettled refugees are arriving. Resettled refugees do not know where they will be resettled and municipalities do not know how to cope with resettled refugees who appeared suddenly in their jurisdiction, since they have not been involved.

Other than the cities of Suzuka and Misato, which accepted refugees in their jurisdiction, there were a few civil groups which expressed interest in providing integration support to refugees in cooperation with local municipalities. One of them is a citizens' movement in Matsumoto City, which is about 200 kilometres from Tokyo and has a population of 200,000. In 2008, a group of citizens with the support of the author, started a campaign to accept resettled refugees and provide integration support. Their activities included several symposiums and seminars on refugees and their integration, a film festival and two music concerts. Several members have visited New York and Maryland in the USA, to learn best practices. Knowing the importance of the support of the local government, the group approached the mayor of the city who expressed strong support. Several possible employers expressed interest in providing employment and a consultative body was established with members from three NGOs, the City of Matsumoto, a university and several employers. Notable for his support was the Mayor of Matsumoto, Dr Sugenoya, who has been supporting victims of the Chernobyl nuclear accident in 1986. He stated that he wants to make Matsumoto a Humanitarian City and one of his policies is acceptance of refugees in the city by internalizing a global concern at the local level.

However, the Government did not show interest in those activities. While the group did as much as possible, lack of interest and information on the refugees from the Government gradually alienated the group from the resettlement programme. Towards the end of 2011 when the problems with the first group of refugees had become an issue, the Government approached the Matsumoto citizens' group and RHQ sent the second group of resettled refugees to Matsumoto to meet possible employers. However, the refugees declined to settle in Matsumoto, citing cold weather and salary levels that are lower than in Tokyo. The group members were also disappointed by the impression that the refugees were quite different from what was expected. The group expected refugees to be hardworking, disciplined and purposive to overcome whatever barriers they may face, and when such expectations were not met (in a brief encounter), members were alienated. The refugees' rejection of Matsumoto as a place to settle was the last straw, and the group withdrew from the resettlement initiative. A precious local initiative ended in disappointment of all concerned.

In designing the integration mechanism, the Government has not fully utilized the expertise of UNHCR and IOM, which have not been allowed to participate in the IMWG. It was not until the beginning of 2012, facing unexpected implementation difficulties that the Government decided to establish the Resettlement Experts Committee under the IMWG and invite both organizations to participate as observers. The Government has been jealously guarding national jurisdiction over refugee issues from a global organization, UNHCR.

The Burmese community has not been able to participate in integration support effectively. One important factor in refugees' choice of destination is the presence of a community of co-ethnics. Friends and family members could provide help to new arrivals by providing information, support and sometimes livelihood. There are an estimated 100 Karen refugees in the Tokyo area;

however, they are divided into three groups. The total number of ethnic, economic, cultural or political groups of Burmese nationals would exceed 100, but their interreactions are limited.[31]

> The refugee community (in Japan) is internally divided. Burmese refugees belong to a bewildering array of various political groups with every acronym imaginable, and deciphering the numerous personal vendettas and political issues that surround the formation of each group would be akin to untangling a scouring pad into a line of thread.
>
> (Banki 2006: 42)

The fact that only Karen refugees are the resettlement target group caused frustration of other ethnic groups. Such splits make their coordinated assistance to resettled refugees difficult. Fragmentation also makes it difficult for the Government to discuss assistance issues with refugees, even if it wished to do so.

The third gap is the incentive gap. For accepting 30 refugees, the Government (MOFA, Agency for Cultural Affairs, Ministry of Labour and Welfare and MOJ) budgeted 162 million Yen (US$1.62 million at the exchange rate of $1 = 100 Yen) for initial assistance, including Japanese language training. The per capita budget is 5.4 million Yen or US$54,000, which is the second highest after New Zealand that budgets US$61,250.[32] A family of five would cost US$270,000. Most of the money was given to RHQ to be spent for the 180 days central training. On the other hand, local municipalities that host resettled refugees do not receive funding although long-term integration support costs the municipalities significantly. Most municipalities, financially and politically, cannot afford to provide additional services to refugees when their own citizens are suffering from economic difficulties. Other than material compensation, ideational incentives and reputational values, such as public acknowledgment of employers or NGOs that provide support, should have been possible but such has not been offered by the Government. The absence of incentives and compensation alienates municipalities and the civil society or individuals from joining local integration support. Production of international/national public goods did not occur at the local level.

In terms of financial incentives, the Government should consider diverting part of the large amount of money contributed to UNHCR for longer term integration support in local towns and villages. For instance, while a reduction in contributions to UNHCR of US$1 million will not have a visible impact on UNHCR, it would be a huge sum of money for local governments and employers. The 'marginal impact' would be much bigger for the same amount of money. It would also introduce more balance between asylum, resettlement and financial contributions. Public acknowledgement by the government of local support initiatives is an inexpensive yet effective incentive to encourage community participation and support. Local employers may find some refugees to be valuable workforce. Finally, the presence of refugees in the local community could well raise citizens' awareness of the grim reality of forced displacement and appreciation of how lucky and privileged they are in the contemporary world.

The way forward

As long as these three gaps remain, refugees will assume that there are fewer future opportunities in Japan than those offered by other resettlement countries. Refugees (and asylum seekers) have penetrating eyes and refugees will exercise a 'reverse selection' and 'deselect' Japan as a country of resettlement. To increase integration support in such a way as to help refugees' transition from a protracted refugee situation to urbanized life in Japan, the three aforementioned gaps have to be filled.

To fill the jurisdictional gap on resettlement at the global level, the government should internalize the global norms into the domestic decision-making process. A consolidated refugee policy that balances the relative weights between asylum and burden sharing (resettlement and financial contributions) is desirable. A perspective of an outward looking Japan is important. Excessive focus on domestic concerns and the 'ability to settle' criterion should be avoided. At the national level, Japan's resettlement needs to be internalized and 'owned' at the local level by municipal governments, employers, NGOs and volunteers who share common understanding of the value and necessity of refugee resettlement. The argument that resettlement of refugees (territorial protection) promotes the 'national interest' of Japan has to be translated into 'local interests'.

To fill the participation gap, the government should actively seek stakeholders' involvement in decision-making and management of the resettlement programme. The Government's role should be limited to deciding the quota, countries of first asylum from where refugees are resettled and resettlement-related budgets. RHQ's historical 'monopoly' needs to be replaced by a consortium of service providers not only in Tokyo but also elsewhere. The Government should identify local governments that are willing to offer integration support and network them into a club of supporters. In particular, the Government should include refugees in the decision-making process as their ideas, interests and behaviour will affect the viability of the resettlement programme. Until now, refugees have been treated as 'objects' of assistance and not as 'subjects' who have agency. Refugees transmit their perceptions to their friends and relatives in the camps and influence the decision-making of other refugees whether or not to join the Japanese programme.

The Government should encourage the Refugee Coordination Committee (RCCJ), which was established in 2010 by refugees with UNHCR support, to connect (resettled) refugees with the Government on the one hand and assistance NGOs on the other. All RCCJ members are refugees who know the culture and capacities of refugees as well as having experience and knowledge of how to survive and adapt to Japan and become self-reliant. It is significant that refugees themselves took an initiative to cooperate with each other to provide support to resettled fellow refugees. In April and June 2012, the RCCJ held two online seminars on resettlement, inviting the Vice-Minister of Foreign Affairs, a prominent parliamentarian and an asylum lawyer. There were thousands of viewers and that exceeded expectations. It was another example of refugee agency.

On the part of civil society, concerned NGOs should develop shared views on the value of resettlement for host communities, and define respective roles and responsibilities, such as advocacy, fund-raising, research and assistance activities. A positive development is that, in 2008, a consortium of NGOs called the Forum for Refugees Japan (FRJ) was created under the guidance of UNHCR. FRJ is a network of 13 NGOs that coordinates their activities for refugees and asylum seekers in Japan. FRJ achieved a milestone in 2012 by reaching an agreement with the MOJ and the Japan Bar Association that all parties will join efforts to improve the RSD process and assistance to asylum seekers, such as finding alternatives to detention. This tripartite agreement is epoch-making and will have a significant impact on the future of the Japan's refugee policy, including the resettlement programme.

UNHCR should provide the government with a more detailed analysis of social and cultural profiles of refugees in the camps. The pilot programme was started with several rather naive assumptions about refugees' aspirations, capacities, social relations and cultural practices, creating misunderstanding and confusions. UNHCR should also help the government identify reasons for the unpopularity of the Japanese resettlement programme by asking refugees who have chosen other destination countries. Asking refugees who have come to Japan why they came would not be meaningful, while asking refugees who deselected Japan why they opted for other countries would reveal a lot about the reasons for 'Japan passing.' Third, UNHCR should provide interested municipal governments with rationales why accepting refugees is in their interest, taking into account local conditions. This is a challenging task and UNHCR should learn more about Japan's local communities in terms of their political, social and economic opportunities and constraints. Finally, UNHCR should persuade the government to use part of the annual financial contributions to UNHCR for the resettlement programme in such a way as to benefit or compensate the municipal governments that accept refugees.

To fill the incentive gap, the Government should increase direct funding to the municipalities that accommodate refugee families. The argument that resettlement is in the national interest of Japan is not sufficient to persuade local people who ask 'why should we accept refugees in our town?' Local interests can be ideational, such as public honouring of employers by the Ministers, which would go a long way in encouraging local actors. Local goodwill should be acknowledged. Considering refugee issues can be highly educational for students, opening their eyes and giving them a better understanding of Japan's position in the global society.

Finally, the Government must improve its information and communication strategy on resettlement, so that a shared view is created among the government, the refugees, the host community and civil society. Good communication is a key to any public policy. The Government's very restrictive information and communication policy at the beginning of the pilot programme has caused confusion, suspicions and doubts on the part of the refugee community and civil society. While this reflects traditional Japanese bureaucratic culture, the restrictive communication policy, lack of transparency and accountability did not promote understanding of the values of resettlement and damaged the credibility of the programme. Only when sustained efforts are made to inform and involve local governments, civil society, the general public and refugees themselves, will resettlement as an international public good be produced at the local level and the resettlement programme will survive.

Conclusion

The Japanese Government's decision to start a pilot resettlement programme was a surprise to the humanitarian community. It was made by a small number of policy elites who realized that by responding to a call from UNHCR, the criticism that Japan is a free-rider on the Global Refugee Regime can be addressed and its national interest in terms of reputational value would be promoted. However, domestic implementation of the programme was much more difficult than many expected. Territorial protection of refugees in the form of resettlement does not necessarily offer human security for refugees unless a robust domestic integration support system is established. Such a system is a 'national public product' to be co-produced by the Government, local municipalities, civil society and the general public. Unfortunately, such a system does not yet exist in Japan as shown by the three gaps (jurisdictional gap, participation gap and incentive gap) which form structural barriers. Due to the poor communication and information strategy adopted by the Government, there are no shared interests among the supporters and the public at the local level. Rather than focusing on the technical 'fixes' like selection criteria, the government should look at the big picture and fill the three gaps mentioned earlier.

The refusal by the three refugee families to come to Japan, or 'Japan passing', was a shock and a turning point for the pilot programme that forced all concerned parties to reconsider the past approaches and re-examine hitherto untested assumptions and mindsets. However, there are also signs of hope. No doubt there are many individuals who have goodwill to help displaced refugees. Refugees and civil society are becoming more active than before. The Government has also changed its approach and increased transparency and communications with the civil society. If these genuine efforts continue, the pilot programme has a chance to overcome the 'pain of birth' and one can even hope that the resettlement programme becomes one of the keys to 'open the doors of Japan to the global society', as Minister Nakagawa proudly mentioned on World Refugee Day, 20 June 2012. (This article covers developments up to January 2013).

Notes

1   When an asylum seeker's request for recognition of refugee status is rejected, he/she can file an appeal to the Minister of Justice, who is required to ask for the opinions of

RACs in making decisions on filed appeals. So far all the RACs' opinions have been respected by the Minister. As of January 2013 there are 56 RACs who are appointed from among academic experts on legal issues or international affairs. RACs are divided into groups of three to examine appeal cases. See www.moj.go.jp/ENGLISH/ IB/ib-01.html (accessed 10 January 2013).

2 Cabinet Office website, www.cas.go.jp/jp/seisaku/nanmin/konkyo.html.
3 Calculated from statistics compiled by the Japan Lawyers Network for Refugees (JLNR) www.jlnr.jp/stat/index.html accessed 13 July 2012.
4 Mukae argues that joining the 1951 Convention was a tactical concession by Japanese bureaucrats to ensure that Japan could still retain its fundamental isolationism by maintaining sovereign control over refugee status determination and its domestic refugee policy.
5 Official records of the 169th National Diet, available at http://kokkai.ndl.go.jp/.
6 www.cas.go.jp/jp/seisaku/nanmin/081216ryoukai.html.
7 MOJ, NPA and the city of Tokyo have been conducting a joint campaign to reduce the number of irregular stayers, by which the number was reduced from some 300,000 in 1993 to 78,000 in early 2012. In the course of the campaign, a large number of asylum seekers (possibly refugees) who did not have regular visa status have been arrested. This is one of the reasons for the recent surge in the number of asylum claims in Japan. On crimes by foreigners, see Okada 2007: 5–19. On crime statistics, see www.moj.go.jp/ nyuukokukanri/kouhou/nyuukokukanri04_00016.html (accessed 10 January 2013).
8 Email comments from DG Inami to the author.
9 Cabinet Understanding of 16 December 2008 on the implementation of a third-country resettlement programme on a pilot basis, available in Japanese at www.cas. go.jp/jp/seisaku/nanmin/081216ryoukai.html (accessed 10 January 2013). Also, see a presentation by a MOJ official in charge of RSD at the second Resettlement Experts Meeting of 19 June 2010.
10 Records of the seventh meeting of the Resettlement Experts Council (REC) of 7 November 2012.
11 A paper presented by the Senior Protection Officer of the UNHCR Office in Japan at the first meeting of REC on 8 May 2012.
12 Confirmed by the former Minister of State (in charge of resettlement), Masaharu Nakagawa (February 2012–October 2012), on 13 October 2012 in his lecture on resettlement at Toyo Eiwa University.
13 Author interview with Ms Marip Seng Bu, Director, Refugees Coordination Committee in Japan (RCCJ), 14 March 2012.
14 In August 2011 the author together with a dozen students visited Mae La camp and met the members of the second group. When the second group arrived at the Narita Airport, some of the students welcomed them there, but there were no visible Burmese community members.
15 RHQ resettlement staff at the sixth meeting of REC of 7 November 2012.
16 UNHCR staff remark at the sixth meeting of REC of 7 November 2012.
17 Ibid.
18 A comment made by a senior staff of RHQ in a press interview with a Burmese newspaper. See Boehler 2012.
19 Res. 2, adopted at the House of Representatives, 17 Nov. 2011/Res. 1, adopted at the House of Councillors, 21 November 2011.
20 Representative of the Cabinet Office at the sixth meeting of REC of 17 November 2012.
21 Minister Nakagawa's comment in 11th meeting of Migrant and Refugee Studies on 13 October 2012 at Toyo Eiwa University.
22 Records of the sixth meeting of REC.
23 Cabinet Understanding of 16 December 2008 on the implementation of a third-country resettlement programme on a pilot basis, available in Japanese at www.cas. go.jp/jp/seisaku/nanmin/081216ryoukai.html (accessed 10 January 2013).

24 UNHCR briefing material submitted to the first meeting of REC of 8 May 2012.
25 Comment of a convention refugee at the seventh meeting of REC of 5 December 2012.
26 Data from Cabinet Office Annual Report 2004 available at www8.cao.go.jp/shoushi/whitepaper/w-2004/html-h/html/g1223120.html (accessed 10 January 2013). 1 US$=90 Yen. Data still relevant.
27 IOM presentation at the second meeting of REC of 19 June 2012.
28 Ibid.
29 Two of RHQ's senior managers are seconded from MOFA.
30 FRJ press release (Japanese), dated 1 October 2012.
31 Comment of Professor Kei Nemoto at the seventh meeting of REC of 5 December 2012.
32 MOFA presentations at the second and sixth meetings of REC of 19 June and 7 November 2012.

References

Asahi Shimbun Digital, 2012. 'Japan unlikely to reach goal for Myanmar refugees', Daisuke Furuta, 17 February 2012.
Banki, S. 2006. 'The triad of transnationalism, legal recognition, and local community: Shaping political space for the Burmese refugees in Japan', *Refuge: Canada's Periodical on Refugees* 23(2): 36–46.
Banki, S and Lang, H. 2008. 'Difficult to remain: The impact of mass resettlement', *Forced Migration Review* 30: 43–44.
Barkin, J.S. 2010. *Realist constructivism: Rethinking international relations theory*. Cambridge: Cambridge University Press.
Baylis, J., Smith, S. and Owen, P. (eds) 2008. *The globalization of world politics: An introduction to international relations* (4th edn). Oxford: Oxford University Press.
Betts, A. 2009. *Forced migration and global politics*. Oxford: Wiley-Blackwell.
Betts, A. and Loescher, G. (eds) 2011. *Refugees in international relations*. Oxford: Oxford University Press.
Boehler, P. 2012. 'Burmese refugees face tough time in Japan'. *The Irrawaddy*, 19 April.
Bowles, E. 1998. 'From village to camp: Refugee camp life in transition on the Thailand–Burma Border'. *Forced Migration Review* 2: 11–14.
Brettell, C.B. and Hollifield, J.F. 2008. *Migration theory: Talking across disciplines* (2nd edn). London: Routledge.
Brooks, H.K. 2004. 'Burmese refugees in Thai border: The provision of humanitarian relief to Karen refugees on the Thai–Burmese border', in Andrzej Bolesta (ed.) *Conflict and displacement: International politics in the developing world*. Libra, pp. 107–110.
Brown, C. and Ainley, K. 2009. *Understanding international relations* (4th edn). London: Palgrave Macmillan.
Castles, S. and Miller, M.J. 2009. *The age of migration: International population movements in the modern world* (4th edn). London: Palgrave Macmillan.
Dean, M. and Nagashima, M. 2007. 'Sharing the burden: The role of government and NGOs in protecting and providing for asylum seekers and refugees in Japan' *Journal of Refugee Studies* 20(3): 481–508.
Flowers, P.R. 2009. *Refugees, women, and weapons: International norm adoption and compliance in Japan*. Stanford: Stanford University Press.
Haddad, E. 2008. *The refugee in international society: Between sovereigns*. Cambridge: Cambridge University Press.

Hatano, Y. 2010. (Former Japanese Ambassador to the UN Organizations in Geneva) Essay in *Ouyuu Kaihou*, No. 97, December 2010, p. 7.

Helton, A.C. 2002. *The price of indifference: Refugees and humanitarian action in the new century*. Oxford: Oxford University Press.

Japan Broadcasting Corporation (NHK), World Network, Myanmar Refugee Resettlement, 26 March 2012.

Japan Forum on International Relations 2010. *Prospects and challenges for the acceptance of foreign migrants to Japan*, November 2010. Available at: www.jfir.or.jp/j/pr/pdf/33.pdf (accessed 20 January 2013).

Japan Lawyers Network for Refugees 2011. Letter to the Minister of Foreign Affairs dated 26 September 2011. Available at: www.jlnr.jp/statements/20110926_mofa.pdf (accessed 20 January 2013).

Kingston, J. 2007. 'Diplomat rues Tokyo's "lack of humanity" to asylum seekers', *Japan Times*, 8 July. Available at: www.japantimes.co.jp/text/fl20070708x1.html (accessed 20 January 2013).

Kaul, I., Grunberg, I. and Stern, M.A. (eds) 1999. *Global Public Goods: International cooperation in the 21st century*. Oxford: Oxford University Press.

Kipgen, N. 2008. 'Japan's action changes Asia's image', *The Korean Times Opinion*, 29 December.

Klotz, A. and Lynch, C. 2007. *Strategies for research*, in Constructivist International Relations series, M.E. Sharpe.

Kubalkova, V., Onuf, N. and Kowert, P. (eds) 1998. *International relations in a constructed world*, M.E. Sharpe.

Liberal Democratic Party 2008. Gaikoku Jinzai Kouryu Suishin Renmei, *Jinzai Kaikoku: Nihongata Imin Seisaku no Teigen (A proposal for a Japanese-style immigration policy)*, 12 June.

Loescher, G. 2001. *The UNHCR and world politics: A perilous path*. Oxford: Oxford University Press.

Loescher, G., Betts, A. and Milner, J. 2008. *The United Nations High Commissioner for Refugees (UNHCR): The politics and practice of refugee protection into the twenty-first century*. London: Routledge.

Loescher, G. and Milner, J. 2008. 'Burmese refugees in South and Southeast Asia: A comparative regional analysis', in Loescher, G., Milner, J., Newman, E. and Troller, G. (eds) *Protracted refugee situations*. Tokyo: United Nations University Press.

Loescher, G., Milner, J., Newman, E. and Troller, G. 2008. *Protracted refugee situations: Political, human rights and security implications*. Tokyo: United Nations University Press.

Ministry of Foreign Affairs (MOFA) 2011. ODA Report 2011. Available at: www.mofa.go.jp/mofaj/gaiko/oda/shiryo/hakusyo/11_hakusho_pdf/pdfs/11_hakusho_0401.pdf (accessed 20 January 2013).

Ministry of Health, Labour and Welfare 2010. National Living Conditions. Available at: www.mhlw.go.jp/toukei/saikin/hw/k-tyosa/k-tyosa10/2-7.html (accessed 20 January 2013).

Ministry of Justice 2012. '*Heisei 23nen Ni Okeru Nannminn Ninteishasu Nadoni Tsuite (Number of refugee status recognitions in 2012)*'. Available at: www.moj.go.jp/nyuu-kokukanri/kouhou/nyuukokukanri03_00085.html (accessed 9 March 2013).

Morowasa, H., Dussich, J.J.P. and Kirshhoff, G.F. (eds) 2012. *Victimology and human security*. Oisterwijk: Wolf Legal Publishers.

Mukae, R. 2001. *To be of the world: Japan's refugee policy*. Florence: European Press Academic Publishing.

National Police Agency 2011. *Statistics on suicide during 2010*. March 2011. Available at: www.npa.go.jp/safetylife/seianki/H22jisatsunogaiyou.pdf (accessed 20 January 2013).

Newman, E. and van Selm, J. (eds) 2003. *Refugees and forced displacement: International security, human vulnerability, and the state*. Tokyo: United Nations University Press.

Okada, K. 2007. *Gaikokujin to Hanzai (Foreigners and Crime)*. Reference, National Dire Library, July.

Oshige, M. 2011. 'Myanmar refugees in Japan find adjusting to new life difficult', *Daily Yomiuri Online*. 6 August 2011. Available at: www.yomiuri.co.jp/dy/national/T110805006342.htm (accessed 20 January 2013).

Takizawa, S. 2011. 'Refugees and human security: A research note on the Japanese refugee policy', *Journal of the Graduate School of Toyo Eiwa University* 7: 21–40. Available at: http://ci.nii.ac.jp/naid/110008426208/ (accessed 20 January 2013).

TBBC (Thailand and Burma Border Consortium) 2011. Programme report.

UNHCR 2003. *Agenda for protection*. Third edition. Available at: www.unhcr.org/refworld/docid/4714a1bf2.html (accessed 3 March 2013).

UNHCR 2004. Protracted Refugee Situations, Standing Committee to the Executive Committee of the High Commissioner's Program, thirtieth meeting, EC/54/SC/CRP.14, 10 June. Available at: www.unhcr.org/refworld/docid/4a54bc00d.html (accessed 20 January 2013).

UNHCR 2009. *A Report on the local integration of Indo-Chinese refugees and displaced persons in Japan*. Available at: www.unhcr.or.jp/protect/pdf/IndoChineseReport.pdf (accessed 20 January 2013).

UNHCR 2011a. *Asylum levels and trends in industrialized countries in 2011*. Available at: www.unhcr.org/4e9beaa19.html (accessed 20 January 2013).

UNHCR 2011b. *UNHCR Resettlement Handbook*. Geneva.

UNHCR 2012. 'Financial updates: Top donors'. Available at: www.unhcr.org/pages/49c3646c27e.html (accessed 6 February 2013).

UN Treasury n.d. 'UN operational rates of exchange'. Available at: http://treasury.un.org/operationalrates/Default.aspx. (accessed 10 January 2013).

USCRI 2011. *Help end human warehousing*. Available at: www.refugees.org/our-work/refugee-rights/warehousing-campaign/ (accessed 20 January 2013).

Weiss, T.G. and Daws, S. (eds) 2007. *The Oxford handbook on the United Nations*. Oxford: Oxford University Press.

Weiner, M. 1995. *The global migration crisis: Challenges to states and to human rights*. London: Harper Collins College Publishers.

Wendt, A. 1999. *Social theory of international politics*. Cambridge: Cambridge University Press.

Wight, C. 2006. *Agents, structures, and international relations: Politics as ontology*. Cambridge: Cambridge University Press.

Yamada, H. (ed.) 2007. *Nihon no Nanmin Ukeire: Kako, Genzai, Mirai* (Japan's Acceptance of Refugees: Past, Present and Future). Tokyo Foundation.

Yamashita, N. 2013. 'Local integration of Myanmar refugees resettled in Japan'. Graduate School of Toyo Eiwa University, unpublished mimeograph.

Zolberg, A.R. and Benda, P.M. (eds) 2001. *Global migrants, global refugees*. New York/Oxford: Berghahn Books.

10 Coping as an asylum seeker in Japan

Burmese in Shinjuku, Tokyo

Koichi Koizumi

Introduction

Japan is a country that has seen relatively few asylum seekers compared to other industrialized countries such as the US and Europe. For various reasons, Japan appears to be a less favoured destination (see Takizawa, this volume). Nevertheless, numbers have been increasing in recent years, from an average of around 50 cases per year until the 1990s, to an all-time high of 1,867 in 2011 (Ministry of Justice 2012, Japan Immigration Association 2012). The largest single group of applicants for refugee status in Japan come from Burma. Indeed, the Burmese form a significant proportion of all asylum seekers in Japan. Many of the Burmese live in the Tokyo Greater Metropolitan area and are particularly concentrated in the Shinjuku Ward of the city, which is the focus of this study.

Research on refugees in Japan has mainly focused on the analysis of legal issues and the application process itself (see for example Honma 2005; Abe 2010). While terms such as 'asylum seeker' and 'refugee' have strict legal definitions, this chapter is not concerned with legal categories as such. The aim of this chapter, by contrast, is to examine the problems encountered by Burmese applicants for refugee status in daily living in Japan. It is a description of how their lives and thoughts (e.g. about work experience, social networking, community participation, connections with friends and family in their home countries, long-term goals and hopes) have changed with the passage of time during their stay in Tokyo/Shinjuku. The uniqueness of Burmese applicants' individual experiences interweaves and creates a clear picture of a small part of their complex social world. The reason Burmese refugees have been chosen is because, as a fixed group of people visible in society, it is easier to discern how they change socially over time and how they cope as a community. The insecure position these people occupy in Japan affects the choices they can make in their daily lives, in both physical and mental terms, which are also affected by many other factors.

The investigative methods used in this study incorporate examination and analysis of the reports produced by private organizations assisting the applicants as well as interviews conducted with leaders and individuals within the refugee applicant community, related private groups, local authorities, governmental institutions and people involved in international organizations. Information on official governmental decisions was obtained from the relevant websites.

Seeking asylum in Japan

It is easy to enter Japan on a short-term visa, and once there to apply for asylum, which is free of charge. Applicants who receive a negative decision may appeal within seven days. Applicants whose appeal is rejected may within the next six months apply for a judicial review, in which they must be legally represented. While going through the application process, applicants may be granted Permission for Provisional Stay, which does not include the right to work. Material assistance is available for those considered eligible by the Refugee Assistance Headquarters (UNHCR 2005).

In 2011, the countries with the largest number of applicants for refugee status in Japan were as follows: Burma (491), Nepal (251), Turkey/Kurdistan (234), Sri Lanka (224) and Pakistan (169). The applicants appealing decisions included 444 from Burma, 231 from Sri Lanka, 213 from Turkey, 191 from Nepal and 142 from Pakistan; Burmese comprise 25.8 per cent of the total. In the same year, 2,999 decisions were made (including those on applications filed previously) and 21 applicants were granted refugee status, a recognition rate of 0.7 per cent, including 14 applicants on appeal. This number comprises 18 refugees from Burma and three from other countries. Meanwhile, 248 applicants were granted residence on humanitarian grounds, bringing the total granted asylum to 269. The number of Burmese granted asylum (i.e. either refugee status, or residence on humanitarian grounds) was 214, approximately 80 per cent of the total, which included applicants from 17 other countries. Thus, comparatively more Burmese are granted asylum; however, the total number of those granted refugee status, including Burmese, is falling (Ministry of Justice 2012) (Figure 10.1).

Applicants for refugee status in Japan face a variety of obstacles. These include navigating the application process, obtaining permission for special residence status, managing work conflicts and handling problems with accessing medical and education facilities. Applicants for refugee status who are not recognized as eligible for assistance must work, regardless of whether they have legal permission to do so. However, their employment status is frequently unstable. As the economy continues to slump, the problems faced by foreign residents generally, such as employment and education, become all the more serious.

Simply applying for refugee status in Japan does not automatically grant one any rights for everyday life. Government assistance depends on having residence status.[1] While the applicants wait for a decision, most cannot work legally and must depend on the government for living and housing expenses. In the case of an illness, the applicants have to pay medical costs out of their own pocket and hope for reimbursement from aid organizations at a later date. Furthermore, many Japanese language education programmes are restricted to those who have already been granted refugee status, and there are limited places available to those still going through the application process. Those who have no official

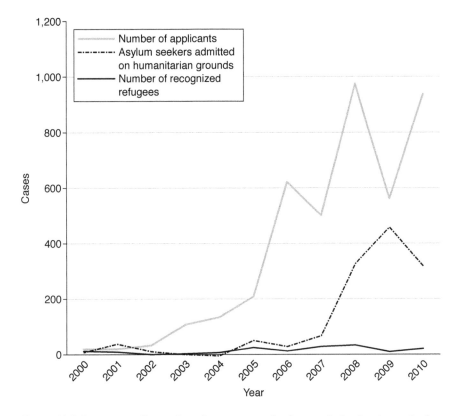

Figure 10.1 Burmese applicants for refugee status and refugees admitted on humanitarian grounds (source: On Refugee Numbers and Others, by the Ministry of Justice, and UNHCR statistics).

status of residence and have been granted provisional release cannot work legally; those who are illegally residing or working in Japan live under constant fear of being discovered by the immigration authorities or police. Living with such a degree of stress causes some to develop mental disorders (interview with Taeko Kimura, Asian Friendship Home).

The situation of the Burmese

To date, there has been no established economic field in Japan for the Burmese and no foundation to help support their culture. Most of them were forced to leave Burma and had very limited options as to where they could go, Japan being relatively easy to enter. At present, most Burmese residing in Japan were born between 1960 and 1989, so are in their twenties through to early fifties. Burmese who have registered as foreign residents are 8,366 in total (as of 2010); however, it is estimated that the actual number of Burmese residents is 12,000 (comment from an official in the Ministry of Justice). The Burmese have various residence statuses and are largely concentrated in the Greater Tokyo Metropolitan Area. Sixty per cent live in Tokyo and the rest in other areas including Aichi, Chiba, Kanagawa Saitama and Shizuoka. In Tokyo, many reside in Takadanobaba, Ike-bukuro, Otsuka, Sugamo and Nakano. In the past, many resided in the Nakai area of Shinjuku ward; however, they subsequently moved to the more easily access-ible Takadanobaba area of Shinjuku ward (see Figure 10.2; Shinjuku Self-governing Creation Institute, Working Group Report on Foreigners 2012 (3): 21).

Burmese citizens comprise 135 ethnic groups recognized by the Burmese Government (Chin, Kachin, Karenni, Shan and so on). The majority ethnic Burmese, although belonging to the same state, do not necessarily identify with members of the ethnic minorities. There is a tradition of strong bonds and mutual assistance within each ethnic minority. While there are at present approximately 30 Burmese democratic organizations, there is not much exchange between dif-ferent ethnicities. The tension between the different ethnic groups in Burma is also perpetuated among the Burmese living in Japan. In other words, there exist differences within the group recognized as Burmese. Wherever necessary, names of individual ethnic groups have been used in this chapter. Where differentiation is not relevant, the general term 'Burmese' has been applied.

The setting of Shinjuku

According to the Ministry of Justice statistics on foreign registration, in October 2010, registered foreigners in Japan numbered 2,134,151; this is a small percentage,

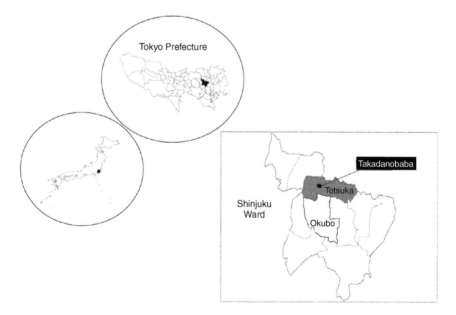

Figure 10.2 Overview of Japan and Shinjuku.

only 0.02 per cent of the Japanese population of 125,820,000. Of these, the Burmese living in Japan numbered 8,366 as of 2010 (Japan Immigrant Association 2010). They were mostly concentrated in Tokyo (5,163 or 61.7 per cent of the total Burmese population), followed by 473 in Aichi, 405 in Chiba, 390 in Kanagawa and 314 in Saitama (ibid.).

The statistical methods differ between the Ministry of Justice and the Tokyo Metropolitan Government, and the statistics for 2010 on Burmese residents differ slightly between the two sources. According to the latter source, there were a total of 5,174 Burmese residents in 2010, 4,926 living in Tokyo's 23 special wards and 247 in the outer metropolitan areas. The Burmese were over-whelmingly concentrated in the central 23 wards, with the top three wards in terms of Burmese immigrant numbers being Shinjuku (1,128), Toshima (933) and Kita (668) (Tokyo Metropolitan Government 2010, Japan Immigration Association 2010).

Compared with the rest of Tokyo and with the country as a whole, Shinjuku is remarkable for the increase in its foreign resident population, both absolutely and as a proportion of the population. Foreign residents in Shinjuku ward have been increasing for the past 30 years and 11 per cent of the total population of 320,000 are foreign nationals. The main reason for this was a wave of foreign immigrants in the late 1980s, particularly 1985–1990. These included foreigners working in the entertainment district of Kabukicho, who moved to the neigh-bourhood of Okubo in order to be within a short walking distance of their work.

Burmese citizens rank third in the numbers of foreigners registered as living in Shinjuku, with 1,153 people as of 1 January 2012 (Shinjuku Self-Governing Creation Institute 2012: 20–21), following (North and South) Koreans and Chinese. Numbering only 11 in 1980, the Burmese significantly increased in the first half of the 1990s and fell slightly in 2005;[2] however, after the immigration reforms in 2007, the numbers rose again. The increase in 1990–1992 was due to democratic activists and minorities leaving Burma because of the military junta taking control in 1988. The rise from 2008 to 2009 was due to the increase in Burmese applicants for refugee status because of large-scale demonstrations in Burma's former capital Yangon in 2007, and subsequent crackdown by the junta. The Burmese population of Shinjuku ward is significantly affected by swings in the immigration system, changes in the refugee control system and the political situation back in Burma (see Figure 10.3).

Estimates in Shinjuku ward (2010) show roughly 60 per cent of Burmese to be refugees (including applicants for refugee status as well as those with refugee or humanitarian status). In total, 373 are employed under the provisions of Designated Activities,[3] and of the other 266, 148 are Special Permanent Resi-dents (Shinjuku Self-governing Creation Institute 2012 (3): 20). Upon being suc-cessfully granted refugee status, they are expected to register as aliens; hence, this number is thought to be close to the number of actual applicants. When the Burmese apply for refugee status, they register in Shinjuku Ward as 'no status of residence'. Many of them work in part-time service jobs at places such as bars or hotels. The Burmese are thought to obtain special residence permission for

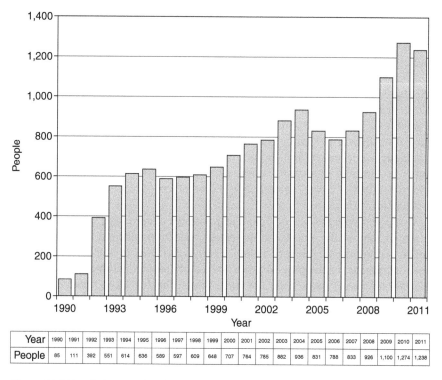

Year	1990	1991	1992	1993	1994	1995	1996	1997	1998	1999	2000	2001	2002	2003	2004	2005	2006	2007	2008	2009	2010	2011
People	85	111	392	551	614	636	589	597	609	648	707	764	785	882	936	831	788	833	926	1,100	1,274	1,238

Figure 10.3 Population trends of Burmese registered aliens in Shinjuku Ward (source: Report on Research 2011, by Shinjuku Self-administered Centre for Research).

humanitarian reasons under the Designated Activities status by way of the refugee application process. This practice began after the Immigration and Refugee Act was revised in 2005.

Through surveys on residential situations and daily life, the administration of Shinjuku ward has a good grasp of the administrative needs of its foreign population; however, in 2010, short-term residents staying for less than five years were estimated to be over 70 per cent of the total, and of these, 30 per cent stayed only one–three years. Medium-term residents, who stayed between five and 10 years, were mainly in their 30s and 40s, comprising less than 70 per cent of the total in this age bracket. Long-term residents, who stayed over 10 years, were in their 40s and constituted about 50 per cent of the total in this group. Compared with residents from other countries, on an average, medium- and long-term residents are young. The Burmese short-term residents have fewer members in their twenties and a proportionally higher number in their thirties (Shinjuku Self-governing Creation Institute 2012 (2): 18–19).

Segregating things by household status, single-person households are just under 90 per cent of the total. Compared with other foreign nationals, the proportion of couple-only or couple-with-child households is slightly higher and single-parent households is proportionally low (ibid.: 18). In terms of gender breakdown, out of the 1,274 total Burmese (Shinjuku Ward, Foreign Alien Resident Registration 2010), 808 are male and 466 are female, amounting to men outnumbering women by almost 2:1. In terms of age breakdown, over 80 per cent are concentrated in the 20–40 demographic, with 668 men and 383 women. By examining both status of residence and gender, 70 per cent of men are thought to be refugee applicants, without residence, Designated Activities or Special Permanent Resident status. More than 40 per cent of women, who are not thought to be undergoing the recognition process, have residence status such as Employment, Exchange Student and Dependant Resident status, which is a higher proportion than men (ibid.: 16). Among those interviewed, a number of men came to Japan by themselves, received refugee status and afterwards married Burmese women who were residing in Japan under Exchange Student Visa status.

As noted earlier, after the 2005 reform of the Immigration and Refugee Act, Designated Activity status began to be given to refugee applicants, including the Burmese; however, when examining their residence status between 2005 and 2010, while those thought to have come as refugees, with Employment Qualification status, Designated Activity status or Special Permanent Resident status increased greatly, there was a large decrease in those coming with temporary visitor status or 'other' (many of these registered as foreigners but have no residence status). The category that saw the greatest increase was the Designated Activities status, in which even applicants turned down for refugee status were given this special residence permission, which allows them to perform designated activities for humanitarian reasons (Figure 10.4). Once certain conditions are met, changing from Designated Activities to Special Permanent Resident is possible (ibid.: 17).[4]

In recent years, the reason that the 'other' category has been decreasing is thought to be because those illegally residing in Japan had obtained Special Permanent Resident or Designated Activity status by applying through the recognition process. Takadanobaba, in the Totsuka area, is witnessing an increase in those with Employment Qualification, Designated Activities and 'other', as noted below.

Takadanobaba neighbourhood

Takadanobaba, located within Shinjuku Ward, is known as Tokyo's 'Little Yangon'. Burmese restaurants can be found around the train station and there are approximately 20 eateries, general stores and beauty establishments run by Burmese. The Burmese number roughly 500 in Takadanobaba alone and, when areas such as Takada in Toshima are added, close to 1,000 of them live in this neighbourhood. Takadanobaba Blocks one through four are particularly representative of this population increase. There has been no change in the residential areas where Burmese live but their population has increased, particularly in central Totsuka and the surrounding areas (ibid.: 19).

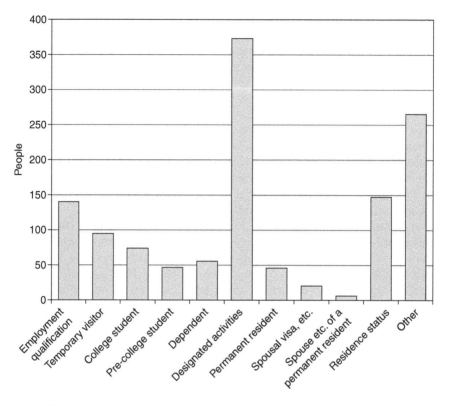

Figure 10.4 Burmese in Shinjuku by residence status 2010 (source: Report on Research 2011, by Shinjuku Self-administered Centre for Research).

In 1990, many Burmese, beginning with democratic activists, began to con-centrate in the Nakai area, also situated in Shinjuku Ward. The 'Burmese created a community centred around community stewards and monks and would all come together at food establishments or beauty parlours, creating an environ-ment where it was easy for them to congregate' (Shinjuku Self-governing Cre-ation Institute 2012 (3): 21). However, after this, their main congregation spot moved to Takadanobaba.

The question is why do they live in Shinjuku ward and why do so many decide to live specifically in Takadanobaba? According to Kawamura (2008: 75–109), Shinjuku ward is the district where the greatest multiculturalization and multiracialization has occurred. It is in the centre of Tokyo and has many inter-national corporations. Many people temporarily stay in the city and come and go every day. Particularly after World War II, this multi-ethnic consciousness was heavily affected by the presence of *zainichi* Koreans (Koreans who settled per-manently in Japan over the course of the twentieth century). Shinjuku's Kabuki-cho district developed as an entertainment centre due to the energy of young people who came to Tokyo from the countryside during the economic boom of the 1950s and 1960s. Moreover, after 1980, people from other Asian countries began coming to Japan, creating new 'ethnic towns'. In the 1990s, the bursting of the economic bubble and housing market decline supported the advance of foreigners. From the middle of the 1990s, managers of apartment buildings were forced to accept foreigners as residents or leave rooms empty (Inaba 2008: 66).[5] Offices and storefronts for lease were the same; real estate companies hired foreign staff and produced multilingual signs and advertisements.

The Okubo area is not historically thought to be an area that invited foreigners; however, it became a place where foreigners resided for the following reasons: (1) there were many temporary visitors who came to Japan with the help of people of the same nationality. These people went through Shinjuku temporarily to eventually reach other nearby prefectures. In other words, Shinjuku became the 'gateway' to Japanese society. (2) There are many Koreans, but numbers of different ethnicities increased in the 1980s and more than 90 nationalities were represented among registered residents. International marriages increased and there were also children without nationalities (stateless) and visa overstayers. (3) Many people who wanted the anonymity of large cities moved so they could keep their circumstances private (Kawamura 2008: 80).

The beginning of the 1990s saw several incidents of intra-foreigner fights and murders in the community. Local area councils asked for help from authorities and police, and efforts were made to create multicultural neighbourhoods. From the side of local community, efforts were made to make the neighbourhoods more friendly to them. The need for attention to this issue at the national level was felt at the beginning of the twenty-first century, when focus shifted from foreigners passing through the country to taking up permanent residence (ibid.: 81).

Roughly 35,000 foreigners from 118 countries are residing in Shinjuku (as of 2010). For this reason, Shinjuku ward, as an active measure to foster the characteristics of a place where many foreigners live, is pushing forward its commitment to 'multi-cultural community building'.[6] As the foreign population increased, the ward had to create policies that treated them the same as Japanese nationals and provided them with necessary administrative services. However, there was no division specializing in refugees or refugee applicants and no appropriate financial aid or welfare support system. In 2005, Shinjuku ward created the 'multi-cultural plaza' and a platform for exchange, teaching Japanese to foreign residents and providing help in Korean, Chinese, English, Thai and Burmese. The 'Foreign Resident Comprehensive Counselling and Aid Centre' was established and provided a one-stop shop for consultations on matters such as status of residence and issues affecting the livelihood of immigrants. In 2009, there were 63 cases of Burmese going for consultations, although the content of these consultations was not disclosed.

During the interviews, various points regarding the large population of Burmese in Takadanobaba came up. First, because it is close to many train lines, such as the JR Yamanote line, Seibu Shinjuku line and Tozai Subway line, it is thought to be easy for people to assemble there from various regions. The networks among Burmese are strong and since they meet frequently, information travels quickly among them. Takadanobaba is home to several municipal centres open to the public where the Burmese are able to gather, such as the Shinjuku Recycling Activity Centre and the Consumer Activity Support Centre. In general, the rent in Takadanobaba is relatively inexpensive compared to elsewhere in Shinjuku and landlords are willing to rent rooms to foreigners. There are many Burmese restaurants; moreover, it is easy for immigrants to commute to them, and they also offer good places for employment. Various places of worship are located in the area including Islamic mosques, Buddhist temples and Christian churches, serving the many ethnicities that make up the Burmese community.

Thus, even if the Burmese dispersed to different regions, it appears likely that they would gather with their fellow countrymen in the area. To the Burmese, Takadanobaba is the central node of their national network.

Multiple problems

What sorts of problems are faced by applicants for refugee status such as the Burmese? What effect does the city environment of Tokyo have on their movements and decisions? The movements of applicants after they immigrate have a serious impact on their lives as refugees.

First, let us examine the consultation services offered by aid groups. In 2009 the Refugee Assistance Headquarters (RHQ), a governmental aid organization, provided 29,180 consultations, of which 16,792 were for refugee applicants and 12,388 for special permanent residents (Foundation for the Welfare and Education of the Asian People 2009: 53). From July 2009 to June 2010, the Japan Association for Refugees, which specializes in supporting applicants for refugee status (including Burmese), handled 11,050 consultation cases, 6,355 legal aid cases and 4,695 livelihood support cases (Japan Association for Refugees 2010). As not all applicants come to these organizations, the numbers are not entirely comprehensive but show a rough picture of the situation.

Many consultations focused on life issues related to poverty and the refugee application process. Legal consultations focused on refugee recognition and status of residence issues. The status of residence or lack thereof is related to work, social and medical insurance and access to all public services. Most problematic was the legal prohibition on employment for those with provisional stay or provisional release status. Specifically, they cannot work, have no social security and cannot benefit from the national health insurance system. All material support for their livelihood is thus cut off.

Here I would like to provide a space to present the voices of applicants concerning the various problems they face in their lives. To the applicants, just as poverty and residence status are connected, the various problems they face are interconnected, multilayered and affect their social connections as well as their individual sense of well-being. These are the voices of the Burmese refugees.

I came with only clothes on my back, was always afraid that my visa would run out and could not speak Japanese at all, but I had to find a way to live. Just finding a place to stay was difficult.

(Male)

In Burmese landlord house, some of the rooms where Burmese stay, have dozens of people in them.

(Male, 20s)

Because of the lack of access to information and economic resources, the Burmese immigrants face difficulties in entering 'foreigner houses' (Shinjuku Regional Culture Department 2008: 108) or renting apartments (because key money, security deposits and agent fees are all high) and have little time to search for property; hence, many rely on the support of Burmese friends or acquaintances. Even if they find a place to stay, in many situations, they cannot stay very long. Fellow Burmese run mediation businesses for apartment hunting and guarantor services.[7] Burmese brokers who charge for consultations on Japan and information about daily needs are also appearing.

This 48-year-old Burmese man found a place to live with the help of a Burmese intermediary, but

[I]t has been 9 months and I have used all the money I had. Four of us live in a 6-tatami mat room. During the midnight or early morning, when I come home to sleep, it is noisy and I wake up. I cannot rest.

In an environment where rest is impossible, he still has to work. His job and living situation may be imagined from the statements by other Burmese:

In the morning, we would go to places where they gathered day-labourers, go to the day's construction site, clean or pick up garbage and then learn Japanese.

(Male, late 40s)

We would go to the place where labourers gathered and then work at construction sites picking up garbage or washing dishes. I wandered from address to address. I could not pick what I did but I had work. Sometimes I would be paid commuting costs, sometimes not.

(Male, 40s)

I worked at a bar franchise's factory in Tsukiji, splitting tuna. It was tough work that wore me out.

(Male, 40s)

When I came to Japan, I worked at the same restaurant as ethnic Kachin and registered as an alien in Kita Ward. However, they wrote that I had no status of residence. I was recognized as a refugee and my visa was changed to Long-Term Resident. Before that, it was Provisional Release and with that you cannot get permission to legally work but I worked very hard. Even if you worked the same amount as a Japanese, there was a difference in the amount you got paid hourly.

(Female, 40s)

These were workplaces that were dirty, dangerous and tough. The recent economic downturn also meant that construction work was hard to find:

People on provisional release had no legal permission to work; however, in actuality, they would go to work at regional farms (due to lack of farm trainees, there was a demand for workers) and factories.

(Male, 20s)

If Burmese wanted to work, there were jobs. They knew the situation and went to the employer. But even when they were working, they were afraid:

Even when faced with serious problems, applicants are scared to go to the ward office. Until my application goes through, it is hard to walk around out in the open. If my husband, who was working, came back 5 minutes late, I would get worried.

(Female, 40s)

When I get called by immigration, it is scary and stressful.

(Male, 20s)

If I do not appear to get my visa renewed at the designated day and time, I will be detained. If I am supposed to come in at 11, I cannot come in at exactly 11, I have to come before. Everyone is afraid of renewal. If they tell us '4th floor,' that means we are going to detention.

(Ibid.)

A few become cynical:

(I was actually afraid but) I decided to walk around even if it meant they would catch me. I had no home to go back to in Burma and my mother is in Canada, so I thought they would not deport me. The deportation centre is better than the one in Thailand.

(Male, 40s)

Because alien registration allows access to identification documentation, it is spread around by word of mouth and most people register:

I work part-time jobs, try not to use any money and save it. I do not do any recreation.

(Male, 20s)

Because there are many who support themselves with part-time work, their livelihood collapses if they get sick or otherwise cannot work. Because of the prolonged recession, late payment or non-payment of wages and accidents at work are also problematic. There are many cases of back pain from lifting heavy objects or lack of exercise, and broken bones from accidents. Foreign workers are not aware of places such as the Labour Administration Office, which provides labour consultation and dispute resolution services.

Many Burmese have no residence status and their rate of entry into the national health insurance system is very low, due partly to language difficulties which mean they fail to obtain accurate information. In 2007, out of the 833 Burmese registered in Shinjuku ward, less than 9.7 per cent were registered with the national insurance (Horiuchi 2007: 229–240). Most applicants without residence status work without insurance and pay for medical bills in multiple payments; go to free or low-cost clinics or simply forgo medical treatment. Some medical facilities will refuse to treat people without insurance, while some Burmese believe that medical facilities will report them to immigration; thus, they avoid them. The Burmese applicants generally believe that it is important not to become sick or injured.

In order to get a job in Japan, it is indispensable to be able to speak Japanese. However, some Burmese believe that 'It is factory work so I do not need to speak Japanese' or 'Even if I do not have a visa, I can get a job through connections with other Burmese'. Some Burmese believe that they work long hours because they have no valid visa, whereas others believe that even with a visa one must work long hours. Some employers compliment their work ethic: 'Burmese do not complain to their superiors and work without ever saying no' (interview with one Japanese restaurant manager). It is possible that employers use their lack of visa or not being able to speak Japanese as an excuse to take advantage of their work ethic.

In some cases, it is not enough for them to rely on fellow Burmese and they must turn to Japanese citizens for assistance: 'Even if I can pay rent, I need a Japanese guarantor or no one will rent an apartment to me'. 'I went to the Hello Work (Public Employment Introduction Office) in many towns, but they turned me down for the sole reason that I could not speak Japanese'. 'I went to Hello Work time after time, but I could not find anything and lost my self-confidence'.

Finding employment and a place to live with an unstable status is very difficult. Because they cannot complete forms in Japanese or explain their situation to those at the counter, it is hard for them to receive public services.[8] This creates significant problems in their daily lives. Eventually they begin to think, 'If I can learn Japanese, I will be able to live by myself'.

Over 90 per cent of the Burmese wish to learn Japanese. However, because they have little economic room to manoeuvre, they must put their livelihood first and have few opportunities to study. According to a survey by Shinjuku ward in March 2008, the problems most Burmese faced in their lives were language-related issues (74.3 per cent). Next was hospital and medical care (37.1 per cent) and then work (28.6 per cent) (Shinjuku Regional Culture Department 2008: 72).

Although they are able to form networks because they share ethnicity and culture, in some cases, meeting with co-ethnics can cause anxiety. Some Burmese cannot talk about their anti-governmental guerrilla activities in gatherings of Burmese, because these groups now contain many people who they require to be careful of, for example, Burmese who have come to Japan with valid passports to make money. If they are known to be guerrillas, people may avoid them. Thus, many Burmese have learnt to be silent, and occasionally report having felt depressed because of this. The applicants are very careful about protecting their personal information.

They are able to speak directly to their families back in Burma, but some have heard the buzz of wiretaps on the telephone line (Comment from a Burmese refugee). Their letters are also opened.

Because Japanese governmental regulations are strict, the Burmese are afraid of being picked up as illegal immigrants and, furthermore, they cannot speak Japanese. Many Burmese confine themselves to their rooms, and the number of people who have become ill has increased. Of particular note have been cases of AIDS. Various factors contribute to a great feeling of unease, including thinking about their families back in Burma, not receiving refugee recognition, economic difficulties of living in Japan, not being able to see a future for themselves in Japan or that their physical condition is deteriorating. Restrictions on their movement and living space have a negative effect on their individual social lives.

Mechanisms for supporting applicants

Since 1983, Japan has instituted certain aid facilities for applicants for refugee status, although these are quite limited both financially and in scope. They target applicants who have difficulties with their livelihood while waiting for recognition of their status. Since 1995, these operations have been entrusted to RHQ. As part of their activities, RHQ give an allowance for living costs to applicants.[9] Since December 2003, they have also provided accommodation to those without housing, although the number of these is limited.

This aid was only given to a very small number of applicants initially and was not given out because of legal obligations but from a humanitarian perspective. However, because of the great increase in applicants, the budget was exhausted and in 2008, the government put a temporary halt to living cost aid. It was restarted afterwards; however, in April 2009, standards for receiving aid were tightened up and in May, nearly 100 people had their aid cut off for budgetary reasons. At present, after assessment, nearly all applicants receive aid (comments from RHQ employees). Living costs require a qualifying examination[10] and there were problems with the time it took for the aid to begin, but this time was somewhat shortened and UNHCR as well as other non-governmental organizations helped to fill the gap with a limited amount of aid.

As noted earlier, public aid covered 3–4 months and RHQ supported applicants, on an average, for 8–9 months; however, this did not meet the actual waiting period, which averaged two years. From July 2004, RHQ created and is currently operating a counter specifically to respond to problems of recognized refugees and applicants (medical, employment-related, educational, living-related, accident-related, legal formality-related, etc.). RHQ gives out Burmese-language versions of *Handbook for Living in Japan* and *Medical Terminology* to refugees and applicants for free (FWEAP 2011b). Now, let us hear from those working at RHQ:

> Every day, a number of applicants will visit RHQ's office asking for living cost allowances. Compared with refugees from Indochina and refugees who have had their status recognised, their living situation is very tough and unstable. They have trouble meeting a minimal standard of living. As the applicants increase, the variety and complexity of their cases increase and it is impossible to solve their problems easily.
>
> ((a RHQ refugee advisor) (FWEAP 2011a: 62))

> In the case of an applicant for refugee status, because their activities and access to social services such as health insurance are restricted on the basis of their status of residence, there is a limit to the services we can provide.
>
> ((Refugee advisor) (FWEAP 2011: 62))

If, as a foreigner or outsider, one is cut off from the public support system, one will face many related problems. Thus, long-term aid is required until employment is possible.

There are also those who come seeking assistance on starting offices or businesses. The Refugee Enterprise Support Fund, formed to support, through enterprise, the economic independence of refugees who have little chance of employment and inadequate access to social services, was approved as a Public Welfare Foundation in March 2012 and was able to begin making investments (Entrepreneurship Support Program for Refugee Empowerment 2012). Its aim is to enable refugees to earn money on their own and to support job creation for other refugees.[11] Because of the sudden increase in aid applicants, the supporters struggle to support the budget and actual costs.

UNHCR takes the position that the responsibility for dispensing aid lies unequivocally with the receiving country, that is, Japan (UNHCR 2012). It recommends that recognized refugees and applicants for refugee status go to Japanese organizations, and does not independently confer refugee recognition. UNHCR makes recommendations on the recognition process and on dealing with refugees and applicants (handling information, detainment issues, poverty aid, children education and medical services), standards for recognition and other issues such as settlement. As part of its consultative function, it expresses legal opinions on how immigration authorities and the court system interpret the Refugee Convention in terms of recognition. In principle, it gives limited financial aid (lump sum payments) to and only to those who apply for living aid from RHQ until their allowance commences. Its connections with individual refugees and their communities are limited. Practical issues (detainment and illness counselling) are entrusted to the International Social Service of Japan, a related non-governmental organization (NGO), as well as the Japan Association for Refugees (JAR).[12] Allowances are immediately available on the basis of the judgment of NGO workers.

Support for applicants is handled by a number of private organizations, some involved full time in refugee assistance and others where the work is handled by one to two employees in addition to their main job. Poverty inquiries are handled by RHQ. From 1995, UNHCR strengthened its ties with various refugee aid organizations – 'Partnership in Action', created a joint Refugee Forum, Japan Forum for UNHCR and NGOs (J-FUN) in 2006. And, furthermore, specifically for refugees and asylum seekers coming to Japan, UNHCR set up Forum for Refugees Japan (FRJ) in 2004 and aligned their respective aid activities.

Coping mechanisms: vertical structures amongst ethnicities

Aid organizations state that Burmese residing in Japan are relatively many, have many organizations and groups, are able to obtain legal services by themselves and are able to solve their problems in their own community. They are able to live in Japan for long periods of time without speaking Japanese if they are in environments where they can work and are close to their fellow Burmese. In Aichi, there is a community of recognized Burmese refugees who help fellow Burmese applicants for refugee status by accompanying them as interpreters for issues such as medical consultations. There are around 27 or 28 Burmese groups. Each group meets once or twice a month and there are web pages for various groups and organizations on the internet. Because there is no powerful public aid system and they have been marginalized in Japanese society, they must, on their own, increase opportunities to amplify their own voices, share limited resources and support and advance their own Burmese social networks.

However, they do not always work in large cohesive groups. One (recognized) Burmese refugee said, 'Some organizations get along and others do not. The only time they all get together is during festivals. Because of what has happened in the past, it is impossible for everyone to get together on the spot'.

NGOs are advancing aid activities that allow refugees to help each other; however, they often end up having to divide activities by ethnic group. In addition, the groups' political activity overlaps with social-cultural activities. Groups engage in a wide range of political activities such as gatherings, propaganda, fund-raising and so on; hence, it is said that aid distribution is problematic. Support from Japanese aid is limited to financial support for issues such as expenses for cultural decorations or leasing activity space (FWEAP 2011).

Burmese gatherings tend to be broken down by organizations and groups: for example, the Association of the United Nationalities (AUN) is an ethnic minority organization that has a political character and at the same time only allows members to assist each other. Each ethnic group has its own leader and opportunities such as jobs are passed around the membership. Jobs are available as long as the seeker has no preference for the type of work, and although there is no large financial fund, they lend small amounts of money to each other. They assist each other on temporary matters but are undeveloped as an aid organization. During emergencies, they give money to one another. They hope to develop connections outside the boundaries of their organization; however, their capability to dispense information and bring Japanese and themselves together is weak. Issues such as jobs, work and living arrangements tend to break down by ethnic group. In addition, foreigners in Tokyo tend to be segregated by country of origin: Takadanobaba has the Burmese, Ikebukuro has Chinese and Shin Okubo has Koreans (Tajima 1996). In the refugee community, the Federation of Workers Union of Burmese Citizens (FWUBC) is said to cross ethnic boundaries.

The Burmese go to Burmese stores, karaoke establishments and restaurants that have bulletin boards displaying various information including rooms for rent and job opportunities. Most Burmese foodstuffs are available in Japan. The spread of Burmese restaurants, supermarkets and rental video stores also means the spread of employment opportunities for Burmese. Around 15 years ago, the first community newsletter was created and there are now a number of free and paid newsletters being printed.

The Burmese gather in Islamic mosques, Christian churches and Buddhist temples on weekends and these become settings for job searches, birthday parties, wedding ceremonies, information exchange and meeting new people.[13] Ethnicity and religion bring the Burmese of Shinjuku together (Shinjuku Self-governing Creation Institute (2012), *Institute Report* 2011, Working Group Report on Foreigners (3): 21). People gather for Buddhist dharma talks and meditation events. It is an important opportunity to gather in groups and focus on well-being for people who are disadvantaged and face constant struggle.

In Burmese communities, not participating in important events or not treating other members with hospitality is seen as selfish and met with strict social penalties, leading to people being ostracized. Hospitality and tolerance are particularly important, and support for the group or ethnicity one belongs to is prioritized over one's individual social obligations (Tanabe 2008).

Visiting, receiving guests and exchanging pleasantries with friends are a central part of daily life. At their core is the thinking that it is important to treat others with hospitality. These visits have the role of creating, strengthening and supporting connections amongst one another. In particular, social exchange is important in maintaining emotional stability in a foreign country.

They are unable to meet their compatriots in other countries such as the United States or the United Kingdom, but send each other news and information about what they have been doing via email.

Individual life strategies: a future with no clear foundation

While in the past some Burmese would only consider applying for refugee status after their visas expired, or despite suggestions would not see any necessity to apply for refugee status at all, one Burmese group leader now commented, 'Now, I do not think there are many overstayers at all'. Overstayers are not left in isolation and live on the periphery of the Burmese community of Japan. The same Burmese group leader continued: those who were less likely to be recognized as refugees exhibited various patterns of thinking such as 'They do not do anything political and just save as much money as they can. If they get caught, they go home' or 'People who have been in Japan for long periods, 10 years or so, have no job if they go back to Burma, so they want to stay here. If they get caught, they will apply for refugee status'.

Some say that even when they receive official recognition of refugee status, they witness no significant change in their life in terms of employment or living situation compared with when they were in the application process (Male, 40s). Even after being recognized, refugees must extend their visas every three years. One advantage of this is that once a refugee travel document is produced, the refugee may travel out of the country, although destinations are restricted.[14]

In time, some Burmese surrender to the government and go home without fulfilling their ambitions, whereas some leave Japan to go to countries such as the United States, Canada or Australia. Some marry Japanese nationals and some bring their families over from Burma. Their motivations vary. Those who were interviewed are limited in number and may not be representative. Let us examine a few.

Case 1: Mr A (late 40s)

A former government employee, after falling out of favour with authorities during the protest struggles, he bought a valid passport for a large sum of money and came to Japan on a tourist visa in 1991. He applied for refugee status and received recognition. He worked for about four years at construction sites and doing electric work. After this, he pooled financial resources with friends and opened a Burmese restaurant. He says that the restaurant allows him to use his own time freely and that in the future, he would like to return to Burma and become an elementary school teacher and teach children.

Mr A's status of residence is long-term resident; however, he is still considering returning to his home country, so he has not applied for permanent residency. He takes care of his fellow countrymen and is in contact with those overseas as well. Those like him who have decided to return home at some time in the future are estimated to account for 20–30 per cent and many are former activists.

Case 2: Mr B (26)

Mr B came to Japan in 2006 with his father. His father is in Nagoya and his mother is in New York. Because his grandfather participated in a democratic demonstration in 1988, they could not continue the family business. His grandfather cut ties with him in a public newspaper, so he was able to obtain Burmese citizenship and a passport. He is currently a student at a Japanese university. He lives in a six-tatami mat apartment with one other Burmese man (an applicant for refugee status).

Mr B visits Burma every other year but does not consider himself to be a refugee and feels safe in being able to return any time. He goes to the Burmese embassy once a month to pay 10,000 Yen and get his passport renewed. If he were to apply to become a refugee, he would attract the eye of the Burmese government and be unable to return, so he does not participate in demonstrations. The embassy staff take pictures of those who do participate in such demonstrations, so he is careful of this and tries not to associate with refugee applicants. In the future, he is thinking of going to the United States. Mr B does not have legal refugee status; however, his background overlaps with those of refugees in many areas.

Case 3: Mr C (48)

In 1962, Mr C's family's factory and movie theatre were nationalized and confiscated by the junta. When he was a student, he participated in democratic demonstrations, was persecuted and fled to Japan. His mother and older brother are in Canada. He bought a passport in Thailand and came to Japan by ferry via Pusan, South Korea. He pursued his refugee recognition all the way to the Supreme Court and won. He learned Japanese in a kitchen by writing characters on the back of receipts and studying on his own. He borrowed small amounts of money from friends and churchgoers; there was no interest charged but a strong belief that he would pay back the money.

Mr C is currently a Master's degree student at a Japanese university and is struggling to pay tuition fees. He met a Burmese woman who came to Japan on a valid passport at a church and married her. Even if her college student visa runs out, because she married Mr C, a recognized refugee, she may remain in Japan as his spouse. In the future, they hope to teach English to refugees but have not decided where they will live.

Case 4: Mrs D (40)

She came to Japan in 1992. She has given birth to four children in Japan and runs a Burmese restaurant. She opened the restaurant with financial help from her sister based in the United States. Thinking of the future of her children's education, she applied for a United States Green Card several years ago. If she is able to get it, she says she will go to the United States and afterwards go back and forth between Japan and the United States.

Before she was recognized as a refugee, because her former address involved a route through Tokyo station (where many people would be able to see her), she was very nervous and moved to an address near her husband's workplace instead. When her oldest daughter was in sixth grade, she asked, 'Am I not Japanese?' She is leaning towards going to the United States.

Mr A has been in Japan for over 20 years but strongly wishes to return to Burma. Mr B and Mr C are unclear and Mrs D, despite living in Japan for 20 years, is thinking of going to the United States. In general, young Burmese have not decided whether to continue living in Japan.

Even if a Burmese participates in political activity during their refugee application period, after recognition they tend to stop and focus on earning money for the purposes of returning to Burma in the future. Their compatriots are disappointed when this happens; however, this is generally accepted if they have a family. The political situation in Burma is fluid and the possibility of being able to return home is on their minds.

From the outset, the Burmese tend to think that once their country is democratized, they will return home and they focus on teaching their children Burmese rather than Japanese. On the other hand, some families think of moving to countries such as the United States, Canada or Australia and push their children to learn English. The idea of living long-term in Japan is not very strong (Shinjuku Self-governing Creation Institute (2012), *Institute Report* 2011, Working Group Report on Foreigners (3), p. 22). Some intend only to come to Japan initially because it is easy to obtain a visa and then go somewhere else, making Japan a sort of 'first stop'. They think of going to a country where English is spoken because they are familiar with the language.

Other factors include the backgrounds of some Burmese living in Japan who are elite intellectuals having attended college, come from relatively affluent backgrounds and have a grasp of global affairs. Many refugees tend to admire the United States as a symbol of freedom. Mrs D, Mr B and Mr C all have family in Canada and the United States. Mr A wishes to return home but originally wanted to go to the United States and only came to Japan because it was easy to obtain a short-term visa.

As stated before, some Burmese who originally come to Japan wanted to go to the United States, but circumstances forced them to reside in Japan. They are very frustrated and feel they are in danger of straying from their life path. It is not hard to imagine that these people, who are thinking of returning home or moving on to another country, experience great stress because of the pressure to assimilate (such as learning Japanese) from distributors of aid. However, the two possibilities of one day relocating to countries such as the United States or returning to Burma present a nuanced dilemma.

A Burmese woman who lost her job as a teacher by participating in anti-government demonstrations, and came to Japan in 1993 on the advice of her father, was recognized as a refugee (in 2006) and married a fellow Burmese in Japan. She said:

> I obtained refugee status but do not know what will happen to me in the future. I have no idea what will happen with my children's education, my job and savings, what will happen when I reach old age, anything. I am not certain that we will always be able to continue living in Japan, on the other hand I do not know whether we can return to Burma.
>
> (Shinjuku Regional Culture Department 2008: 280–281)

Although they cannot paint a clear picture of their future, the Burmese continue to move forward to long-term and permanent resident status. Even if they wish to return to Burma, an estimated 60 per cent consider remaining in Japan for reasons such as the fact that their children only speak Japanese (one Burmese group leader). Less than 40 per cent of Burmese arrived in Japan at the end of the 1980s to the first half of the 1990s, are now aged 40 and older and fear losing their access to benefits in the future when they have grown old (Shinjuku Self-governing Creation Institute, Working Group Report on Foreigners 2012 (3): 22).

Some Burmese children are essentially stateless because their parents could not notify the Burmese embassy of their birth. In 1985, Japan revised its nationality law to recognize the nationality of both mother and father; however, children born to parents who are stateless remain stateless. Moreover, because over half of them are refugees, they have no money to spare to make donations to establish schools. Even with aid from the Burmese government, it is hard to attend such schools because children who go there are the offspring of those who fought against the government (ibid.). So far, there have been no Burmese children in Japan who have become professionally qualified in Japanese society. The best we have seen are fashion models or designers.

As their children grow and become fluent in Japanese, will the parents be able to overcome the feeling that they are guests (a foreigner consciousness) and will Japan, which has become their country of residence, ever become their home country? Although those wishing to progress to undergraduate and graduate education have increased, children have difficulty reaching levels beyond mandatory education. Although the Ministry of Education has a policy to promote special treatment for refugees, this does not penetrate schools. UNHCR has partnered with three universities to create a higher education programme for refugees to send children to university free of charge, but only accepted four people among an applicant pool of dozens.

Changing attitudes among Burmese and Japanese

Has there, however, been a transformation in local communities? Even though some Japanese oppose a further increase in the foreign population of Shinjuku ward, according to a ward survey, negative reactions are declining. Amongst those acquainted with Burmese, there is little feeling of opposition and contrasts between the Japanese and Burmese immigrants are beginning to fade. There are no incidents where Burmese stand out as troublemakers and it is felt that they observe the rules of daily society. While some come into contact with the Burmese through their children, however, without a specific catalyst it is not common for Japanese to become acquainted with them in their daily lives (Shinjuku Regional Culture Department, Culture International Section, 2008: 97).

In the previous alien registration system, cards were issued indicating 'no status of residence'; however, with the new basic resident registration system, there is a fear that those without residence status will slip through the cracks when dealing with local authorities. If this happens, it is possible that unregistered foreigners will gather in greater numbers in Shinjuku ward, which could lead to unexpectedly increased concern among the local population.

In the wake of the earthquake and tsunami on 11 March 2011, refugees from Uganda, Turkey (mainly Kurdistan), Burma, Nepal and Ethiopia volunteered to work clearing away rubble in the disaster-affected areas. The day after the quake, refugee support groups received telephone calls from refugees 'hoping to head to the disaster-affected areas to help'. In one month, the Burmese donated 1.5 million Yen to the Japanese Red Cross for victims of the Tohoku earthquake. On 30 April 2011, 95 Burmese went to Ishinomaki and Tagajyo in Miyagi prefecture to distribute rice and rake mud. On 18 June of the same year, 150 Burmese went to Ishinomaki (*Chunichi* Newspaper 6 June 2011). A Japanese acquaintance asked, 'Why are the Burmese doing so much to help?' The answer was, 'When something horrible happens in the place they live, they want to help as much as possible. That is how people live' (Comment from one Burmese leader). The applicants for refugee status, whose movements are normally restricted, received temporary travel permits from immigration authorities and joined the efforts. Refugees who took part in the volunteer efforts were driven by a sense of gratitude towards the Japanese society, which had given them shelter. This was a desire to help the society in which they were currently living. While grateful for the support during their application process, their feeling was that this alone was not enough; they wanted to pull their weight, pay taxes and contribute to these efforts. Prior to this, when Shinjuku ward surveyed the registered foreigners in general about activities they wished to undertake in the area, 42.9 per cent of all the Burmese answered 'participation in volunteer activities' and 34.3 per cent answered 'sharing Burmese cuisine and culture with the people of Japan' (Shinjuku Regional Culture Department 2008: 108).

Meanwhile, the incomes of refugees working in factories and the food industry were dealt a serious blow by the planned blackouts that followed after the collapse of the Fukushima nuclear reactor. Some Burmese lost their jobs as a result of the effects of the earthquake, tsunami and nuclear accident, and their incomes decreased due to shorter working hours, placing great strain on their livelihoods. Refugees' families who remained in Burma also experienced anxiety over the possible effects of the nuclear accident on their family members in Japan. Some foreign students returned to Burma. Lacking clear information, some Burmese deliberated over where to evacuate to and whether to return to Burma. Those with resident status in Japan had the ability to temporarily leave the country. Directly after the earthquake, foreigners left Japan in excess of twice the numbers of the same period the previous year, totalling 316,000. Among these, short-term visitors (staying for 90 days or less) were the greatest in number; however, they were followed by foreign students, families of foreign residents and spouses of Japanese citizens. After the disaster, the Burmese reacted in the following ways: (1) out of fear of radiation, approximately 100 Burmese who had overstayed their visas appeared before immigration authorities and headed home; (2) in light of the new political regime in Burma, others held off on applying for refugee status. However, Burmese who were already in the process of applying for refugee status were unable to leave Japan without rescinding their applications.

Conclusion

The Burmese refugees are a populace seeking protection under the law, facing transformations and adaptations in daily life after their escape from Burma. As they try to rebuild their lives while their applications are pending, the range of options available to these refugees is limited and the results are greatly determined by outside factors. During and after their applications, the greatest problem the Burmese face is not a lack of money, but the lack of social support and the difficulty of their labour situation amidst these changes in their daily lives. The workplace is their one point of contact with Japanese society. What is important is not acquiring a special residence permit with its accompanying restrictions, but acquiring the definitive status of refugee, receiving Japanese language tuition and being able to bring their families over from Burma. The logical step is to set a fixed time period and to permit work after that period for all applicants (not just those who applied while they still had a valid visa, as at present) even if application procedures are taking longer. This will also prove useful after refugee status is granted, enabling self-sufficiency at an early juncture.

Jacobsen and Landau (2005: 52) advocate for UNHCR to promote the right to work for refugees and those who have applied for refugee status.[15] Given that many applicants for refugee status do not have residency status, it is important to give due consideration to the interests of the applicants. The expenses municipalities face in terms of public facilities, multilingual services and personnel costs are by no means insignificant. The potential efforts by local staff and volunteers are also great. The ability to work would reduce the financial burden on the hosts, promote self-sufficiency and prevent alienation from Japanese society. The problem of residency status has been a source of misunderstanding and criticism of refugees. By distancing the applicants from society and placing them on the periphery, the status quo causes the refugee issue to be discussed in the same context as 'illegal residents'. Under the expansion of the Basic Resident Registration Law to include foreign residents, collaboration is to be undertaken between the Ministry of Justice and local municipalities; by the same token, this collaboration must also include efforts on Japanese language education and work training for refugees and applicants. Though the connection between the national government's refugee policy and local municipalities' general foreign resident policies (multicultural policies) was previously weak, it may now be possible to tie them together on a regional basis.

In Shinjuku ward, the sharing of prompt and accurate information with foreign residents, including the Burmese, is becoming an important issue. Communication technologies should be used to regularly communicate information to community groups about services, events, new laws and refugee-related human rights. A key element in such cases is not only to ensure that information is shared but also that individuals are actually able to receive the services. Support groups can be formed within the Burmese community that enable volunteers to look after health-care issues, mental health and children separated from their parents. In the future, true courtesy and perseverance will be required.

Moreover, in order to provide appropriate support, manpower and money will be needed.

It is possible to position the evaluation period for refugee status as a preparatory period for permanent residence and for Japanese-language training and work training to be undertaken. Though some worry that granting permission to work will pose difficulties for prompt returns to countries of origin, rationally delineating the time periods and enabling work will enable applicants to become financially self-sufficient and return home if their applications are rejected (Azuma 2005: 6). A support system for daily life is a prerequisite for proper refugee status evaluation procedures and must be prepared in tandem with the status evaluation system.

In regional society, Burmese are still 'guests' and are mainly noticed during special events, and are still unable to become independent societal actors. In addition, there are difficulties in breaking away from the ethnic group and forming connections that transcend this group. All this led to the formation of the Refugees Coordination Committee Japan (RCCJ) in 2009 to enable refugees in Japan to express themselves and assert their rights. RCCJ is an organization run by and for the refugees.

Improvements have been seen in recent years in comprehensive governmental and civil society initiatives to address refugee issues. One Burmese refugee notes, '[In the past,] I used to try to avoid getting sick, but Japan has begun to change recently'.

In 2011, the Diet marked the sixtieth anniversary of the Refugee Convention and the thirtieth anniversary of Japan's accession to the convention by passing a unanimous resolution pledging Japan's continuous support for the protection of refugees and the resolution of refugee issues. This was the first such resolution adopted in the world. However, social and legal issues still remain. Action has not yet progressed to the point where refugees' unique situation and needs are fully grasped.

The Refugee Convention tasks member countries with protecting refugees as well as all forms of their rights, as they relate to daily life. Nations have the sole authority to determine who enters the country, under what conditions and with what qualifications. This is an essential assumption for the independence and sovereignty of nations. Immigration law is responsible for creating a legal basis for this structure. However, while sovereignty may rest with a nation, it is ceasing to be the case that all refugee immigration and emigration falls within the nation's discretion. The recognition of refugee status can be understood as an element of state sovereignty and exclusive national jurisdiction; however, this view has come under criticism in light of the advancement of international law.[16]

More than 90 per cent of Japan's foreign population comes from Asia. Through special residency permits, many irregular foreign residents have been given regular status. Regionally and in local municipalities, Burmese have seen long-term stays and permanent residency expand; however, at the same time, the unstable political conditions of East Asia ensure that population movements will increase. It is important that Japan continues to fulfil its role in sheltering the people of Burma in the days ahead.

Notes

1 When persons with foreign student and long-term resident visas have financial problems, they can receive public assistance from the government.

2 In the 'Action plan to enforce a society tough on crime' determined in the December 2003 government cabinet meeting for measures against crime, a goal was set to eliminate half of the illegal residents within five years (Ministry of Justice 2010: 13). Strict immigration pre-examinations, individual identification (fingerprints and photographs) and the reinforcement of the identification of false documentation were instituted. The number of refugee applicants detained also increased. The management of long-term stay became difficult and the number of Burmese in Japan, which once exceeded 10,000, decreased. They were granted amnesty by the Burmese Embassy in Japan, who paid for their return home (comment from Burmese refugees).

3 Even when not recognized as a refugee, in the event that their visa status is specially approved, 'Designated Activities' approval is given. Approved refugees became long-term residents and immigrants with special residential status were granted designated activities. Unlike designated activities permission, long-term residents are not limited in work activities and are able to apply for public welfare.

4 According to the documentation (Shinjuku Self-governing Creation Institute, *Institute Report* 2011, Working Group Report on Foreigners (2) 2012: 17), as of December 2009, when the refugee application legal counsel applied to change the status of 37 Burmese living in Japan who had received 'Designated Activities' approval based on humane considerations, the status change standards put forward by the Tokyo Immigration Bureau were (1) stay of 10 years or more, regardless of legality and (2) stay of three years or more with designated activities status.

5 Around 70–80 per cent of real estate clients are foreigners. There are about 152,000 apartment-type residences in Shinjuku, accounting for close to 90 per cent of the total. Among these residences, the numbers of condominiums (purchased or rented) are about 103,000. The majority of residents live in condominiums. Compared with other wards, Shinjuku is on the expensive side (Shinjuku Self-governing Creation Institute (2011), *Institute report* 2010, No. 3, Condominium Working Group Report (1), p. 1 and 20).

6 In Shinjuku ward, in 1991, a 'cooperative system' was created through real estate agents, to set an example for the rest of the country in ending discrimination against residents who are citizens of other countries. Each self-governing entity observes how Shinjuku ward is advancing. Real estate agents sprang up to cater to foreign customers who hired foreigners, prepared contracts and guides in many languages and put up 'Foreigners Welcome' signs. Condominium associations also began to operate in different languages and worked to avoid possible troubles. Neighbourhood associations also actively worked in order to assure the coexistence of multiple cultures. NGO activities are in full swing and there is an especially large number of foreigner support NGOs.

7 A guarantor is required to rent a residence, but there are Japanese people who will agree to be a guarantor for a fee of 150,000 Yen (comment by a Burmese refugee).

8 In the Tokyo metropolitan area Hello Work offices, services such as the foreigner corners are available and English interpretation services are also available. In addition, there are public service facilities that offer employment information for foreigners, such as the Tokyo Employment Service Centre for Foreigners. However, many people are not aware of this. There are also cases in which the applicants can only speak Burmese and do not have sufficient English ability for communication.

9 As a rule, welfare is distributed for four months in the form of a maximum of 85,000 Yen per month for a single person, with living expenses of 1,500 Yen per day, accommodation expenses of 40,000–60,000 Yen per month and actual medical costs. Because there are many cases in which labour is not authorized, the funds are

provided even if visa status is lost. Because medical costs are paid after the fact, the request for pre-payment is strong. The average number of recipients for one month in 2011 was 387. There were reports of fraudulent claimants in the media; however, it is said that RHQ does not have the power to investigate.

10 Out of those who applied as refugees, based on the RHQ survey, the Ministry of Foreign Affairs gives aid to those in need, but there are a number of conditions. (1) Before making a refugee application, application to receive funds cannot be made, (2) as a rule, the welfare period is four months, (3) refugee applicants are no longer eligible after the second time and (4) use of shelter is only possible in the event that the time period since landing is short.

11 The maximum loan of 1 million Yen is a small amount; however, it is financing that allows refugees to operate restaurants.

12 The JAR mainly offers legal services and social counselling. It also does important work in livelihood support. The organization has entered into a collaborative business plan with UNHCR and carries out refugee registrations on UNHCR's behalf, offers phone support, financial support, etc. Because its funds are limited, its aim is to strengthen network building and create a system where problems may be referred to other groups (a JAR manager).

13 For example, the NPO Myanmar Culture and Welfare Association (Itabashi ward) purchased a three-storey building, making the first floor an assembly room, second floor an altar room and third floor as quarters for a Buddhist priest invited from Burma. The association was run by volunteers with no full-time employees and people were meant to pay what they could.

14 One may use refugee travel documents to go to the US and the UK; however, Thailand does not accept such documents, so they use re-entry permits to Japan. 'The official in South Korea had never seen the re-entry permit from Japan so I had to explain it' (Comment from a Burmese refugee).

15 Because it is the self-governing entities who are responsible for providing refugees and refugee status applicants with proper documentation, including travel documents, permission for labour activities and photo ID, making the regional self-governing entities and organizations realize their responsibility to refugees as well as responsibilities for health, residence, education, refugees should be included in this. Social responsibility can be achieved by organizations and public groups employing refugees.

16 Kouki Abe, presentation at the 2011 Peace Studies Association of Japan.

References

Abe, K. 2010. *Kokunaihou no Bouryoku o Koete* (Beyond the Violence of International Law). Tokyo: Iwanami Publishing.

Azuma, Y. 2005. *Kokusai Kihan kara Mita Nihon no Nanmin Seisaku: Nanmin Shinseisya he no Safty Net Kouchiku ni Mukete* (Japanese Refugee Policy in view of International Law: to set up Safety Net for asylum seekers). Unpublished mimeograph.

Cabinet Secretariat 2004. *Nanmin Nintei Shinseisya he no Shien ni tuite* (Assistance to Applicants for Refugee Status). Available at: www.cas.go.jp/jp/seisaku/nanmin/040708sien.html (accessed on 28 April 2012).

Chunichi Newspaper 2011. *Nanninra ga Gareki Hiroi de Ongaesi* (Debris cleaning up by refugees and asylum seekers as a form of gratitude). Available at: www.chunichi.co.jp/s/article/2011061690101654.html (accessed 19 June 2011).

Entrepreneurship Support Program for Refugee Empowerment 2012. News release dated 1 March 2012. Available at: http://espre.org/2012/03/authorization/ (accessed 9 March 2012).

FWEAP (Foundation for the Welfare and Education of the Asian People) 2001. *Nanmin Sinseisya tou ni taisuru Seikatu Jyoukyou Chousa* (Livelihood Survey on asylum seekers). Available at: www.fweap.or.jp (accessed 9 March 2012).

FWEAP 2002. *Nanmin Sinseisya no Jyuu Kankyou ni kansuru Jyoukyou Chousa* (Housing Survey on asylum seekers).

FWEAP 2009. *Ai* (Love), No. 33.

FWEAP 2011. *Nihon de Kurasu Nanmin Teijyuusya* (Refugee Residents living in Japan). Available at: www.rhq.gr.jp/japanese/know/kur/02_22.htm (accessed on 29 May 2011).

FWEAP 2011a. *Sousyuuhen* (compilation), No. 35.

FWEAP 2011b. *Livelihood Handbook revised version.*

Honma, H. 2005. *Kokusai Nanminhou no Riron to sono Kokunaiteki Tekiyou* (Theory of International Law and its Domestic Practice).

Horiuchi, Y. 2007. 'Koureikasuru Gaikokujin no Shakai Hosyou: Sono Genzai to Mirai Shinjuku-ku no Data kara' (Social Security of Ageing Foreign Residents: Present and Future from the Shinjuku Ward Data), in Kawamura, C. and Son (eds) *Ibunkakan Kaigo to Tabunka Kyousei: Dare ga Kaigo o Ninauno ka* (Health Care for people with different cultures and multicultural symbiosis: who will take responsibility for Health Care?). Tokyo: Akashi Publishing.

Immigration Bureau of Japan, Ministry of Justice 2006. *Guide to Refugee Status Determination.*

Inaba, Y. 2008. 'Uketsugareteiku Shinjuumin no Machi no Idenshi' (The passed down genes of the town of new residents), in Kawamura, C. (ed.) *'IminkokkaNihon' to Tabunka Kyouseiron* ('The Immigrant Nation, Japan' and the multicultural symbiosis theory). Tokyo: Akashi Publishing.

Jacobsen, K. and Landau, L. 2005. 'Recommendations for urban refugee policy', *FMR* 23, Refugee Studies Centre, University of Oxford, p. 52.

Japan Association for Refugees 2008, 2009, 2010. *Annual Report.*

Japan Association for Refugees 2011. *Newsletter.* Vol. 5 and Vol. 6.

Japan Association for Refugees 2011a. *Aratana Nanmin Ukeire to Shinjuku-ku: Daisangoku Teijyuu ni atatte Watashitachi ga dekirukoto o Kangaeru* (Refugee renewed acceptance and Shinjuku ward: Let's think about what we can do for refugee resettlement), International Symposium report.

Japan Immigration Association 2010. *Zairyuu Gaikokujin Toukei, Heisei 22 nen edition* (Statistics of Registered Foreigners in Japan, 2010).

Japan Immigration Association 2012. *Kokusai Jinryuu* (International Human Mobility), No. 299, No. 301 and No. 302.

Kawamura, Chizuko (2008). *'Iminkokka Nihon' to Tabunka Kyousei Riron* ('The Immigrant Nation, Japan' and the Multicultural Symbiosis Theory). Tokyo: Akashi Publishing.

Kawamura, C. and Son, W. 2007. *Ibunkakan Kaigo to Tabunka Kyousei: Dare ga Kaigo o Ninauno ka* (Health Care for people with different cultures and multicultural symbiosis: who will take responsibility for Health Care?). Tokyo: Akashi Publishing.

Ministry of Foreign Affairs 2009. *Refugee Acceptance in Japan.* Available at: www. mofa.go.jp/mofaj/gaiko/nanmin/main3.html (accessed 28 April 2012).

Ministry of Justice 2010. *The Basic Plan for Immigration Control.* 4th Edition.

Ministry of Justice 2012. *Heisei 23nen ni okeru Nanmin Ninteisyasuutou ni tuite* (The number of People admitted to Refugee Status in 2011). Available at: www.moj.go.jp/ nyuukokukanri/kouhou/nyuukokukanri03_00085.html (accessed 5 April 2012).

Morisita, Y. n.d. *Nihon de Seikatusuru Nanmin/Higo Kibousya no Iryou/Kenkou Mondai* (Medical and Health Problems of Refugees and Asylum Seekers in Japan), unpublished mimeograph.

Shinjuku Regional Culture Department, Culture International Section 2008. *Heisei 19 nendo Shinjuku-ku Tabunka Kyousei Jittai Chousa Houkokusyo* (Shinjuku ward Report of the Survey done on Multicultural Symbiosis in 2007).

Shinjuku Self-governing Creation Institute 2010, 2011, 2012. *Institute Report*, 2010 No. 1, No. 2 and No. 3, 2011 No. 2, No. 3, 2012 No. 1.

Shinjuku Self-governing Creation Institute 2012. *Institute Report* 2011, Working Group Report on Foreigners (2) and (3).

Shinjuku Ward 2012. *Foreign Alien Resident Registration in Shinjuku*. Available at: www.city.shinjuku.lg.jp/kusei/36toukei.html (accessed 11 April 2013).

Tajima, J. 1996. 'Toshi Chiiki Syakai to Ajiakei Gaikokujin', in *Tosi to Toshika no Shakaigaku* (Urban Community and Asian Foreigners in Sociology relating to Cities and Urbanization). Iwanami Lecture Series *Modern Sociology*, Vol. 18, pp. 151–169. Tokyo: Iwanami Publishing.

Tanabe, H. 2008. *Makeruna! Zainichi Birumajin* (Don't be beaten! Burmese Residents in Japan). Tokyo: Nashinoki-sha Press.

Tokyo Metropolitan Government 2010. *Statistics of Registered Foreigners in Tokyo*. Available at: www.toukei.metro.tokyo.jp/gaikoku/ga-index.htm (accessed 20 January 2013).

UNHCR 2005. *Information for Asylum Seekers in Japan*. Available at: www.unhcr.or.jp/protect/pdf/info_seekres_e.pdf (accessed 20 January 2013).

UNHCR 2012. *Refugee Protection handled by UNHCR Office in Tokyo*. Available at: www.unhcr.org.jp/protect/u_protection/index.html (accessed 28 April 2012).

Postscript

Gerhard Hoffstaedter

This edited volume brings together researchers from across the world to discuss fieldsites as disparate as The Gambia is from Japan. Yet all chapters in this volume highlight the void that exists between protection that legal documents such as the 1951 Convention relating to the Status of Refugees and its 1967 protocol want to afford to refugees and the reality of life for most refugees. The UNHCR remains the main agency charged with determining refugee status and safeguarding refugees from persecution and harm. As such it continues to perform the most difficult of tasks in often very challenging environments, as the preceding chapters attest. The move from refugee camps to urban settings presents new challenges to the UNHCR, host countries and refugees themselves.

The chapters in this volume provide candid insights into the theory/policy of refugee protection and its realities on the ground.

Refugees in urban areas face a range of challenges, such as access to health care, education and a livelihood. Urban refugees are often not afforded host state protection. They often live in legal limbo and occupy spaces of illegality or marginality in cities across the globe. Living on the margins of both society and the law brings other more nefarious and dangerous threats to already vulnerable populations.

Urban refugees demonstrate extraordinary resilience in making new lives for themselves in difficult situations and settings. Their experiences as recounted in this book have brought to the fore issues such as culture, diversity and the divergent needs of different refugee populations in a major city. In some cases urban settings allow refugees to blend in, work and participate in some ways in daily life.

What became very clear in the chapters covering a vast range of countries is that local particularities often have grave and far-reaching effects for the lives of refugees in urban settings. Global norms may offer a discourse to make sense of how international law should be applied and how urban refugees in countries with already weak rule of law or non-signatories to the UN refugee convention could be integrated. However, as the case studies show, even in signatory countries integration is challenged and difficult and the UNHCR as well as refugees have to find ways around local particularities outside of the easy grasp of global norms and discourses.

The chapters in this volume complicate the continued homogenization of refugee experiences that often still focus on the camp as a locus of being a refugee or the processes of being granted refugee status and resettlement. Chapters here focus on the multiple processes of seeking asylum as well as seeking means to live or sustain themselves in urban environments. The process of granting asylum thus becomes just another, albeit important, step in the living of life away from one's home, social and kin bonds. Thus all chapters add to the complexity of the refugee experience. They also interrogate the UNHCR's role in protecting refugees and the role host societies and states play in this process.

In conclusion this volume, with the variety of field locations covered, as well as the range of refugee populations described, demonstrates how urban refugees find ways to make livelihoods for themselves, navigate the precariousness of their legal status, access the UNHCR registration process and access services such as health, education and shelter in order to survive, and in some cases prosper, as urban refugees.

Index

Page numbers in **bold** denote figures.

274   *Index*

Japan Association for Refugees (JAR) 231,
250, 256
Japanese resettlement programme 206–38;
announcement of 213–17; difficulties
224–33; implementation 217–24;
incentive gap 233, 235; jurisdictional
gap 226–8, 234; participation gap
228–33, 234–5; selection criteria
217–18, 223–4, 226–8, 231
JAR *see* Japan Association for Refugees
(JAR)
job-grants, Italy 129, 130
jurisdictional gaps 207, 209, 226–8, 234

Kakuma Refugee Camp, Kenya 98, 100,
101, 112, 113n1
Kampala, Uganda 76–95; communal
action and independence 91–3;
institutional mistrust 84–91; numbers of
refugees 80, 81; refugee policies 80–1;
social mistrust and insecurity 81–4
Karen National Union (KNU) 224
Karen refugees *see* Japanese resettlement
programme
Karen Women's Organization 166, 167,
175, 180
Kawamura, Chizuko 248
Kenya *see* Nairobi, Kenya
Kibreab, G. 49
Kituo Cha Sheria, Kenya 103, 107
Kobelinsky, C. 123
Komeito, Japan 216
Kosovo 118
Kuala Lumpur, Malaysia 187–203;
approach to urban refugees 190–2;
assistance and durable solutions 200–1;
community outreach and partnerships
194–7; current urban refugee programme
192–4; education 200; health care 198–9;
housing 197–8; legal status of urban
refugees 189; numbers of refugees 188
Kurdish refugees, Italy 121

Landau, L. 263
Landau, L.B. 78
landmines 46, 73n6
language courses: Italy 128–9, 130; Japan
218, 219, 229, 230, 242
Latham, A. 78
legal aid organizations: Kenya 107
legal assistance, India 144
LGBTI (Lesbian, Gay, Bisexual,
Transgender and Intersexual) refugees
89

Liberal Democratic Party (LDP), Japan
216
Liberian refugees, Gambia 48, 50, 52
Libyan refugees, Egypt 32
livelihood access: Egypt 20–3; India 146,
147–8, 157; Japan 251–3; Kenya 109;
Thailand 177–80, 225; *see also*
employment access
local integration: Africa 49; Egypt 20–2;
Gambia 49–50, 52, 67; Italy 116–17,
127–31; Japan 217–24, 228–33;
Malaysia 189, 201

Mae La Refugee Camp, Thailand 217–19,
220–1, 226–7
Mae Tao Clinic, Mae Sot, Thailand 177
Malaysia *see* Kuala Lumpur, Malaysia
Malaysian Social Research Institute
(MSRI) 199
Malkki, Liisa 123
malnutrition: India 155; Thailand 177
Mandate Refugee Certificates 101, 105,
110
Manipur State, India 144
Mapendo International 103
Matsumoto City, Japan 221, 232
Maung, Cynthia 177
Médecins Sans Frontières 192, 199
medical services *see* health care access
Mercy Malaysia (Malaysian Medical
Relief Society) 199
MFDC *see* Mouvement des Forces
Démocratiques de Casamance (MFDC)
micro-credit: Gambia 51; Malaysia 197
Ministry of Foreign Affairs (MOFA),
Japan 212, 214, 215, 220, 231
Ministry of Internal Affairs and
Communications, Japan 215
Ministry of Justice (MOJ), Japan 212,
213–14, 215, 244–5
Misato City, Japan 220, 221, 232
Mizoram State, India 140, 144
MOFA *see* Ministry of Foreign Affairs
(MOFA), Japan
MOJ *see* Ministry of Justice (MOJ), Japan
Mori, Yoshiro 215–16
Moro refugees, Malaysia 189, 191
Mouvement des Forces Démocratiques de
Casamance (MFDC) 45–6, 47, 73n4
MSRI *see* Malaysian Social Research
Institute (MSRI)
murders of refugees, Egypt 24, 25, 26
Mustafa Mahmoud park, Cairo, Egypt 13,
17–30, **17**, **21**, **24**, **27**

Milton Keynes UK
Ingram Content Group UK Ltd.
UKHW031145141024
449569UK00024B/1062